SEMICONDUCTORS AND SEMIMETALS

VOLUME 22

Lightwave Communications Technology

Part B

Semiconductor Injection Lasers, I

Semiconductors and Semimetals

A Treatise

Edited by R. K. WILLARDSON
CRYSCON TECHNOLOGIES, INC.
PHOENIX, ARIZONA

ALBERT C. BEER
BATTELLE COLUMBUS LABORATORIES
COLUMBUS, OHIO

SEMICONDUCTORS AND SEMIMETALS

VOLUME 22

Lightwave Communications Technology

Volume Editor

W. T. TSANG

AT&T BELL LABORATORIES
HOLMDEL, NEW JERSEY

Part B

Semiconductor Injection Lasers, I

1985

ACADEMIC PRESS, INC.
(Harcourt Brace Jovanovich, Publishers)

Orlando   San Diego   New York   London
Toronto   Montreal   Sydney   Tokyo

ACADEMIC PRESS, INC.
Orlando, Florida 32887

United Kingdom Edition published by
ACADEMIC PRESS INC. (LONDON) LTD.
24–28 Oval Road, London NW1 7DX

LIBRARY OF CONGRESS CATALOG CARD NUMBER: 65-26048

ISBN 0–12–752151–8

PRINTED IN THE UNITED STATES OF AMERICA

85 86 87 88   9 8 7 6 5 4 3 2 1

Contents

Chapter 1   Mode Locking of Semiconductor Lasers

J. P. van der Ziel

Chapter 2   High-Frequency Current Modulation of Semiconductor Injection Lasers

Kam Y. Lau and Amnon Yariv

Chapter 3   Spectral Properties of Semiconductor Lasers

Charles H. Henry

Chapter 4 Dynamic Single-Mode Semiconductor Lasers with a Distributed Reflector

Yasuharu Suematsu, Katsumi Kishino, Shigehisa Arai, and Fumio Koyama

Chapter 5 The Cleaved-Couple-Cavity (C³) Laser

W. T. Tsang

List of Contributors

Numbers in parentheses indicate the pages on which the authors' contributions begin.

SHIGEHISA ARAI, *Department of Physical Electronics, Tokyo Institute of Technology, Tokyo 152, Japan* (205)

CHARLES H. HENRY, *AT&T Bell Laboratories, Murray Hill, New Jersey 07974* (153)

KATSUMI KISHINO,* *Department of Physical Electronics, Tokyo Institute of Technology, Tokyo 152, Japan* (205)

FUMIO KOYAMA, *Department of Physical Electronics, Tokyo Institute of Technology, Tokyo 152, Japan* (205)

KAM Y. LAU, *Ortel Corporation, Alhambra, California 91803* (69)

YASUHARU SUEMATSU, *Department of Physical Electronics, Tokyo Institute of Technology, Tokyo 152, Japan* (205)

W. T. TSANG, *AT&T Bell Laboratories, Holmdel, New Jersey 07733* (257)

AMNON YARIV, *California Institute of Technology, Pasadena, California 91125* (69)

J. P. VAN DER ZIEL, *AT&T Bell Laboratories, Murray Hill, New Jersey 07974* (1)

* Present address: Department of Electrical and Electronics Engineering, Sophia University, Tokyo 102, Japan.

Treatise Foreword

This treatise continues the format established in the books of Volume 21, in which a subject of outstanding interest and one possessing ever-increasing practical applications is treated in a multivolume work organized by a guest editor of international repute. The present series, which consists of five volumes (designated as Volume 22, Parts A through E) deals with an area that is experiencing a technological revolution and is destined to have a far-reaching impact in the near future — not only in the communications and data-processing fields, but also in numerous ancillary areas involving, for example, control systems, interconnects that maintain individual system isolation, and freedom from noise emanating from stray electromagnetic fields.

That the excitement engendered by the rapid pace of developments in lightwave communications technology is universal is borne out of the large number of contributions to this series by authors from abroad. It is indeed fortunate that W. T. Tsang, who is most highly knowledgeable in this field and has made so many personal contributions, has been able to take the time to put together a work of the extent and excellence of the present series. The treatise editors are also greatly indebted to Dr. Patel and the other colleagues of Dr. Tsang at AT&T Bell Laboratories, without whose understanding and encouragement this group of books would not have been possible.

R. K. WILLARDSON
ALBERT C. BEER

Foreword

Lightwave technology is breaking down barriers in communications in a manner similar to the way barriers in computing came down thanks to semiconductor integrated circuit technology. Increased packing densities of components on integrated circuit chips made possible a phenomenal amount of information processing capacity at continually decreasing cost. The impact of lightwave technology on communications is quite similar. We are reaching a point where an exponentially increasing transmission capacity is resulting in our capability to provide vast amounts of information to the most distant reaches of the world at a nominal cost. This revolution in information transmission capacity is engendered by the rapid developments in lightwave communications.

Along with the very large transmission capacity, predicted in the late fifties when the laser was invented, have come a number of additional advantages. Of these advantages, I single out those arising from the nonmetallic nature of the transmission medium. These fall under the broad category of what may be called an immunity from unanticipated electromagnetic coupling. The following rank as very important benefits: freedom from electromagnetic interference, absence of ground loops, relative freedom from eavesdropping (i.e., secure links), and potential for resistance to the electromagnetic pulse problems that plague many conventional information transmission systems utilizing metallic conductors as well as satellite and radio technology. Each one of these benefits arises naturally from the medium through which the light is propagated and is, therefore, paced by the progress in optical fibers.

However, what we take for granted today was not so obvious for many decades following the first practicable use of light for communications by Alexander Graham Bell in 1880. The use of heliographs in ancient Greece, Egypt, and elsewhere and the smoke signaling by various American Indian tribes notwithstanding, Bell's experiments on the use of sunlight for transmitting spoken sounds over a distance of a few hundred meters was undoubtedly the first step toward practical optical communications, since it represents a quantum jump in the increase in the bandwidth used for information transmission. The excitement he felt is keenly expressed in his words:

I have heard articulate speech produced by sunlight. I have heard a ray of sun laugh and cough and sing. I have been able to hear a shadow, and I have even perceived by ear the passing of a cloud across the sun's disk.

The results of his experiments were presented at a meeting of the American Association of Scientific Persons in Boston, Massachusetts. But the generally favorable reaction to Bell's photophone in the popular press was tempered with some skepticism. The following paragraph is taken from an article that appeared on the editorial pages of the August 30, 1880, issue of the *New York Times,* which reported on Bell's results.

What the telephone accomplishes with the help of a wire the photophone accomplishes with the aid of a sunbeam. Professor Bell described his invention with so much clearness that every member of the American Association must have understood it. The ordinary man, however, may find a little difficulty in comprehending how sunbeams are to be used. Does Professor Bell intend to connect Boston and Cambridge, for example, with a line of sunbeams hung on telegraph posts, and, if so, of what diameter are the sunbeams to be, and how is he to obtain them of the required size? . . .

Bell reported optical communication through free atmosphere, but the reporter, unintentionally, seemed to have foreseen the time when optical-fiber cables would be strung from pole to pole or buried underground.

A unique set of circumstances and a host of advances resulting from extensive interdisciplinary efforts have fueled the revolution in lightwave communications and the acceptance of this new technology. The tremendous progress in lightwave communications is a result of necessity as well as of the response of the scientists and engineers to the formidable challenges. The large bandwidth possible with lightwave communications is a direct result of the very high carrier frequency of electromagnetic radiation in the optical region. This advantage was recognized at least as early as the late fifties and early sixties. Yet almost fifteen years elapsed before lightwave communications technology became economically viable. Two primary components of the communications technology paced this development: the light source and the transmission medium. A third component, the receiver, is also important but was not the pacing one in the early years of development of lightwave systems.

The laser was invented in 1958, and within a very few years laser action was demonstrated in a variety of solids, liquids, and gases. The semiconductor injection laser, the workhorse of contemporary optical communications, was invented in 1962, but its evolution to a practical transmitter in a lightwave system took another eight years. In 1970 Hayashi and Panish (and, independently, Alferov in the Soviet Union) demonstrated the first continuous wave (cw) room-temperature-operated semiconductor laser. The potentials of small size, high reliability, low cost, long life, and ability to modulate the light output of the semiconductor laser at very high rates by merely modulating the drive current were recognized early in the game. With the demonstration of the cw room-temperature operation the race was on to exploit all these advantages.

Again, while laser light propagation through the atmosphere was considered in the mid-sixties, everyone recognized the limitations due to unpredictable and adverse weather conditions. To avoid these limitations, propagation in large hollow pipes was also studied, but again practical difficulties arose. It was the development of optical fiber technology to reduce transmission losses to acceptable levels that has led to the practical implementation of lightwave communications. While light transmission through very small-diameter fibers was demonstrated in the early fifties, it was a combination of theoretical advances by Kao and inventive experimentation by Maurer in the late sixties that resulted in the realization of 20-dB/km fiber. Additional fuel was thus provided to speed up the revolution.

Today, new records are continually being set for the longest and the highest-capacity lightwave communications system. Yet these records are thousands of times below the fundamental bandwidth limits set by the carrier frequency of the optical radiation on the rate of information transmission. Furthermore, from very fundamental considerations of light-transmitting materials, there is no reason why the currently achieved lowest losses for optical fibers, in the region of 0.1 dB/km at 1.55 μm, will not be considered too high in the future. It is not inconceivable that fiber losses as low as 10^{-4} dB/km may someday be achieved. It does not take a great deal of imagination to realize the impact of such development.

This is where we are. What future developments will pace the exploitation of lightwave communications? The five-volume mini-treatise on lightwave communications technology aims both to recapitulate the existing developments and to highlight new science that will form the underpinnings of the next generation of technology. We know a lot about how to transmit information using optical means, but we know less than enough about how to switch, manipulate, and process information in the optical domain. To take full advantage of all the promise of lightwave communications, we have to be able to push the optical bits through the entire communications system with the electronic-to-optical and optical-to-electronic interfaces only at the two ends of the lightwave communications system. To achieve this, we will need practical and efficient ways of switching, storing, and processing optical information. This is a must before lightwave communications is able to touch every single subscriber of the present telephone and other forms of communications technology.

We have come a long way since Bell's experiments of 1880, but there is a lot more distance ahead. That is what the field of lightwave communications is all about — more challenges, more excitement, more fun for those who are the actors, and a greater opportunity for society to derive maximum benefit from the almost exponentially increasing information capacity of lightwave systems.

AT&T Bell Laboratories   C. K. N. PATEL
October 9, 1984

Preface

When American Indians transmitted messages by means of smoke signals they were exploiting concepts at the heart of modern optical communications. The intermittent puffs of smoke they released from a mountaintop were a digital signal; indeed, the signal was binary, since it encoded information in the form of the presence or absence of puffs of smoke. Light was the information carrier; air was the transmission medium; the human eye was the photodetector. The duplication of a signal at a second mountaintop for the transmission to a third served as signal reamplification, as in today's electronic repeater. Man had devised and used optical communications even long before the historic event involving the "photophone" used over a hundred years ago (1880) by Alexander Graham Bell to transmit a telephone signal over a distance of two hundred meters by using a beam of sunlight as the carrier. It was not until 1977, however, that the first commercial optical communications system was installed. Involved in the perfection of this new technology are the invention and development of a reliable and compact near-infrared optical source that can be modulated by the information-bearing signal, a low-loss transmission medium that is capable of guiding the optical energy along it, and a sensitive photodetector that can recover the modulation error free to re-treat the information transmitted.

The invention and experimental demonstration of a laser in 1958 immediately brought about new interest and extensive research in optical communications. However, the prospect of practical optical communications brightened only when three major technologies matured. The first technology involved the demonstration of laser operation by injecting current through a semiconductor device in 1962 and the achievement of continuous operation for over one million hours in 1977. The second technology involved the attainment of a 20-dB/km doped silica fiber in 1970, the realization that pure silica has the lowest optical loss of any likely medium, the discovery in 1973 that suitably heat-treated, boron-doped silica could have a refractive index less than that of pure silica, and the recent achievement of an ultralow loss of 0.12 dB/km with Ge-doped silica-based fibers. The third technology is the development of low-noise photodetectors in the 1970s, which made possible ultrahigh-sensitivity photoreceivers. It is the simulta-

xv

neous achievement of reliable semiconductor current-injection lasers, low loss in optical fibers, and low-noise photodetectors that thrusts lightwave communications technology into reality and overtakes the conventional transmission systems employing electrical means.

Since optical-fiber communications encompasses simultaneously several other technologies, which include the systems area of telecommunications and glass and semiconductor optoelectronics technologies, a tremendous amount of research has been conducted during the past two decades. We shall attempt to summarize the accumulated knowledge in the present series of volumes of "Semiconductors and Semimetals" subtitled "Lightwave Communications Technology." The series consists of seven volumes. Because of the subject matter, the first five volumes concern semiconductor optoelectronics technology and, therefore, will be covered in "Semiconductors and Semimetals." The last two volumes, one on optical-fiber technology and the other on transmission systems, will be covered in the treatise "Optical Fiber Communications," edited by Tingye Li and W. T. Tsang.

Volume 22, Part A, devoted entirely to semiconductor growth technology, deals in detail with the various epitaxial growth techniques and materials defect characterization of III–V compound semiconductors. These include liquid-phase epitaxy, molecular-beam epitaxy, atmospheric-pressure and low-pressure metallo-organic chemical vapor deposition, and halide and chloride transport vapor-phase deposition. Each technique is covered in a separate chapter. A chapter is also devoted to the treatment of materials defects in semiconductors.

In Volume 22, Parts B and C, the preparation, characterization, properties, and applications of semiconductor current-injection lasers and light-emitting diodes covering the spectral range of 0.7 to 1.6 μm and above 2 μm are reviewed. Specifically, Volume 22, Part B, contains chapters on dynamic properties and subpicosecond-pulse mode locking, high-speed current modulation, and spectral properties of semiconductor lasers as well as dynamic single-frequency distributed feedback lasers and cleaved-coupled-cavity semiconductor lasers. Volume 22, Part C, consists of chapters on semiconductor lasers and light-emitting diodes. The chapters on semiconductor lasers consist of a review of laser structures and a comparison of their performances, schemes of transverse mode stabilization, functional reliability of semiconductor lasers as optical transmitters, and semiconductor lasers with wavelengths above 2 μm. The treatment of light-emitting diodes is covered in three separate chapters, on light-emitting diode device design, its reliability, and its use as an optical source in lightwave transmission systems. Volume 22, Parts B and C, should be considered as an integral treatment of semiconductor lasers and light-emitting diodes rather than as two separate volumes.

Volume 22, Part D, is devoted exclusively to photodector technology. It includes detailed treatments of the physics of avalanche photodiodes; avalanche photodiodes based on silicon, germanium, and III–V compound semiconductors; and phototransistors. A separate chapter discusses the sensitivity of avalanche photodetector receivers for high-bit-rate long-wavelength optical communications systems.

Volume 22, Part E, is devoted to the area of integrated optoelectronics and other emerging applications of semiconductor devices. Detailed treatments of the principles and characteristics of integrable active and passive optical devices and the performance of integrated electronic and photonic devices are given. A chapter on the application of semiconductor lasers as optical amplifiers in lightwave transmission systems is also included as an example of the important new applications of semiconductor lasers.

Because of the subject matter (although important to the overall treatment of the entire lightwave communications technology), the last two volumes will appear in a different treatise. The volume on optical-fiber technology contains chapters on the design and fabrication, optical characterization, and nonlinear optics in optical fibers. The final volume is on lightwave transmission systems. This includes chapters on lightwave systems fundamentals, optical transmitter and receiver design theories, and frequency and phase modulation of semiconductor lasers in coherent optical transmission systems.

Thus, the series of seven volumes treats the entire technology in depth. Every author is from an organization that is engaged in the research and development of lightwave communications technology and systems.

As a guest editor, I am indebted to R. K. Willardson and A. C. Beer for having given me this valuable opportunity to put such an important and exploding technology in "Semiconductors and Semimetals." I am also indebted to all the contributors and their employers who have made this series possible. I wish to express my appreciation to AT&T Bell Laboratories for providing the facilities and environment necessary for such an endeavor and to C. K. N. Patel for preparing the Foreword.

CHAPTER 1

Mode Locking of Semiconductor Lasers

J. P. van der Ziel

AT&T BELL LABORATORIES
MURRAY HILL, NEW JERSEY

I. Introduction

An aspect common to all laser work has been the generation of short pulses. In most cases the thresholds that were attainable, at least initially, were sufficiently high to allow only pulsed operation at low duty cycles. Pulsed operation enables one to generate high peak powers that have a wide range of application, examples of which are the studies of optical nonlinearities and time evolution of luminescence following pulsed excitation. It has proven to be extremely important and fruitful to further extend the regime of short pulse generation to shorter pulse widths and to higher repetition rates.

The shortest pulses have been obtained by mode locking the laser in an external cavity. The mode locking of semiconductor lasers is the topic of this chapter. The experimental studies of semiconductor mode locking have been briefly reviewed by Au Yeung and Johnston (1982a) and Inaba (1983). The initial report of mode locking resulted in 23-psec-long pulses at 3-GHz repetition rates from an AlGaAs-type laser (Ho *et al.*, 1978a,b). Mode locking of an InGaAsP-type laser at 1.21 μm yielded pulses as short as 18 psec at a 2.1-GHz repetition rate (Glasser, 1978). Refinements in the technique and in laser technology have resulted in 0.5–0.6-psec-long pulses (van der Ziel *et al.*, 1981b,c; Yokoyama *et al.*, 1982). The reduction in pulse widths by a factor of approximately 40 is a remarkable achievement in such a short time span. Further large reductions in pulse widths will be more difficult to

1

achieve. Since the mode-locked pulses are transform limited, shorter pulses are accompanied by an increasing spectral bandwidth. The large dispersion of the group velocity in the semiconductor laser and the bandwidth of the gain results in a natural limitation of the pulse width.

Prior to 1978, a number of attempts had been made to achieve mode locking of semiconductor lasers. Optical modulation corresponding to the round-trip transit time was observed in a number of cases (Morozov *et al.*, 1968; Mohn, 1969a,b; Broom, 1969; Broom *et al.*, 1970; Paoli and Ripper, 1970; Harris, 1971; Morikawa *et al.*, 1976; Risch and Voumard, 1977; Risch *et al.*, 1977; Ikushima and Maeda, 1978; Salathe, 1979). By introducing a dispersive Fabry–Perot element in the external resonator, the emission spectrum was reduced to a single mode and the self-pulsing behavior was eliminated (Voumard, 1977). The noise properties induced by the reflected waves of an external cavity have also been studied (Hirota and Suematsu, 1979). The mode-locking work suffered from the relatively poor quality of lasers available at that time, as well as the lack of capability for accurately measuring pulses of less than 200-psec duration by second-harmonic auto-correlation and streak camera techniques.

The short pulse widths currently available have outstripped the present needs for optical communication. The mode-locking technique, when coupled with external modulators and a multiplexer, appears to be a promising source of lightwave radiation for future systems that require short pulse widths and high bit rates. There are several techniques that yield somewhat longer pulses. Direct modulation of the gain by sinusoidal microwave or short pulse current injection results in relaxation oscillations if the total applied current passes through the threshold. By operating at a sufficiently high frequency or short pulse width, and over a limited current range, single pulses of 20–50-psec duration can be obtained. These techniques are discussed in Chapter 2. The theory of active and passive mode locking is described in Part II of this chapter, and the experimental work is discussed in Part III.

II. Theory of Semiconductor Laser Mode Locking

1. THE MODE-LOCKING PROCESS

There exists a considerable amount of earlier theoretical work on the generation of short pulses by mode locking of solid-state and dye lasers that is applicable to semiconductor laser mode locking. Two distinct, but complementary, approaches have been used to analyze mode-locking behavior. The models correspond to the spectral and time analysis of the mode-locking process, respectively. In the spectral analysis, an ensemble of standing wave

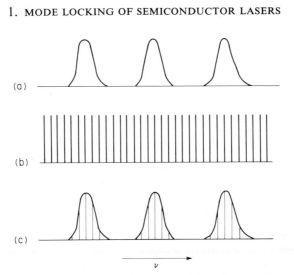

FIG. 1. Laser modes with an external cavity. (a) Laser Fabry–Perot modes spaced at Δv_L, (b) equivalent external cavity modes spaced at Δv_m, and (c) composite cavity modes.

cavity modes of the laser and the external cavity are considered (Fig. 1). The frequency spacing of modes v_m and v_{m+1} (where m is the longitudinal mode index of the composite cavity) is

$$\Delta v_m = c \left/ \left(2 \sum_i n_i l_i \right), \right. \tag{1}$$

where c is the velocity of light and $\sum_i n_i l_i$ the total optical length of the composite cavity with regions of length l_i and refractive index n_i.

A nonlinearity such as a resonant time-dependent gain or a saturable absorption induces polarization sidebands spaced Δv_m from the original modes. The presence of a polarization amplitude at the frequency corresponding to the mode spacing produces a transfer of energy between the modes and induces a fixed; coherent phase relationship between the modes. The locking of the phases extends over a finite spectral full width at half-maximum (FWHM) of Δv corresponding to the locking of

$$N = \Delta v / \Delta v_m \tag{2}$$

modes of the composite cavity.

The duration of the pulses results from the interference between the standing wave cavity modes. The pulse FWHM is the transform limit of the locked modes

$$\tau_p = a(\Delta v)^{-1}, \tag{3}$$

where a is a numerical factor that is somewhat dependent on the pulse shape.

In the second approach to mode-locking theory, the mode-locked standing waves of the cavity are decomposed into traveling waves. The interference of these modes results in a radiation pulse of width τ_p propagating in the cavity. The theory describes how the pulse envelope is modified as it propagates through the cavity. In the steady state, the envelope is replicated after each transit. The losses due to mirror transmission and internal absorption and scattering are made up by the gain. The dispersive effects resulting from the finite bandwidth and dispersion are compensated for by a time delay resulting in phase shifts within the pulse.

The mode-locking process is a mode of operation quite different from that obtained if the lasers are directly modulated by a sinusoidal or pulsed current. In the latter excitation scheme, the dynamic response of the laser photon and carrier density equations produces relaxation oscillations that build up from the spontaneous emission level and can be as short as 30–50 psec. When the laser is mode locked, it is the internal radiation pulse circulating in the cavity that, on its return to the active medium, causes an intense burst of stimulated emission from the inverted population. Thus, the pulse needs only to be amplified from the initial value reduced by the round-trip losses. The laser photon and carrier density equations are less important in determining the pulse widths. This is clearly illustrated for the case of passive mode locking with a slow, saturable absorber in which there is a brief period of net gain resulting from the delicate balance of the depletion of the gain and the saturation of the absorber, which depends on the intensity of the optical pulse as it returns to the active medium. The observation of subpicosecond pulses from passively mode-locked lasers indicates that these processes can be two orders of magnitude shorter than the relaxation processes present in the photon and carrier density equations.

2. MODEL FOR MODE LOCKING

Several theories of active and passive mode locking of semiconductor lasers have been reported (Haus and Ho, 1979; Aspin and Carroll, 1979; Glasser, 1980; Haus, 1980a,b, 1981; van der Ziel, 1981b; Au Yeung, 1981; Au Yeung and Johnston, 1981). The origins of the coupling leading to pulse shortening include sinusoidal modulation of the carrier density (and, hence, the gain), the depletion of the gain owing to stimulated emission by the reinjected pulse, and the presence of a region of saturable absorption.

A model for semiconductor mode locking that is based on a pulse propagating in a cavity is developed here. The optical gain is derived from the carrier density equations of the laser. Expressions for the pulse widths are obtained by expanding the gain and loss over time around the peak of the pulse. The theory is in good agreement with the experimental observations presented in Part III.

The pulse repetition rate for semiconductor mode locking is typically in the 0.5–2-GHz range. This rate is not arbitrary but is dictated by the time response of the gain recovery during the interval between the pulses. In the absence of intense stimulated emission, the gain recovery is determined by the carrier injection rate and the electron lifetime. The latter is typically ≈ 2 nsec, and significant but incomplete recovery of the gain is achieved during the interval between the pulses. At very low repetition rates, the gain builds up to a high level that results in a high level of spontaneous emission and self-pulsing yielding wide pulses. At very high repetition rates, the gain has insufficient time to recover to a sufficiently high value and the resulting pulses are relatively weak and broad. This limit is relaxed when the injected current is much greater than the threshold current.

The saturable absorption present in semiconductor lasers has a time response that is fast compared with the pulse repetition rate but is assumed to be long compared with the pulse width. The latter defines the absorber as a slow, saturable absorber (New, 1972, 1974; Haus, 1975a) in contrast to a fast, saturable absorber (Haus, 1975b), in which the absorber can recover in a short time compared with the pulse width. This is particularly advantageous since mode locking with a slow absorber is much less critical than with a fast absorber and results in shorter pulses.

The theory of active mode locking is derived from the work of Haus (1981) and van der Ziel (1981b), which in turn followed the theory of synchronously mode-locked dye lasers (Auschnitt and Jain, 1978; Auschnitt *et al.,* 1979; Haus, 1975a–c, 1976). The theory of passive mode locking is derived from the work of Haus (1981).

a. The Pulse Envelope Equation

The mode-locking process is described by the time dependence of the pulse envelope function. For simplicity, a ring cavity geometry, shown in Fig. 2, is assumed with a single pulse propagating in one direction only (Haus, 1981). The optical electric field of the pulse, after m passes through

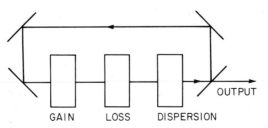

FIG. 2. Schematic diagram of the effect of the material parameters on a unidirectional pulse propagating in an external ring cavity resonator.

the cavity, is

$$E_m(t) = v_m(t) \exp(j\omega_0 t) + \text{c.c.,} \tag{4}$$

where j is the imaginary number, ω_0 the center frequency, $v_m(t)$ the time-dependent envelope function and c.c. the complex conjugate. The ring cavity contains a region of gain $G(t)$ of length l_g and a lossy region of absorption described by $L(t)$ of effective length l_l consisting of a fixed loss, a spectrally dependent loss and a saturable loss. Dispersion effects are included in a phase factor ϕ. The change in the pulse envelope occurring in a single transit can be expressed in terms of the operator

$$v_{m+1}(t) = D \exp(Gl_g - Ll_l - j\phi)v_m(t) \tag{5}$$

acting on the mode envelope. The operator D delays the pulse by a round-trip transit time T_R. The net change in the pulse envelope is assumed to be small; hence, the exponential function in Eq. (5) can be expanded to the first order yielding

$$v_{m+1}(t) \approx D(1 + Gl_g - Ll_l - j\phi)v_m(t) \approx D(1 + O)v_m(t), \tag{6}$$

where

$$O = Gl_g - Ll_l - j\phi \tag{7}$$

is the operator describing the changes induced in $v_m(t)$ by the medium in one transit.

At equilibrium the pulses must be replicated after each transit. The dynamic properties of the active medium in the cavity produce a reshaping of the pulse. Consequently, the round-trip transit time is changed slightly from the transit time in the absence of pulse shaping, which is defined by T_R, by a small time difference δT_R. The time shift enters as an unknown parameter and is present in both active and passive mode locking. With the time shift, the replication requirement can be written as

$$v_{m+1}(t) \approx Dv_m(t - \delta T_R) \approx D[1 - \delta T_R(\partial/\partial T)]. \tag{8}$$

Equating Eqs. (5) and (8) yields the operator equation describing the effect of mode locking on the mode amplitude

$$[O + \delta T_R(\partial/\partial t)]v_m(t) = 0. \tag{9}$$

The operator O is a time-dependent function. For both active and passive mode locking, O contains the internally induced time dependence owing to the reduction in gain by the stimulated emission and the bleaching of the saturable absorption. The operator also contains the explicit time dependence of the gain from the gain recovery and the external modulation. Equation (9) has been solved for the case of active mode locking resulting in

Hermite–Gaussian pulses (Haus, 1981; van der Ziel, 1981b) and for passive mode locking assuming a slow, saturable absorber yielding $sech^2(t/t_p)$-shaped pulses (Haus, 1981). In the following sections, the case of gain modulation is treated in the presence of a saturable absorber. Both the gain and loss terms have a finite spectral bandwidth. In the following treatment, it is assumed that the loss term contains the smallest bandwidth-limiting term. This is clearly the case when there is residual Fabry–Perot reflection from the interior laser–external cavity interface or if there is an additional bandwidth-limiting element inserted in the external cavity.

b. The Loss Function

The loss is considered to consist of three components:

(a) A fixed time-independent loss L_0 describing the mirror losses, the internal scattering losses in the laser and external cavity, and the imperfect coupling to the external cavity. The loss defines the quality factor of the cavity Q as

$$L_0 = \omega_0 T_R/2Q. \tag{10}$$

(b) A spectrally dependent loss describing the effect of a filtering element extending over an effective length l_s. A Lorentzian loss profile in the frequency domain centered at ω_0 is assumed such that

$$L_s l_s = L_{s0} l_s \left(1 - \frac{1}{1 + j(\omega - \omega_0)/\omega_s}\right), \tag{11}$$

where ω_s is the bandwidth. The loss operates on the Fourier transform of $v_m(t)$, which is $v_m(\omega - \omega_0)$. For small frequency deviations about ω_0, the expansion of Eq. (11) yields the parabolic dependence on angular frequency to be

$$L_s l_s = L_{s0} l_s \{j[(\omega - \omega_0)/\omega_s] + [(\omega - \omega_0)/\omega_s]^2\}. \tag{12}$$

Hence, in the time domain the spectrally dependent loss can be expressed in terms of the derivative operator

$$L_s l_s = L_{s0} l_s \left(\frac{1}{\omega_s}\frac{d}{dt} - \frac{1}{\omega_s^2}\frac{d^2}{dt^2}\right), \tag{13}$$

where the first derivative term corresponds to a time shift of the pulse and the second derivative is a diffusion term causing a spreading of the pulse over time.

(c) An optically intensity-dependent loss term describing the saturable absorption L_a that extends over a length l_a. For an equivalent two-level system, the saturable loss due to a photon density P is given by

$$dL_a/dt = (L_{a0} - L_a)/\tau_a - c\sigma_a L_a P, \tag{14}$$

where the relaxation time of the absorber τ_a is assumed to be long relative to the pulse width but short compared with T_R so that at the arrival of the pulse the absorber has fully recovered. The absorption cross section is σ_a, and L_{a0} is the low power loss of the absorber.

The photon density is given by

$$P(t) = |v(t)|^2/Ac, \tag{15}$$

where A is the area of the beam in the laser and c the velocity of light. The total loss can be expressed as

$$\frac{\omega_0 T_R}{2Q}\left[1 + L_s\left(\frac{1}{\omega_s}\frac{d}{dt} - \frac{1}{\omega_s^2}\frac{d^2}{dt^2}\right) + L_a(t)\right], \tag{16}$$

where

$$L_s = L_{s0}l_s(2Q/\omega_0 T_R) \tag{17}$$

is the spectrally dependent loss and

$$L_a = L_a(t)l_a(2Q/\omega_0 T_R) \tag{18}$$

is the saturable absorption loss.

c. The Gain Function

The relationship between the gain of the laser, the injected current density, and the optical pulses is shown schematically in Fig. 3 for the case of sinusoidal current modulation. The optical pulses are short compared to the repetition rates, and it is convenient to choose the origin of the time base at the peak of one of the pulses. A typical cycle begins at the end of a pulse, i.e., at approximately $-T_R$ at which point the carrier density has been reduced by stimulated emission of the pulse to a level below the threshold. During the relatively long period of time from $t \approx -T_R$ to $t \approx 0$, there is negligible stimulated emission, and the gain increases by the direct and rf current injection.

The carrier density is given by

$$dn/dt = (1/eV)[I_0 + I_{rf}\cos(\omega_m t + \theta)] - (n/\tau_e) - \zeta^{-1}G(t)P(t), \tag{19}$$

where I_0 and I_{rf} are, respectively, the direct and rf currents injected into the active region of volume V, ω_m the rf angular frequency, θ the phase angle of the rf current relative to the center of the pulse, and τ_e the electron lifetime.

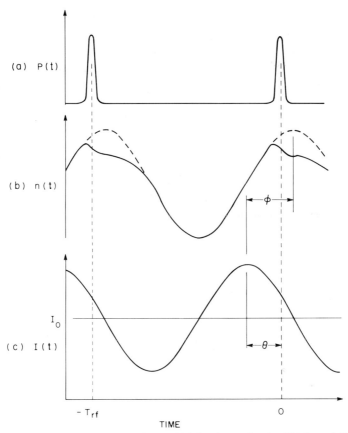

FIG. 3. Schematic of the time dependence of (a) the photon density $P(t)$ through the active medium, (b) the carrier density $n(t)$, and (c) the applied current $I(t)$. The decrease in $n(t)$ due to stimulated emission is represented as the difference between the dashed and solid lines in curve (b). The time origin ($t = 0$) is selected to be at the peak of the current pulse; θ is the phase shift of the peak of the current density relative to $t = 0$, and ϕ is the phase lag of the carrier density relative to the current density. [From van der Ziel (1981b).]

The last term gives the effect of the stimulated emission. The gain is given by

$$G(t) = c\sigma_g[n(t) - n_g]\zeta, \tag{20}$$

where for typical lasers $\sigma_g \approx 1.3 - 1.6 \times 10^{-16} \ \mathrm{cm}^2$ is the cross section for stimulated emission and $n_g \approx 1.5 \times 10^{18} \ \mathrm{cm}^{-3}$ is the carrier density required to overcome the internal losses. The factor

$$\zeta = (l_g/c)(2Q/\omega_0 T_R) \tag{21}$$

relates the round-trip gain to the time-independent round-trip loss.

d. The Phase Factor

The phase factor ϕ in Eq. (7) is frequency dependent and can be expanded in the frequency domain to yield

$$\phi = \phi_0 + \frac{d^2k}{d\omega^2} l_\phi (\omega - \omega_0)^2. \tag{22}$$

Here ϕ_0 is the frequency-independent phase shift, and the second term gives the dispersion of the group velocity in a medium of wave vector k and length l_ϕ. Unless additional strongly dispersive elements such as a grating are introduced in the external cavity, the most dispersive element is the laser itself. For this case $l_\phi = l_g$. The first derivative of ϕ yields a time delay and has been absorbed in other terms. The Fourier transform of Eq. (22) yields the phase as a time-dependent operator

$$\phi = \phi_0 - \frac{d^2k}{d\omega^2} l_\phi \frac{d^2}{dt^2}. \tag{23}$$

e. The Mode-Locking Equation

Using the expressions for the gain, loss, and phase shift, the operator equation [Eq. (9)] acting on the mode amplitude and written in terms of the normalized quantities is

$$\left[G(t) - 1 - L_s \left(\frac{1}{\omega_s} \frac{d}{dt} - \frac{1}{\omega_s^2} \frac{d^2}{dt^2} \right) - L_a(t) \right.$$
$$\left. - j \frac{2Q}{\omega_0 T_R} \left(\phi_0 - \frac{d^2k}{d\omega^2} l_\phi \frac{d^2}{dt^2} \right) - \frac{\delta}{\omega_s} \frac{d}{dt} \right] v(t) = 0, \tag{24}$$

where the time-delay term [Eq. (8)] is given by

$$\delta = -\omega_s (2Q/\omega_0 T_R) \delta T. \tag{25}$$

Equation (24) is qualitatively similar to the expression given by Haus (1981) even though the bandwidth-limiting element and the modulation term have different origins.

The solution of Eq. (24) is complex in the case of strong active and passive mode locking due to the time-dependent gain and loss terms, both of which change significantly during the emission of the pulse. An approximate solution, obtained from a series expansion in time about the center of the pulse, is obtained in Section 2,g, and 2,h. Considerable insight into the mode-locking process is obtained for the situation in which the gain change is small and the saturable absorption term is absent. This effect is associated with weak mode locking and is treated in Section 2,f.

f. Weak Mode Locking

The limit of weak mode locking is obtained when the stimulated emission during the pulse does not significantly deplete the gain (Haus, 1981). The pulse width-limiting effect is due to modulation rather than gain depletion by the stimulated emission. The stimulated emission term in Eq. (19) can then be neglected, and the carrier density is obtained by integration. For direct and rf current excitation at the radial modulation frequency ω_m

$$n(t) = (\tau_e/eV)[I_0 + I_{rf}(1 + \omega_m^2\tau_e^2)^{-1/2}\cos(\omega_m t + \theta - \psi)]. \quad (26)$$

The carrier density has a sinusoidal time dependence that lags the injected current density by the phase angle ψ given by

$$\tan\psi = \omega_m\tau_e. \quad (27)$$

The maximum carrier density occurs at $t = 0$ when $\theta = \psi$. By using Eq. (26) in Eq. (20), the gain can be written as

$$G(t) = g\{1 - 2M[1 - \cos(\omega_m t)]\}, \quad (28)$$

where g is the peak gain at $t = 0$ and M the modulation depth. Expanding the cosine term for a small time interval about $t = 0$ yields

$$G(t) = g(1 - M\omega_m^2 t^2). \quad (29)$$

The first-derivative terms in Eq. (24) cause the pulse to shift in time. For resonant modulation, the first-derivative terms cancel; that is, the time shift of the pulse owing to the loss is compensated for by the time-delay term. From Eq. (24) the resonance modulation condition is

$$L_s + \delta = 0 \quad (30)$$

or

$$\delta T_R = (\omega_0 T_R/2Q)(L_s/\omega_s). \quad (31)$$

The second-derivative terms may be combined to form a complex frequency factor

$$\frac{1}{\omega_D^2} = \frac{L_s}{\omega_s^2} + j\frac{2Q}{\omega_0 T_R}\frac{d^2\mathbf{k}}{d\omega^2}l_\phi \quad (32)$$

that gives the dispersive contribution to the pulse. With these simplifications the mode-locking equation becomes

$$\left(g - 1 - j\frac{2Q}{\omega_0 T_R}\phi_0 + \frac{1}{\omega_D^2}\frac{d^2}{dt^2} - \frac{L_s + \delta}{\omega_s}\frac{d}{dt} - gM\omega_m^2 t^2\right)v(t) = 0. \quad (33)$$

The solutions are Hermite–Gaussian functions. When the first-derivative term is nonzero, the solutions are shifted in time from $t = 0$. It can be shown that the Hermite–Gaussian functions of order higher than zero have a lower gain and are, in general, not excited (Haus, 1981). The lowest-order solution is the Gaussian

$$v(t) = \exp(-\tfrac{1}{2}\omega_p^2 t^2 + \alpha t), \tag{34}$$

where ω_p and α are obtained by substituting the solution into Eq. (33) yielding three eigenvalue equations that correspond to the t^0, t^1, and t^2 terms

$$g - 1 - j\frac{2Q}{\omega_0 T_R}\phi_0 = \frac{\omega_p^2}{\omega_D^2} + \left[\frac{(L_s + \delta)}{2}\frac{\omega_D}{\omega_s}\right]^2, \tag{35}$$

$$\alpha = \frac{\omega_D^2}{2}\left(\frac{L_s + \delta}{\omega_s}\right), \tag{36}$$

and

$$\omega_p = (gM)^{1/4}(\omega_m \omega_D)^{1/2}. \tag{37}$$

The pulse envelope has a Gaussian time dependence and the FWHM is given by

$$\tau_p = 2\sqrt{\ln 2}/\omega_p. \tag{38}$$

The pulse width varies inversely with the fourth root of the gain modulation depth product and inversely with the square root of the product of the effective bandwidth and modulation frequency. In the presence of dispersion, ω_D is complex, and the imaginary component of ω_p in Eq. (37) produces a frequency chirping during the pulse. The detuning of the modulation frequency from the resonance conditions [Eq. (30)] produces a nonzero first-derivative term in Eq. (33) and introduces the time shift given by Eq. (36) away from the peak of the gain curve. The gain required for mode locking is obtained from the real part of Eq. (35),

$$g - 1 = \mathrm{Re}\left[\left(\frac{\omega_p}{\omega_D}\right)^2 + \left(\frac{L_s + \delta}{2}\frac{\omega_D}{\omega_s}\right)^2\right]. \tag{39}$$

The net gain has a minimum value at resonance

$$g - 1 = (gM)^{1/2}\omega_m\,\mathrm{Re}(1/\omega_D) \tag{40}$$

and increases with detuning. The phase shift is given by the imaginary part of Eq. (35),

$$\phi_0 = -\frac{\omega_0 T_R}{2Q}\,\mathrm{Im}\left[\left(\frac{\omega_p}{\omega_D}\right)^2 + \left(\frac{L_s + \delta}{2}\frac{\omega_D}{\omega_s}\right)^2\right] \tag{41}$$

and is zero in the absence of dispersion corresponding to real ω_D.

The spectral width of the Gaussian pulse amplitude [Eq. (34)] corresponds to a Gaussian spectral intensity of

$$E(\omega)^2 = \exp\left[-\left(\frac{\omega - \omega_0}{\omega_p}\right)^2 + \frac{\alpha^2}{\omega_p}\right]. \tag{42}$$

The spectrum has a FWHM intensity of

$$\delta\omega = 2\sqrt{\ln 2}\,\omega_p. \tag{43}$$

The transform limit of the product of the bandwidth and pulse FWHM is

$$\Delta\nu\,\tau_p = (2/\pi)\ln 2 = 0.441. \tag{44}$$

A comparison of the experimental value of $\Delta\nu\,\tau_p$ with Eq. (44) is frequently cited as a measure of how close the observed pulse approaches the transform limit. The transform limits of the bandwidth-pulse width product are a function of the pulse shape. The products for several pulse shapes common to mode locking are given in Table I. In addition, the relationship between the pulse FWHM and the FWHM of the second-order autocorrelation function obtained by second-harmonic generation is also given.

TABLE I

SECOND-ORDER AUTOCORRELATION WIDTHS AND
WIDTH PRODUCTS FOR FIVE TRANSFORM-LIMITED
PULSE SHAPES[a,b]

$P(t)$	$\tau_p/t_{1/2}$	$\Delta\nu\,\tau_p$
Gaussian		
$\exp(-4\ln 2\,t^2/\tau_p^2)$	0.7071	0.4413
Hyperbolic sech		
$\mathrm{sech}^2(1.7627t/\tau_p)$	0.6482	0.3148
One-sided exponential		
$\exp(t\ln 2/\tau_p),\ t \geq 0$	0.5	0.1103
Symmetric two-sided exponential		
$\exp\left(-\dfrac{2\,\lvert t\rvert\ln 2}{\tau_p}\right)$	0.4130	0.1420
Lorentzian		
$[1 + (2t/\tau_p)^2]^{-1}$	0.500	0.2206

[a] Data from Ippen and Shank (1977) and Sala et al. (1980).
[b] t_p and $\tau_{1/2}$ are the FWHM of $P(t)$ and the autocorrelation, respectively, and $\Delta\nu$ is the FWHM of the frequency spectrum.

g. *Intense Mode Locking with Saturable Absorption*

As the current is increased to a level well above the threshold value, the stimulated emission owing to the propagating pulse produces a significant reduction in the carrier density and gain as the pulse passes through the laser. The relationship between the optical pulse, gain, and sinusoidal current is shown schematically in Fig. 3. For illustrative purposes, the situation is assumed in which the gain recovers fully to its equilibrium value in the interval between the pulses. The electron lifetime τ_e is 2–3 nsec, and the recovery of the gain is incomplete for modulation rates is excess of several hundred megahertz. It will be shown later that at higher frequencies the stimulated emission produces a reduction in the average gain, leading to an approximate clamping of the gain. The time-dependent gain during the emission of the pulse is shown schematically in Fig. 4. There is a reduction in the gain of ΔG during the pulse. The sinusoidal dependence [Eq. (20)] is

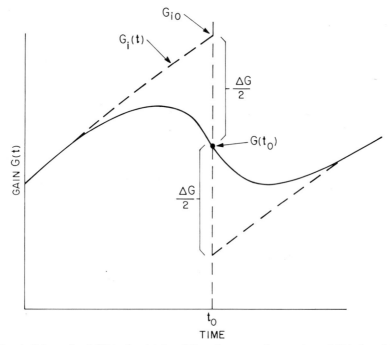

FIG. 4. Schematic of $G(t)$ in the vicinity of the pulse centered at t_0, where $G_i(t)$ is the gain in the absence of stimulated emission and ΔG the reduction in gain due to the emission of the pulse. [From van der Ziel and Logan (1982). Generation of short optical pulses in semiconductor lasers by combined dc and microwave current injection. *IEEE J. Quantum Electron.* **QE-18,** 1340. © IEEE 1982.]

clearly inadequate for describing the gain. Under these conditions, the trailing edge of the pulse is terminated by the reduction in the gain to a value below the threshold owing to the stimulated emission rather than by the trailing edge of the sinusoidal gain. Thus Eq. (29) is not applicable, and the pulse widths will be reduced from the values obtained from Eq. (37).

An approximate solution to the mode-locking equation is obtained by expanding the gain and loss in a second-order power series over time about the center of the pulse, which is defined to be at $t = 0$ (van der Ziel, 1981b). The solution takes into account the dynamic effect of the reduction in gain and the saturation of the absorption and is valid near the center of the pulse.

The widths of the pulses are assumed to be short compared to T_R. The carrier density in the absence of stimulated emission near $t = 0$ but including the effect of the previous pulses at $-mT_R$ (where m is an integer) is obtained by integrating Eq. (19) from $-T_R$ to $t \simeq 0$ yielding

$$n(t) = (\tau_e/eV)[I_0 + I_{rf}(1 + \omega^2\tau_e^2)^{-1/2} \cos(\omega t + \theta - \psi)]$$
$$- \Delta n \exp(-t/\tau_e)/[\exp(T_R/\tau_e) - 1]. \tag{45}$$

Equation (45) differs from Eq. (26) by the last term. This term gives the effect of the depletion of the gain by previous pulses and its incomplete recovery in the time between pulses.

A Gaussian mode amplitude [Eq. (34)] will be assumed, and the center of the pulse is defined at $t = 0$, hence $\alpha = 0$. The phase shift now accounts for the shift of pulse position with respect to the carrier and current densities. The optical photon density is

$$P(t) = P_0 \exp(-\omega_p^2 t^2), \tag{46}$$

and half of the integrated photon density is

$$E = \tfrac{1}{2}\sqrt{\pi} P_0/\omega_p. \tag{47}$$

The time dependence of the gain, obtained from Eqs. (19) and (20), is

$$dG/dt = (dG/dt)_i - c\sigma_e GP, \tag{48}$$

where $(dG/dt)_i$ denotes the change in the gain due to the current injection, spontaneous decay, and the effect of the previous pulses in reducing the gain. The last term gives the effect of stimulated emission of the pulse near $t = 0$. The power series of the gain over time, obtained by integrating Eq. (48), is

$$G(t) = G_{i0}(1 - c\sigma E) + (G'_{i0} - c\sigma G_{i0}P_0)t$$
$$+ \tfrac{1}{2}(G''_{i0} - (G'_{i0} - c\sigma_e G_{i0}P_0)c\sigma_e P_0)t^2 + \cdots, \tag{49}$$

where the primes denote differentiation with respect to time and the subscript 0 denotes evaluation at $t = 0$. The derivatives are

$$G_{i0} = \xi c \sigma_e \{(\tau_e/eV)[I_0 + I_{rf}(1 + \omega^2 t_e^2)^{-1/2} \cos(\theta - \psi)] - n_g$$
$$- \Delta n/[\exp(T_R/\tau_e) - 1]\}, \tag{50}$$

$$G_{i0}' = \xi c \sigma_e \{-(I_{rf} \omega \tau_e/eV)(1 + \omega^2 \tau_e^2)^{-1/2} \sin(\theta - \psi)$$
$$+ (\Delta n/\tau_e)/[\exp(T_R/\tau_e) - 1], \tag{51}$$

and

$$G_{i0}'' = \xi c \sigma_e \{-(I_{rf} \omega^2 \tau_e/eV)(1 + \omega^2 \tau_e^2)^{-1/2} \cos(\theta - \psi)$$
$$- (\Delta n/\tau_e^2)/[\exp(T_R/\tau_e) - 1]\}. \tag{52}$$

The power series of the saturable loss, obtained by integrating Eq. (14) near $t = 0$, is

$$L_a(t) = L_{a0}(1 - c\sigma_a E) - c\sigma_a L_{a0} P_0 t + \tfrac{1}{2} L_{a0}(c\sigma_a P_0)^2 t^2. \tag{53}$$

Complete recovery of the saturable absorption in the interval between the pulses is assumed; hence, substituting Eqs. (49) and (53) into Eq. (24) and equating the equal powers over time yields three equations:

$$G_{i0}(1 - c\sigma_e E) - 1 - L_{a0}(1 - c\sigma_a E) - j\frac{2Q}{\omega_0 T_R}\phi_0 l_\phi = \frac{\omega_p^2}{\omega_D^2}, \tag{54}$$

$$-(G_{i0}' - c\sigma_e G_{i0} P_0) - c\sigma_a L_{a0} P_0 = \omega_p^2\left(\frac{L_s + \delta}{\omega_s}\right), \tag{55}$$

and

$$-G_{i0}'' - \omega_p^2\frac{L_s + \delta}{\omega_s}c\sigma_e P_0 + L_{a0}c^2\sigma_a(\sigma_a - \sigma_e)P_0^2 = \frac{2\omega_p^4}{\omega_D^2}. \tag{56}$$

The simultaneous solution of Eqs. (54)–(56) yields the pulse width ω_p, the peak power P_0, and the effective phase shift of the pulse relative to the carrier density ϕ_0 and the detuning effect $L_s + \delta$. The solution of Eqs. (54)–(56) is, in general, quite complicated. However, a qualitative understanding of the pulse behavior can be made by noting the similarity of these equations to Eqs. (35)–(37). The derivative terms of G_{i0}' are due to the time dependence of the gain and carrier relaxation. For intense modulation, these terms can have considerable magnitude. The power-dependent terms give the effect owing to gain reduction and saturation of the absorber by the optical pulse. The net gain − loss at $t = 0$ is obtained from

$$G_{i0}(1 - c\sigma_e E) - 1 - L_{a0}(1 - c\sigma_a E) = \text{Re}[(\omega_p/\omega_D)^2], \tag{57}$$

which illustrates that the net gain − loss at the center of the pulse is positive. The phase shift ϕ_0 is obtained from

$$\phi_0 = (-\omega T_R/2Q)\,\text{Im}[(\omega_p/\omega_D)^2].\qquad(58)$$

If the net detuning effect is taken to be zero, then Eqs. (55) and (56) become decoupled, and Eq. (55) gives an expression for the peak power at $t = 0$:

$$P_0 = G'_{i0}/c(\sigma_e G_{i0} - \sigma_a L_{a0}).\qquad(59)$$

Thus, the saturable absorber enhances the peak power. From Eq. (56), the pulse width is given by

$$-G''_{i0} + L_{a0}c^2\sigma_a(\sigma_a - \sigma_e)P_0^2 = 2\omega_p^4/\omega_D^2.\qquad(60)$$

From Eq. (52) both the rf term (for $\theta - \psi < \pi/2$) and the population-reduction term are less than zero, indicating that $-G''_{i0} > 0$, corresponding to pulse formation. The population-reduction term thus enhances the rf current term and gives rise to the additional pulse-shortening effect described at the beginning of this section. The saturable absorption effect contributes to pulse narrowing when $\sigma_a > \sigma_e$. The effect of frequency detuning can be qualitatively understood from Eq. (56). A negative frequency detuning corresponds to $f_{rf} < f_R$, where f_R is the round-trip frequency and produces $\delta T > 0$ and $\delta < 0$. Thus ω_p will be increased, and the pulse width will be decreased by the second term in Eq. (56). Conversely, a positive detuning ($f_{rf} > f_R$) results in pulse broadening. From Eq. (55) the negative frequency detuning appears to decrease the peak power P_0. However, the detuning also causes a phase shift that modifies $G_{i0}(t)$, and hence the dependence of P_0 on detuning is not directly evident. It has been shown experimentally that the total pulse energy is fairly independent of detuning (van der Ziel, 1981b). Hence, for negative frequency detuning, the peak power actually increases as the pulse width decreases. The effect of the frequency-limiting element and the dispersion are included in the ω_D^2 term. Decreasing the frequency bandwidth in Eq. (56) results in a smaller ω_p and, hence, lengthens the pulses.

h. Passive Mode Locking

The conditions for passive mode locking in the Gaussian mode amplitude approximation are also contained in Eqs. (54)–(56). For direct current excitation, $G_i(t)$ from Eq. (49) contains only the steady-state gain and the term describing the recovery of the gain following the emission of the pulse. The gain recovery term is time dependent and is expressed as $1 - \exp(-t/T_e)$. Hence G'_{i0} and G''_{i0} are relatively small in comparison to the gain saturation terms for short pulse widths and will be neglected in the following discussion.

From Eqs. (54)–(56), the mode-locking equations are

$$G_{i0} - 1 - L_{s0} - (\sigma_e G_{i0} - \sigma_a L_{a0})cE = (\omega_p/\omega_D)^2, \tag{61}$$

$$(\sigma_e G_{i0} - \sigma_a L_{a0})cP_0 = \omega_p^2(L_{s0} + \delta)/\omega_s, \tag{62}$$

and

$$(-\sigma_e^2 G_{i0} + \sigma_a^2 L_{a0})cP_0^2 = 2\omega_p^4/\omega_D^2. \tag{63}$$

Since $\omega_p > 0$ for mode locking, Eq. (61) shows that the net gain at the center of the pulse is positive. Similarly from Eq. (63), the saturable absorber must bleach more easily than the gain is depleted. The description given here requires a time shift with $(L_{s0} + \delta) < 0$ in order for mode locking to occur. The time delay δ has a somewhat different definition in the passive mode-locking case than for the case of active mode locking with sinusoidal modulation. The bleaching of the saturable absorber occurs sooner than the depletion of the gain, and this results in a shortening of the pulse periods and an increase in the frequencies. Substituting Eq. (62) into Eq. (61) yields the net gain before the arrival of the pulse

$$G_{i0} - 1 - L_{s0} = \omega_p^2[(1/\omega_D^2) + ((L_{s0} + \delta)/\omega_s)(\sqrt{\pi}/2\omega_p)], \tag{64}$$

which can be shown to have a negative value. Thus, as shown schematically in Fig. 5, the net gain is negative before the arrival of the pulse, becomes positive at the peak of the pulse, and can be shown to be negative after the pulse. It is thus the transient period of positive net gain, determined by the relative saturation of the absorber and the depletion of the gain, that causes the short-pulse formation in passive mode locking rather than the other material parameters of the laser.

The shape of the mode-locked pulse is sensitively affected by the relative saturation of the absorber and depletion of the gain. Computer calculations of mode locking indicate that the pulse shape is more closely approximated by an asymmetric, exponential pulse shape (New, 1974). In agreement with these calculations, the experimental results described later indicate that the autocorrelation function has exponential shape.

The series expansion of the gain over time that led Haus to the Gaussian pulse envelope is most accurate near the center of the pulse. It will deviate from the actual mode amplitude in the wings of the curve where the series expansion loses its validity. The passive mode-locking equations were simplified by neglecting the G_i' and G_{i0}'' terms that correspond to a constant pump-and-recombination term $G_{i0}(t) = G_{i0}$. Thus with this approximation, both the gain and the saturable absorber effectively recover completely in the interval between pulses. This approximation can be applied directly to Eq. (48) by neglecting the dG_{i0}/dt term, thus obtaining a more accurate description of the gain depletion. Integrating Eq. (48) yields

$$G(t) = G_{i0} \exp(-E(t)/E_g), \tag{65}$$

where G_{i0} is the initial value of the gain before the arrival of the pulse, E_g the integrated photon density required for gain depletion, and

$$E(t) = \int_{-\infty}^{t} P(t') \, dt' \tag{66}$$

the integrated photon density of the pulse at time t. Integrating Eq. (14) for the saturable loss yields

$$L_a(t) = L_{a0} \exp(-E(t)/E_a), \tag{67}$$

where L_{a0} is the initial value of the saturable loss and E_a the integrated photon density required for loss saturation.

The mode-locking equation [Eq. (24)] with this form of $G(t)$ and $L_a(t)$ has

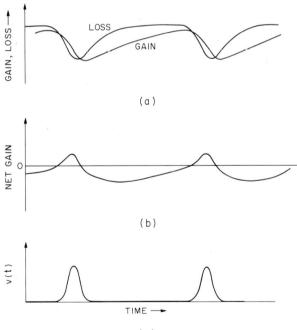

(a)

(b)

(c)

FIG. 5. Passive mode locking. (a) The saturable loss bleaches more readily than the gain is depleted. (b) This results in a short period of net gain that amplifies the pulse amplitude $v(t)$ in (c). At other times the loss attenuates $v(t)$.

mode amplitude solutions (Haus *et al.*, 1975; Haus, 1981)

$$v(t) = V_0 \operatorname{sech}(t/\tau_{\mathrm{p}}), \tag{68}$$

where V_0 is the peak amplitude and τ_{p} the pulse width. The wings of the pulse envelope are exponential, and near $t \approx 0$, the pulse envelope has a Gaussian shape described by $\exp(-t^2/2\tau_{\mathrm{p}}^2)$. Substitution of Eq. (68) into Eqs. (65) and (67) yields the time-dependent gain and saturable loss

$$G(t) = G_{i0} \exp\{-(E_0/E_{\mathrm{g}})[1 + \tanh(t/\tau_{\mathrm{p}})]\}, \tag{69}$$

$$L_{\mathrm{a}}(t) = L_{\mathrm{a}0} \exp\{-(E_0/E_{\mathrm{a}})[1 + \tanh(t/\tau_{\mathrm{p}})]\}, \tag{70}$$

where E_0 is half of the total integrated photon density in the pulse.

Expanding Eqs. (69) and (70) to second order and equating the terms of equal powers of $\tanh(t/\tau_{\mathrm{p}})$ yields

$$G_{i0}\left[1 - \frac{E_0}{E_{\mathrm{g}}} + \frac{1}{2}\left(\frac{E_0}{E_{\mathrm{g}}}\right)^2\right] - 1 -$$

$$L_{\mathrm{a}0}\left[1 - \frac{E_0}{E_{\mathrm{a}}} + \frac{1}{2}\left(\frac{E_0}{E_{\mathrm{a}}}\right)^2\right] - j\frac{2Q}{\omega_0 T_{\mathrm{R}}}\phi_0 l_\phi = \frac{1}{\omega_{\mathrm{D}}^2 \tau_{\mathrm{p}}^2} \tag{71}$$

$$G_{i0}\left[\frac{E_0}{E_{\mathrm{g}}} - \left(\frac{E_0}{E_{\mathrm{g}}}\right)^2\right] - L_{\mathrm{a}0}\left[\frac{E_0}{E_{\mathrm{a}}} - \left(\frac{E_0}{E_{\mathrm{a}}}\right)^2\right] = \frac{L_{\mathrm{s}} + \delta}{\omega_{\mathrm{s}} \tau_{\mathrm{p}}}, \tag{72}$$

$$-\frac{1}{2}G_{i0}\left(\frac{E_0}{E_{\mathrm{g}}}\right)^2 + \frac{1}{2}L_{\mathrm{a}0}\left(\frac{E_0}{E_{\mathrm{a}}}\right)^2 = \frac{2}{\omega_{\mathrm{D}}^2 \tau_{\mathrm{p}}^2}, \tag{73}$$

which are eigenvalue equations for the $\operatorname{sech}(t/\tau_{\mathrm{p}})$ mode amplitude solution and exhibit qualitative similarities with the Eqs. (54)–(56) corresponding to the Gaussian mode envelope.

From Eq. (73), the formation of passive absorber pulses requires that the saturable absorber bleach more readily than the gain is depleted. Equation (71) indicates that the gain is positive at the peak of the pulse. The net gain before the start of the pulse is obtained by adding Eqs. (71) and (72) and substracting Eq. (73), yielding

$$G_{i0} - 1 - L_{\mathrm{a}0} = -\frac{1}{\omega_{\mathrm{D}}^2 \tau_{\mathrm{p}}^2} + \frac{L_{\mathrm{s}} + \delta}{\omega_{\mathrm{s}} \tau_{\mathrm{p}}}, \tag{74}$$

and is negative. Similarly, the gain after the passage of the pulse is obtained by substracting Eq. (72) and (73) from Eq. (71),

$$G_{i0}\left[1 - \frac{2E_0}{E_{\mathrm{g}}} + \frac{1}{2}\left(\frac{2E_0}{E_{\mathrm{g}}}\right)^2\right] - 1 -$$

$$L_{\mathrm{a}0}\left[1 - \frac{2E_0}{E_{\mathrm{a}}} + \frac{1}{2}\left(\frac{2E_0}{E_{\mathrm{a}}}\right)^2\right] = -\frac{1}{\omega_{\mathrm{D}}^2 \tau_{\mathrm{p}}^2} - \frac{L_{\mathrm{s}} + \delta}{\omega_{\mathrm{s}} \tau_{\mathrm{p}}}, \tag{75}$$

and is also negative.

The gain medium thus exhibits a transient period of net positive gain that causes amplification of the pulse and a loss at other times during which the pulse is attenuated. The approximation made here is that the time response of the gain and the saturable absorber is slow compared to the pulse width but sufficiently rapid to recover in the period between the pulses. The pulse shape will be modified at high frequencies when the gain recovery is not complete and the derivatives of the gain are significant. These conditions, however, have been neglected here.

III. Review of Experimental Work

3. ACTIVE MODE LOCKING

a. *Mode Locking in the Gaussian Pulse Approximation*

The first active mode locking of a cw AlGaAs-type double heterostructure laser in an external cavity that resulted in short pulses was reported in 1978 (Ho *et al.,* 1978a). The external cavity (Fig. 6) consisted of a silvered spherical mirror with a 5-cm radius. The initial direct current threshold of the laser was 190 mA and was reduced to 140 mA after alignment of the spherical mirror. The laser was positioned at the radius of curvature and was driven by a 3-GHz dc microwave current through a bias network.

The autocorrelation was obtained from the second harmonic generated in a LiIO$_3$ crystal and is shown in Fig. 7. Assuming a Gaussian pulse shape, the

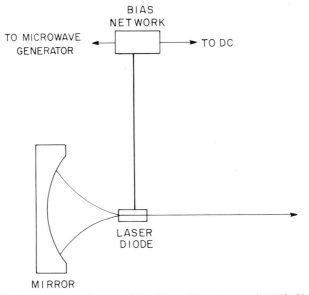

FIG. 6. Schematic of a mode-locked laser. [From Ho *et al.* (1978a,b).]

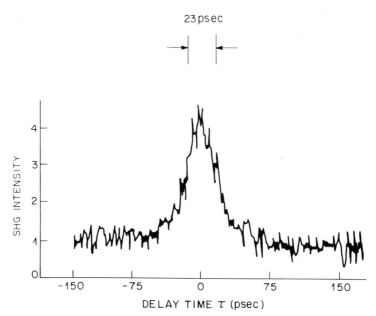

FIG. 7. Intensity autocorrelation trace obtained from second-harmonic generation. [From Ho *et al.* (1978a,b).]

autocorrelation FWHM of 32.6 psec corresponds to a pulse FWHM of 23 psec. The width of the autocorrelation envelope was found to be approximately constant for direct currents between 150 and 200 mA. The autocorrelation also contains a series of sharp spikes having a 6.8-psec separation that is equal to the round-trip transit time through the laser and corresponds to the wavelength spacing of the longitudinal modes.

Similar results have been observed by using an $In_{0.79}Ga_{0.21}As_{0.48}P_{0.52}$-type laser operating at 1.21 μm (Glasser, 1978). By using a spherical mirror with a 3.488-cm radius, pulses with a FWHM of 18 psec were obtained. The mode locking occurred at 2.1 GHz, corresponding to a single cavity 7 cm long. Doubling the cavity length indicates the presence of a second optical reflection within the cavity.

Strip-buried heterostructure GaAlAs-type lasers with an external cavity consisting of a collimating lens and a mirror were also mode locked (van der Ziel and Mikulyak, 1980). The strip-buried heterostructure active region of these lasers consists of a narrow $Al_{0.03}Ga_{0.97}As$-type active strip loading an $Al_{0.15}Ga_{0.85}As$-type optical waveguide (Tsang and Logan, 1979). Consequently, the optical wave extends well beyond the strip, yielding a large optical spot, and hence the devices are capable of high-output emission before damage occurs. The angular spread of the beam is small. This facili-

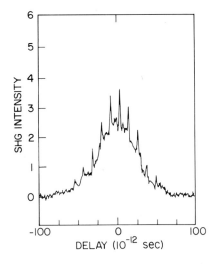

FIG. 8. Second-harmonic-generation auto-correlation measurement of the externally mode-locked pulses. [From van der Ziel and Mikulyak (1980).]

tates the collection of the emission by the lens and returns the reflected radiation back into the laser cavity.

The second-harmonic autocorrelation (Fig. 8) consists of a relatively broad spectrum upon which is superimposed a series of extremely sharp coherence spikes. As shown in Fig. 9, the broad second-harmonic autocorrelation is closely approximated by a Gaussian with a FWHM of 50.5 psec, implying a Gaussian pulse FWHM of 36 psec. These results confirm the mode-locking theory of Section 2,g, corresponding to relatively weak saturable absorption, which results in relatively long Gaussian pulses.

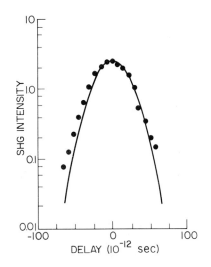

FIG. 9. Second-harmonic autocorrelation measurement showing the Gaussian dependence on the delay time. ● represents the data of Fig. 8. Solid curve is a plot of $I(2\omega) = 2.5$ exp $- (0.033 \, \Delta\tau)^2$. [From van der Ziel and Mikulyak (1980).]

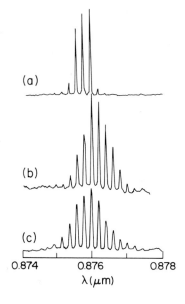

FIG. 10. Optical spectrum of laser emission at 3.5-mW output power: (a) Laser alone, (b) free-running mode-locked emission at 1.9 GHz, and (c) actively mode-locked emission at 1.9 GHz. [From van der Ziel and Mikulyak (1980).]

Figure 10 shows the optical spectra for the laser by itself and the laser free running at 1.9 GHz and the laser actively mode locked at 1.9 GHz. The longitudinal modes are broadened to ≈ 0.65 Å when the laser is mode locked. Substracting the estimated shift to longer wavelength during the pulse of 0.17 Å due to the increase in the carrier-dependent refractive index and deconvoluting the instrumental width of 0.3 Å yields a pulse spectral width of 0.37 Å. Hence, $\Delta\nu\tau_p$ is 0.52, in reasonable agreement with the transform limit value of $\Delta\nu\tau_p = 0.441$ for Gaussian pulses.

The envelope of the optical mode spectrum (Fig. 10c) is replotted in Fig. 11 and also exhibits Gaussian behavior with a FWHM of 9.5 Å. The sharp coherence peaks in Fig. 8 have a FWHM ≈ 1.6 psec, corresponding to Gaussian components of 1.13 psec. Hence, $\Delta\nu\tau_p = 0.42$ for the spectral envelope and the temporal sharp spikes. This is close to the transform-limit value indicating that the sharp spikes are due to interference of the spectral modes. The mode spacing corresponding to a 1.9-GHz repetition rate is 0.049 Å, and so the spectral width of a single longitudinal mode group contains approximately eight intense modes of the composite cavity. The autocorrelation indicates that it is these modes that are locked. The interference of the different mode groups results in the sharp spikes in the autocorrelation. The modes of the different longitudinal mode groups, however, are not locked. It is the combined effect of these mode groups that produces the observed autocorrelation pulse width.

It has been proposed that the failure to lock the modes of the different

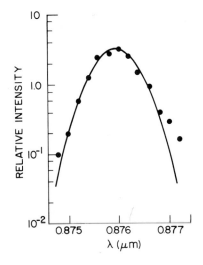

FIG. 11. Intensity of the longitudinal modes of the externally mode-locked emission, which shows the Gaussian dependence of the mode envelope on wavelength. ● represents the data of Fig. 10(c). Solid line is a plot of $I(\lambda) = 3.31 \exp - 0.031(\lambda - 8759.4)^2$. [From van der Ziel and Mikulyak (1980).]

mode groups is due to the spontaneous emission noise of the laser (Haus and Ho, 1979). Since the spikes in the autocorrelation result from the interference of different mode groups, limiting the laser bandwidth to a single mode group will eliminate the structure in the autocorrelation and improve the coherence of the laser beam. The bandwidth limitation has been obtained by inserting a Fabry–Perot etalon consisting of a dielectrically coated quartz plate in the external cavity (Ho, 1979a). An AlGaAs-type stripe geometry laser that had one facet that was antireflection (AR) coated with a $\frac{1}{4}$ layer of SiO was used. When driven by 190-mA dc and 0.5-W rf currents at 433 MHz, a Gaussian autocorrelation was obtained that was free of sharp spikes. This is consistent with the optical spectrum that consisted of a single band ≈ 1 Å wide. The Gaussian pulses had a width of 63.6 psec. Removing the etalon yielded pulse widths ≈ 40 psec, and the multimode group emission resulted in spikes in the autocorrelation.

A transverse junction stripe geometry laser has also been used in a similar experiment (Ito *et al.,* 1980). The external cavity length was ≈ 60 cm, corresponding to a 254.5-MHz repetition rate. The lasers were not AR coated, and an etalon consisting of a 140-μm-thick glass plate dielectrically coated to yield a finesse of 10 was inserted in the external cavity and limited the emission to a single, longitudinal mode group. The lasers were dc biased to the threshold, and a sinusodial microwave current was superimposed. The autocorrelation FWHM was 60 psec and, taken to be Lorentzian, corresponded to a Lorentzian laser pulse with a FWHM of 30 psec. A similar pulse width was obtained by using a fast streak camera. The assumption of a Lorentzian profile is quite different from most other mode-locking work.

b. Comparison with Theory for the Case of Short Pulse Generation in the Presence of Saturable Absorption

(1) *Numerical Values for the Pulse Width.* The operating parameters of mode-locked lasers that yield short pulses have been studied experimentally (van der Ziel, 1981b) and have been compared with the theory of Section 2.9. The mode-locking equations [Eqs. (54)–(56)] in the Gaussian approximation are simplified for the case of net zero detuning. The peak power is then given by Eq. (55) and the pulse width by Eq. (56). The peak power may be approximated by

$$P_0 = \frac{T_R \omega_p t_e}{\sqrt{\pi}} \left[\frac{I_0 + I_{rf} \cos(\theta - \phi) - I_{0T}}{eV} \right], \tag{76}$$

where I_{0T} is the effective direct current threshold in the absence of $I_{rf} \cos(\theta - \phi)$.

For numerical values, we use for the active region $n_g \sigma_g = 200$ cm^{-1}, $n_g = 1.5 \times 10^{18}$ cm^{-3}, $\tau_e = 2 \times 10^{-9}$ sec, $\tau_p = 3 \times 10^{-12}$ sec, and $V = 12 \times 0.15 \times 0.038 \times 2 \times 10^{-8}$ cm^{-3}, where the additional factor 2 in V accounts for the current spreading outside the stripe. The saturable absorber is assumed to be given by $\sigma_a = 2 \times 10^{-16}$ cm^{-2}, $n_a = 10^{16}$ cm^{-3}, and $l_a = 10^{-3}$ cm.

The value of ω_c is determined by the bandwidth of the etalon and other frequency-limiting elements such as the modulation of the transmission resulting from the residual reflectivity of the laser. The measured FWHM of the etalon is 6 Å and $\omega_c = 2.2 \times 10^{12}$ sec^{-1}. Assuming Gaussian pulses of width τ_p, the autocorrelation width is $\tau_{1/2} = 2^{1/2} \tau_p$. Equation (56) can be written as

$$\tau_{1/2}^{-4} = 4.3 \times 10^{-8} \left\{ 0.6[I_0 - I(-T_{rf})] + 12.6 I_{rf} \cos(\theta - \phi) \right.$$
$$+ 800 \frac{\delta f}{\tau_{1/2}^3} [I_0 + I_{rf} \cos(\theta - \phi) - I_{0T}]$$
$$\left. + 250 \frac{[I_0 + I_{rf} \cos(\theta - \phi) - I_{0T}]^2}{\tau_{1/2}^2} \right\}. \tag{77}$$

The units are τ (psec), I (mA), and δf (MHz). The value of $I(-T_{rf})$ is the equivalent current corresponding to $n(-T_{rf})$. The first term derives from the G_{i0}'' term, the δf term describes the frequency detuning, and the last term gives the effect of saturable absorption that is distinguished from the other terms by the quadratic dependence on device current. The relative importances of the terms is obtained from the different current dependences.

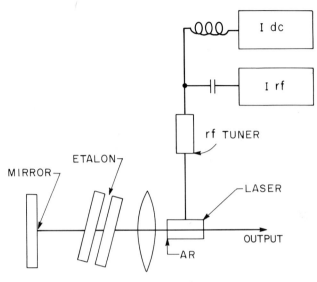

FIG. 12. Experimental arrangement used to study the characteristics of external cavity mode locking. [From van der Ziel (1981b).]

(2) *Laser Geometry.* The external cavity arrangement is shown in Fig. 12. The lasers are standard proton-bombarded, stripe-geometry, 380-μm-long double heterostructure devices with a 0.15-μm-thick by 12-μm-wide active region. An antireflective $\lambda/4$ coating of SiO was evaporated on the laser facet facing the external cavity. The residual reflectivity was less than 1%. The light emitted through the AR-coated facet was collimated by a lens and passed through a Fabry–Perot etalon consisting of two dielectrically coated, $\lambda/20$, flat glass plates spaced by 75 μm. The etalon acts as a double-pass optical filter limiting the lasing bandwidth to a small fraction of the bandwidth obtained without the etalon. This shortens the pulse width. Wavelength tuning of the etalon was obtained by changing the spacing and by tilting the etalon so that its resonant transmission coincides with a Fabry–Perot peak of the laser. The emitted radiation was reinjected into the active region by careful adjustment of the lens and the mirror. It is extremely important to avoid extraneous reflections, either from the external cavity or from the output circuit, back into the active region. Even weak, spurious reflections severely affect the mode-locking behavior and result in long pulses.

The residual reflectivity of the AR-coated and uncoated facets of the laser forms a Fabry–Perot mode selector that limits the bandwidth and hence the pulse width. For a 380-μm-long laser, the mode spacing is ≈ 2.1 Å. With a 1-GHz repetition rate, corresponding to a 15-cm-long optical cavity, the mode spacing is 0.022 Å. Because of the Fabry–Perot of the laser cavity, the composite cavity has transmission peaks at the modes of the composite cavity that are modulated by the transmission characteristics of the laser cavity that typically has a spectral width of ≈ 0.2 Å (Fig. 1). The modes of the composite cavity thus appear as small groups (≈ 10) of narrowly spaced modes separated by the mode spacing of the laser cavity. In active mode locking, the external cavity modes belonging to such a mode group are locked, but the modes corresponding to distinct mode groups are, in general, not locked. Locking of different mode groups has been achieved in passive mode locking.

(3) *Light versus Current Dependence.* The dependence of the light output on direct current is shown in Fig. 13. Without feedback and rf current,

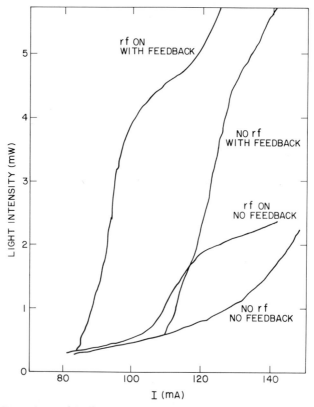

FIG. 13. Dependence of the light power with and without rf current and external feedback. $f = 970$ MHz. [From van der Ziel (1981b).]

the threshold of the AR-coated laser is 130 mA. With rf power, the threshold is reduced to 112 mA with a kink at 120 mA. Below the kink, the external quantum efficiency is similar to the curve without rf current. The introduction of feedback lowers the direct current thresholds to 114 and 82.5 mA for the case without and with rf current, respectively.

(4) *Exponential Pulse Shape.* The second-harmonic autocorrelation of the shortest pulse shape obtained (Fig. 14) is remarkably peaked at zero delay. As shown in the semilog plot, the autocorrelation is closely approximated by a symmetric exponential function. For a single-sided, exponential pulse with a FWHM of τ_p, the autocorrelation shape is described by $\exp(-|t|\ln 2/\tau_p)$, and the exponential slope of the curve yields $\tau_p = 5.3$ psec. The exponential wings of the autocorrelation indicate that the net gain changes rapidly during the emission of the pulse and, together with the short pulse widths, is consistent with the presence of appreciable saturable absorption. A second contribution to the exponential, second-harmonic, autocorrelation tails is the presence of a statistical spread in pulse widths (Van Stryland, 1979). The harmonic generation enhances the shortest pulses having the highest peak power. This is reflected in an autocorrelation that is sharply peaked at $\tau = 0$. For relatively simple pulse width distributions, it has been shown that the autocorrelation has an approximately double-sided exponential shape. If such a distribution of pulse widths is assumed, the shortest pulses in the distribution have widths less than the width obtained from the exponential autocorrelation.

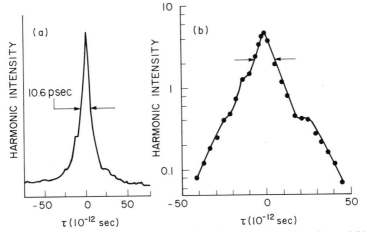

FIG. 14. (a) Second-harmonic autocorrelation function at optimum detuning and (b) semilog plot of the intensity from (a) showing the exponential dependence on delay time. Assuming a single-sided exponential pulse shape, the FWHM is 5.3 psec. [From van der Ziel (1981b).]

(5) *Dependence on Direct Current.* The FWHM of the second-harmonic autocorrelation curves, denoted by $\tau_{1/2}$, as a function of direct current is shown in Fig. 15. Narrower pulses are obtained by slightly reducing the applied frequency. The squares in Fig. 15a indicate the minimum value of $\tau_{1/2} = 10.6$ psec obtained at the detuning that minimizes the pulse width, and the corresponding autocorrelation trace is shown in Fig. 14. The widths were measured for optimum second-harmonic generation; thus, there is a contribution due to detuning. An approximate dependence on current is obtained by retaining only the direct, rf, and saturable absorption terms. The comparison of Eq. (77) with experiment is further complicated by the non-linear dependence of the light output on current. A linear interpolation between the threshold and the high-current light output of Fig. 13 yields an

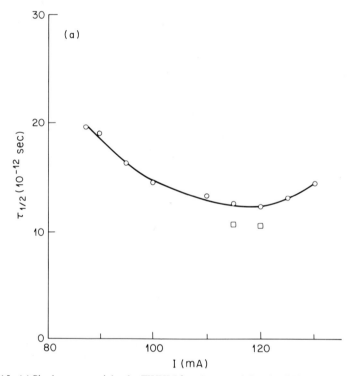

FIG. 15. (a) Single exponential pulse FWHM for a laser modulated at 970 MHz. ○ represents the FWHM for optimum second-harmonic-generation intensity, and □ represents the minimum widths obtained at optimum frequency detuning. $f_0 = 970$ MHz and $I_{th} = 82.5$ mA. (b) Dependence of $\tau_{1/2}^{-4}$ as a function of $(I_{0eff} - 82.5)^2/\tau_{1/2}^2$ from the data in (a). The linear dependence results from the rf current and saturable absorption contribution and is given by Eq. (78). The decrease at higher currents is attributed to the kink in the $L-I$ curve. [From van der Ziel (1981b).]

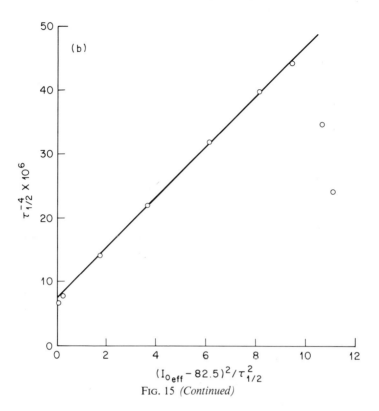

FIG. 15 *(Continued)*

effective current dependence. The light output at a measured current was extrapolated to the effective value $I_{0\text{eff}}$. Figure 15b shows the values of $\tau_{1/2}^{-4}$ versus $I_{0\text{eff}}$ obtained from Fig. 15a. The linear region is given by

$$\tau_{1/2}^{-4} = 8 \times 10^{-6} + 4 \times 10^{-6}[(I_{0\text{eff}} - 82.5)^2/\tau_{1/2}^2] \tag{78}$$

and is in excellent agreement with the data. The domination of the quadratic current term illustrates the importance of the saturable absorption.

In Eq. (77), the first term, which is linear in current, has a negligible value relative to the other terms. From Fig. 13, the reduction in the current threshold due to the rf current is 27.5 mA. Taking this value for $I_{\text{rf}} \cos(\theta - \phi)$, Eq. (77) for the rf current and saturable absorption contributions yields

$$\tau_{1/2}^{-4} = 14.9 \times 10^{-6} + 10.75 \times 10^{-6}[(I_0 - 82.5)^2/\tau_{1/2}^2], \tag{79}$$

which is in good agreement with Eq. (78).

(6) *Dependence on RF Current.* The experimental dependence of $\tau_{1/2}$ on rf current is shown in Fig. 16. The power was varied by means of fixed and

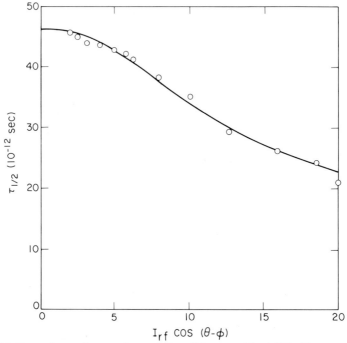

FIG. 16. Dependence of $\tau_{1/2}$ on rf current. $I_0 = 120$ mA, and $f = 1$ GHz. [From van der Ziel (1981b).]

variable attenuators. For these data, the I_{OT} was 105 mA, and since the dc threshold current with the rf on was 85 mA, the maximum effective rf current $I_{rf}\cos(\theta - \phi)$ was 20 mA. The direct current was 120 mA. In obtaining the rf current dependence, the detuning effect is neglected, and the data are fitted to the solid curve in Fig. 16, which is a plot of

$$\tau_{1/2}^{-4} = 3.3 \times 10^{-8} - 2.3 \times 10^{-8} I_{rf} \cos(\theta - \phi)$$
$$+ 1.7 \times 10^{-6}\{[15 + I_{rf}\cos(\theta - \phi)]^2/\tau_{1/2}^2\}. \tag{80}$$

Using the values of the currents in Eq. (77) yields

$$\tau_{1/2}^{-4} = 38.7 \times 10^{-8} + 54.2 \times 10^{-8} I_{rf}\cos(\theta - \phi)$$
$$+ 10.75 \times 10^{-6}\{[15 + I_{rf}\cos(\theta - \phi)]^2/\tau_{1/2}^2\}. \tag{81}$$

As was found for Eq. (78), the saturable absorption term dominates in Eq. (80). The magnitude of the saturable absorption term is a factor of 2.3 smaller than in Eq. (78) and is also considerably smaller than the corresponding term in Eq. (81). The difference is attributed to the slightly different operating conditions of the laser for the two sets of data.

(7) *Frequency Detuning.* From Eq. (77), the pulse width is a strong function of the mismatch between the applied frequency and the round-trip frequency. For negative detuning, this term adds to the remaining terms and results in shorter pulse widths. This is illustrated in Fig. 17, which shows the FWHM of the autocorrelation as a function of detuning. A reduction in the autocorrelation width of a factor of two is observed at -12 MHz. For large, negative detunings, the lasers are no longer properly mode locked. The frequency detuning is accompanied by large phase shifts of the pulse relative to the rf current. The measured dependence is shown in Fig. 18. When the applied frequency is lower than the round-trip frequency, the pulse repetition time fixed by T_{rf} is longer than T_R. The laser partially compensates for this time difference by advancing the phase of the pulse. As indicated in Fig. 17, this is associated with pulse shortening. The detuning effect does not greatly modify the average output power; hence, with the shorter pulse width, the peak power increases. This effect reaches its limit when the phase advances to the point at which mode locking is no longer possible. Modulation detuning effects have also been described by Goodwin and Garside (1983).

(8) *Dependence on Pulse Repetition Rate.* The dependence of the pulse width on repetition rate is shown in Fig. 19. The external cavity was approximately 32 cm long, corresponding to a fundamental frequency of 472 MHz.

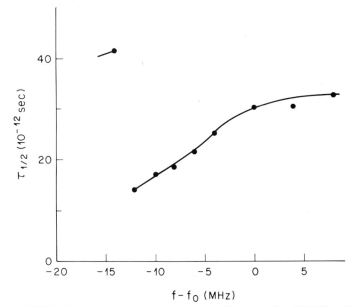

FIG. 17. FWHM of the autocorrelation as a function of detuning. $f_0 = 980$ MHz. [From van der Ziel (1981b).]

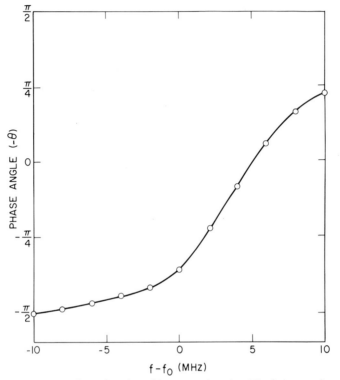

FIG. 18. Phase angle $-\theta$ as a function of frequency detuning. The frequency for optimum second-harmonic generation f_0 is $1 GHz$, and $I_0 = 120$ mA. [From van der Ziel (1981b).]

Higher repetition rates are obtained by driving the rf current at multiples of the fundamental (round-trip) frequency. At the jth harmonic there are j pulses circulating in the cavity. From the mode-locking point of view, the jth multiples of the modes of the composite cavity are locked. At 472 MHz, the lifetime of the carriers is of the same order as the repetition rate. Consequently, the coherence of the carrier density is partially lost in the interval between the pulses. The resulting pulse width is broad and is limited by the spontaneous emission rate. The minimum pulse width occurs approximately in the range 0.9–1.5 GHz. The pulse width increases at 1.879 MHz because there is insufficient time between the pulses to build up the carrier density to a high value, and the peak value of the optical pulse power that can be extracted from the gain medium is reduced. Consequently, the optical pulse cannot reduce the stimulated gain as rapidly as at longer pulse intervals, and the pulse width increases owing to the gain limitation. The pulse width decreases with increasing current.

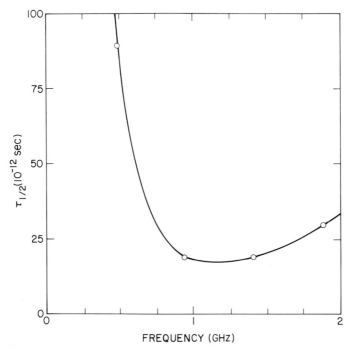

FIG. 19. FWHM of the autocorrelation as a function of frequency. $I_0 = 120$ mA, and $f_0 = 472$ MHz. [From van der Ziel (1981b).]

c. Subpicosecond Pulses from a Ring Laser

The ring laser geometry supports two oppositely directed pulse trains that interact in the saturable absorption region. This process has been termed colliding-pulse-mode locking (CPM) and has resulted in sech²-type pulses with FWHM as short as 0.09 psec in a passively mode-locked dye laser containing a thin (≈ 10-μm) dye jet stream with a saturable absorber (Fork et al., 1981; Shank et al., 1982). The two counter-rotating pulses create a transient grating that synchronizes, stabilizes, and shortens the pulses. The requirements for effective collision pulse mode locking are that the pulses overlap in the saturable absorbing medium. This condition occurs because a minimum amount of energy is lost from the pulses when the collision occurs in the saturable absorber. In addition, the saturable absorber must be sufficiently thin in order that the optical path in the absorber be less than the optical pulse width. Thus, thin saturable absorption regions are required. For example, a pulse FWHM of 0.1 psec corresponds to an optical length of 30 μm, and with a refractive index $n \approx 1.5$, the absorber should be less than 20 μm long.

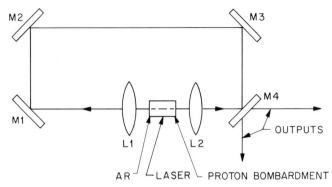

FIG. 20. Schematic diagram of the mode-locked ring laser. [From van der Ziel *et al.* (1981c).]

Subpicosecond pulses have been obtained by active mode locking a GaAs-type laser in an external ring cavity, as shown in Fig. 20 (van der Ziel *et al.*, 1981c). The cavity round-trip length corresponded to a 625-MHz repetition rate. The GaAs-type laser contained a region of saturable absorption introduced by proton bombardment with $10^{15}/cm^2$ protons at 600 keV and a subsequent thermal annealing at 500°C for 5 min. The ring geometry contained a number of losses such that insufficient radiation was fed back into the laser to allow passive mode locking. Injection of a 625-MHz current yielded the autocorrelation shown in Fig. 21, which indicates the presence of

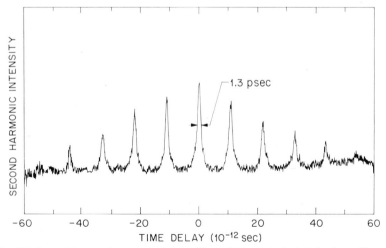

FIG. 21. Second-harmonic autocorrelation of the actively mode-locked ring laser. The peaks have equal FWHM and are spaced at multiples of the 11-psec round-trip transit time through the laser. $I_0 = 115$ mA, $P_{rf} = 0.2$ W, and $f = 625$ MHz. [From van der Ziel *et al.* (1981c).]

a series of approximately five pulses separated by the 11-psec round-trip time through the laser. The absence of an underlying broadband autocorrelation indicates that the external cavity modes of the different laser Fabry–Perot mode groups are mode locked. The central peak of the autocorrelation is an exponential function of the delay time (Fig. 22). Assuming a single-sided exponential line shape, the FWHM of the pulses is 0.56 sec.

It was also confirmed that the pulses collide within the proton-bombarded region. From the measured pulse width and the lack of a pulse from the clockwise rotating beam in Fig. 20 (which is reflected at the exit facet), it was shown that the collision occurs within 20 μm of the proton-bombarded exit facet. The region damaged by the 600-keV protons extends 8–10 μm from the mirror, which is well within the pulse collision limit.

Active mode locking with linear and ring cavity geometries was studied and compared (Olsson and Tang, 1981). In order to reduce the effect of laser cavity reflection, both facets were AR coated, and the laser was placed in the

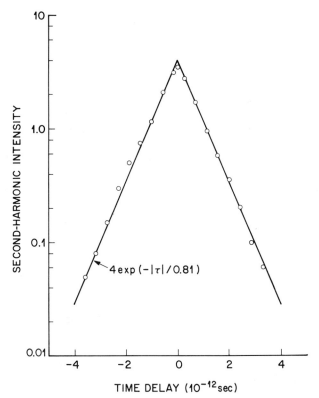

FIG. 22. Semilog plot of the central autocorrelation peak of Fig. 21 showing the exponential dependence on delay time. [From van der Ziel *et al.* (1981c).]

center of the linear cavity. Due to the small residual reflectivity of the facets, there was an approximately 20% amplitude modulation of the output power when the laser was spectrally tuned. With the linear geometry, pulses of 7.8-psec FWHM were obtained. When the cavity was converted to the ring geometry of 16-psec FWHM, long pulses were obtained. Since different lasers were used in the two experiments, the results were not directly comparable. The shorter pulse widths obtained in the linear geometry were attributed to the presence of a larger saturable absorption in the laser used in that experiment.

d. Effect of Laser Geometry

The mode locking of 23 lasers having buried heterostructure, channel substrate, constricted double heterostructure, and oxide stripe geometries have also been studied (Figueroa, 1981; Figueroa et al., 1981). Mode locking with detector-limited pulses (< 60 psec) was observed with lasers that exhibited either self-pulsations or narrow-band noise. Mode locking was not observed with lasers having a broadband noise spectrum. These results indicate that the presence of a nonlinearity, such as saturable absorption or electron traps in the laser, is required to obtain short pulses. Similar results were also observed by using an multimode optical fiber as a resonator (Figueroa et al., 1980b). The frequency of the self-pulsations increases nonlinearly with drive current (van der Ziel, 1979; Dixon and Joyce, 1979). Constructive interference occurs when harmonic relationships exist between the self-pulsation frequency and the cavity frequency (Lau et al., 1980; Figueroa et al., 1980a,b). For a fixed external cavity length, the amplitude of the pulsations exhibits an initial resonance when the two frequencies are equal. A second resonance occurs at higher currents when the pulsation frequency is approximately twice the external cavity frequency and corresponds to the presence of two pulses within the cavity.

Similar intensity changes are observed when the current is kept at a constant value and when the length of the external cavity is varied (Lau et al., 1980; Figueroa et al., 1980a,b). The deviation of the external cavity length from the resonance frequency results in a pulling of the self-pulsation frequency. The pulsation frequency decreases with increasing cavity length. A second resonance is observed at approximately twice the cavity length of the first resonance, and the corresponding length dependence of the frequency shift is somewhat smaller than for the first resonance. The pulsations are quenched at intermediate cavity lengths, in which there is destructive interference of the two frequencies. The pulsations are also quenched for external cavities shorter than 10 cm.

These results appear to be characteristic of non-AR-coated lasers containing a saturable absorber that induces self-pulsations. The feedback in these studies was relatively weak and perturbs the self-pulsations. Increasing the feedback and decreasing the reflectivity of the laser – external cavity interface suppresses the self-pulsations and results in short-pulse mode locking.

e. Reduction of Mirror Reflectivity

The longitudinal mode resonances of the laser can be avoided by angling the active-region stripe at 5 μm from the usual normal to the facet direction (Holbrook *et al.,* 1980a – c; Bradley *et al.,* 1981). When the laser facets are AR coated and the lasers are placed in a 30-cm-long external cavity consisting of 5-cm radius of curvature spherical mirror on one end, a 0.85-NA microscope objective, and a reflecting mirror on the other end (Fig. 23), the spontaneous emission does not exhibit longitudinal laser modes. Above the threshold, the laser emitted in a single, broad mode. Short-pulse operation was obtained by the addition of an aperture and a Fabry – Perot interferometer that restricted the laser bandwidth (Holbrook *et al.,* 1980b). The mode-locked pulses were measured by using a streak camera and were estimated to be of 16-psec duration and 1-W peak power. The measured laser bandwidth was 0.56 Å, and without taking into account the spectral shift during the pulse (which would yield a smaller bandwidth), the time bandwidth product was 0.36, which is somewhat smaller than the transform limit for Gaussian pulses.

The requirement of an AR coating on the intracavity facet to suppress the laser mode structure has been eliminated by using lasers where the oxide isolated stripe angle is increased to 15° from the facet normal (Chen *et al.,* 1982). By using an arrangement similar to Fig. 23, the angled stripe lasers with 20-μm-wide oxide stripes resulted in 11-psec FWHM pulses at a 285.5-MHz repetition rate.

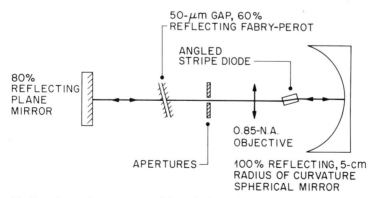

FIG. 23. Experimental arrangement of the actively mode-locked diode laser with an angled stripe in order to reduce the Fabry – Perot reflection. [From Holbrook *et al.,* (1980a,b).]

f. Transient Behavior

The turn-on and decay of an actively mode locked laser has been studied by using a pulsed microwave driving current (Au Yeung et al., 1982). The laser had the proton-bombarded stripe geometry with an initial threshold of 81 mA. The laser was placed in an external cavity ≈ 30 cm long that consisted of a collimating lens, a dielectrically coated Fabry – Perot etalon, and a highly reflecting rear mirror. The laser was driven by a dc bias and a 500-MHz microwave current source that delivered 13.6-dB power into a 50-Ω load. A dividing circuit followed by a electronic pulse was used to drive a diode switch in synchronism with the microwave current. The microwave current was modulated into 600-nsec-long pulses at 1-MHz repetition rates. The diode switch had a 10-nsec turn-on time and a 60-nsec switch-off time.

The onset of the mode locking was fairly slow. The emission was only weakly modulated for the first 10 nsec during which the modulation current was established. At later times, the pulse amplitude increased and the pulse width decreased. The pulses reached the limiting width of 30 – 35 psec after 60 nsec, corresponding to 30 round trips. After termination of the current pulse, the laser energy decreased and the width increased, corresponding to the slow switch-off time of the rf current pulse.

The transient behavior has also been studied by using computer solutions of the rate equations including a feedback term (New, 1974; Aspin and Carroll, 1979). New (1974) showed that the combined section of the amplifier and the saturation of the absorber leads to rapid pulse compression even when the duration of the pulse is far shorter than the relaxation time of the absorber. The initial condition assumed by Aspin and Carroll (1979) contained two equal pulse amplitudes inside the laser cavity. The assumptions of this theory have been criticized by Haus (1980b). About 150 round-trip transits were required before the solution returned to a single pulse. This is larger than the number of transits required in the earlier experimental study.

g. High-Repetition-Rate Mode Locking

The active mode-locking work reviewed in the preceding subsection has emphasized repetition rates in the 0.3 – 5-GHz range. Mode locking at higher repetition rates is experimentally considerably more difficult. However, using mode-locked semiconductor lasers at higher repetition rates (> 5 GHz) may prove to be important for future high-capacity data transmission systems. At these high rates, a significant requirement for the pulsing behavior of the laser is that the injected current must achieve a high level of inversion in the short time interval between pulses. This requires lasers with efficient carrier injection and low threshold currents. In addition to using high-frequency circuitry, the capacitance of the laser chip, including the capacitance due to the blocking junctions or oxide-blocking junctions, must be kept small. This implies the use of small laser chips.

The natural resonance frequency of a laser is

$$f \approx \frac{1}{2\pi} \left[\frac{(1 + ca\tau_{ph})}{\tau_e \tau_{ph}} \left(\frac{I}{I_{th}} - 1 \right) \right]^{1/2}, \tag{82}$$

where τ_e and τ_{ph} are the electron, and photon lifetimes, respectively, and I_{th} is the current at threshold (van der Ziel et al., 1979). The gain given by Eq. (20) can be written as $G = ca(n/n_g - 1)$, where $a = \sigma_g n_g$ is the gain coefficient. Usually the $ca\tau_p$ term is not included in the resonant frequency term. For typical values $a = 200$ cm^{-1}, $c = 0.75 \times 10^{10}$ cm in the medium, and $\tau_{ph} = 10^{-12}$ sec, this factor equals 1.5 which is larger than the unity term.

Improved, high-frequency response can be obtained by using a low-threshold laser and operating well above the threshold, thus resulting in large I/I_{th}. Alternatively, the effective electron lifetime can be reduced by introducing saturable absorption or changing the doping level. This is, in general, counterproductive since it also raises the threshold current. With a large $ca\tau_{ph}$ term, the photon lifetime is less significant in determining the resonant frequency. However, decreasing the cavity length (which decreases τ_{ph}) usually also reduces I_{th} and, thus, enhances the high-frequency performance. Increasing the repetition rate also requires a corresponding reduction of the pulse width in order for the pulses to remain distinguishable.

These concepts have been applied to produce very short pulses at high repetition rates from a self-pulsing laser (van der Ziel and Logan, 1982). Short, 125-μm-long buried optical guide lasers were used to reduce the photon lifetime. The lasers had a 10-mA dc threshold and were slightly degraded to introduce saturable absorption. By operating at 3.5 times the threshold, self-pulsations with less than 19-psec FWHM were observed with 30-psec pulse separation. These are the shortest pulses and the highest repetition rates observed thus far from self-pulsing lasers.

A mode-locking experiment at 7.04 GHz has been performed by using a transverse junction stripe (TJS) GaAs-type laser with an AR-coated facet in a 2.1-cm-long external cavity consisting of a 0.25-NA lens and an aluminum-coated mirror (Suzuki and Saito, 1982). No bandwidth-limiting element was used. The laser threshold was 25 mA before the AR coating was applied. When driven by combined direct and microwave currents, the laser oscillated in a single mode of the diode cavity and in six modes of the external cavity. Transform-limited Gaussian pulses as short as 27 psec were obtained. Characteristic of the emission being confined to a single laser mode group, the autocorrelation curve had a smooth envelope.

The short external cavities required at high repetition rates suggest alternatives to the usual microscope objective and mirror combination. In addi-

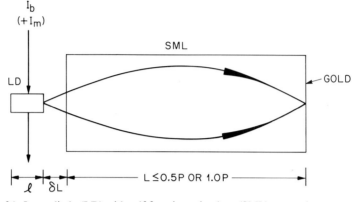

FIG. 24. Laser diode (LD) with self-focusing microlens (SML) external resonator. [From Akiba *et al.* (1981).]

tion to monolithic integration of a laser and waveguide (Johnson, 1980), short optical waveguides, optical fibers (Tucker *et al.*, 1983), and graded refractive index lenses (Akiba *et al.*, 1981) can be used. Such cavities enable one to make a very compact external resonator that will have increased optical stability as well as simplifying the alignment of the reflected beam.

A GRIN lens used as an external cavity has been used to obtain mode locking of an InGaAsP/InP-type laser operating at 1.55 μm (Akiba *et al.*, 1981). The laser had a 15-μm-wide active region defined by a stripe removed from the oxide layer, was 160 μm long, and had a 135-mA threshold (Akiba *et al.*, 1979); no AR coating was used. The experimental configuration is shown in Fig. 24. The best mode-locked operation was obtained by using a full-pitch GRIN lens, which resulted in active mode locking at 5.3 GHz with pulses having 29-psec FWHM. Passive mode locking with FWHM pulses of $\approx$ 30 psec was also observed. With half-pitch GRIN lenses, repetition rates of 8.1 and 9.4 GHz were obtained. Figure 25 shows the autocorrelation and cross-correlation at 9.4 GHz. The FWHM of both correlation functions is 42 psec, corresponding to a Gaussian pulse width of 30 psec. The sharp structure in the autocorrelation indicates the presence of several longitudinal-mode groups in the emission spectrum.

Active mode locking of 1.3-μm ridge-waveguide lasers using an optical fiber external cavity has also been observed (Tucker *et al.*, 1983). The ridge laser is capable of high-frequency performance since the rf current is injected solely into the narrow active region and is blocked by the SiO$_2$ layer over the remaining part of the chip (Kaminow *et al.*, 1983). The SiO$_2$ has a low capacitance relative to the back-biased $p-n$ junctions that are frequently used for current blocking. In addition, the passive current-shunting paths

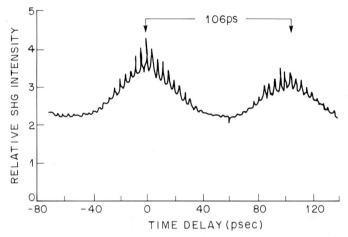

FIG. 25. Second-harmonic auto- and cross-correlation with an 8.0-μm-long, $\frac{3}{4}$-pitch GRIN lens. The direct current is 170 mA. The driving frequency is 9.4 GHz. [From Akiba *et al.* (1981).]

and nonradiative traps at the edges of the active layer are reduced for this geometry.

In the mode-locking experiment, the output from an AR-coated facet was injected into a 3.1-cm-long multimode fiber having a lensed input and a highly reflective rear facet. The fiber length corresponds to a 3.33-GHz, round-trip resonance. The laser was driven at the fundamental, second-, third-harmonic frequencies, corresponding to one, two, and three pulses circulating in the cavity, respectively. In each case, pulses with 29-psec FWHM were obtained from the deconvolution of the picosecond pulses and the known response of the high-speed detector. These relatively long pulses indicate that the lasers did not contain saturable absorption. This is consistent with the more general observation that the quaternary active layers are substantially less sensitive to optical damage and dark-line and regional degradation than GaAs-type active layers.

h. Short-Current Pulse Excitation

The use of short-current pulses from a step recovery diode (comb generator) as an excitation source has also been studied (Au Young and Johnston, 1982b). Stripe geometry lasers defined by proton bombardment were used. The lasers had an initial 81-mA threshold, which increased to 97 mA after AR coating and was lowered to 80 mA when the external cavity was aligned. The external cavity had a length of 30 cm corresponding to a 2-nsec round-trip time. The external cavity length was tuned to the resonant frequency of the comb generator, which emitted 60–90-psec-long pulses.

The resulting optical pulses had single-sided exponential wave forms with FWHM of 9.4 psec. When compared with sinusoidal modulation, the current pulse excitation results in the suppression of multiple pulses owing to the recovery of the gain after the emission of the first pulse and thus allows for a higher dc bias. As its name implies, the comb generator emits a series of frequencies at multiples of the repetition rate. Because the frequency response of the laser is limited to a few gigahertz, the effect of the higher harmonics in the comb output is lost. Thus, the comb generators are expected to be most useful for mode locking at frequencies under 500 MHz and for lasers having a high-frequency response. At higher mode-locking frequencies, at which the laser can respond to only a few of the lowest-frequency teeth of the comb output, the comb generator and cw current excitation are expected to yield nearly similar pulse widths. This was confirmed in a current modulation experiment at 940 MHz in which pulses of 27 and 29 psec were obtained by using comb and cw current excitation, respectively (van der Ziel and Logan, 1982).

i. Time-Division Multiplexing

Multiwavelength operation of a mode-locked laser has been obtained by inserting an electro-optic turner in the external cavity (Olsson and Tang, 1982). The actively mode-locked laser is driven by a dc bias and a train of short-current pulses having a repetition rate that is twice the 50-MHz round-trip frequency so that two pulses circulate in the cavity. The electro-optic tuner was placed near the output mirror and was driven with a 50-MHz rf signal that is phase locked to the 100-MHz pulse train. The tuner phase is electronically adjusted so that the pulses pass through the tuner at the extreme of its variation in optical length. The wavelength separation then corresponded to the full peak-to-peak variation of the optical tuner and was observed to be approximately 6 Å. The pulse widths, as detected by a fast photodiode, were approximately 10^{-9} sec. Similar techniques can be applied to a ring external cavity. The number of wavelengths can be increased by driving the laser at a higher harmonic of the round-trip frequency.

j. Optical Pumping

Picosecond pulses have also been generated from a bulk GaAs crystal by using synchronous two-photon optical pumping by radiation from a mode-locked 1.06-μm Nd–glass laser (Cao et al., 1981a,b; Vaucher et al., 1982). The experimental arrangement is shown in Fig. 26. The $4 \times 5 \times 10$-mm^3 GaAs crystal was AR coated, positioned in a dewar with glass windows, and cooled to approximately 97 K. The crystal was positioned near the high-reflecting end of the external cavity. The output from the crystal was focused on the 75%-reflecting output mirror by the image relay system. Bursts of

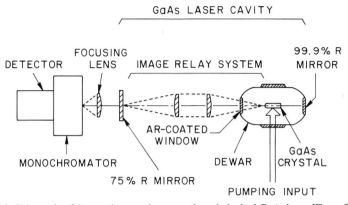

FIG. 26. Schematic of the synchronously pumped mode-locked GaAs laser. [From Cao *et al.* (1981a).]

picosecond pulses from amplified spontaneous emission have also been obtained by removing the image relay system in Fig. 26 (Cao *et al.*, 1981c).

The crystal was synchronously pumped by a pulse train of approximately 100 pulses of 1.06-μm radiation spaced by 6 nsec with each pulse having a single-sided exponential pulse shape with a 7-psec FWHM. Since the cross section for two-photon excitation is relatively small, pump thresholds of 115 mW/cm^2 were required to attain the threshold. The excited volume is larger by a factor of $10^4 - 10^6$ than that for a semiconductor laser; hence, very high output powers were attained.

The pulse widths were obtained from the autocorrelation of the 1.06-μm radiation, the cross-correlation of the GaAs emission with the 1.06-μm radiation using phase matching in an angle-tuned potassium dehydrogen phosphate (KDP) crystal. For 7-psec, 1.06-μm-FWHM excitation, the shortest pulse widths from the GaAs crystal were less than 7-psec FWHM. Mismatching the cavity length of the pump source and the mode-locked GaAs-type laser significantly reduced the GaAs crystal output power. The reduction is consistent with a carrier lifetime of 100 psec, which is in good agreement with the expected value for the doping level used.

Mode locking has also been observed at 1.1 and 1.2 μm by using thin InGaAsP-type layers longitudinally pumped by a mode-locked Kr$^+$ laser (Putnam *et al.*, 1982). The experimental arrangement is shown in Fig. 27. The active layer consists of a 2–3-μm-thick InGaAsP-type platelet grown by liquid-phase epitaxy on an InP-type substrate and is followed by a 5–6-μm-thick InP-type cladding layer. The InP-type substrate is then removed by using a HCl-selective etch, and the two remaining layers were attached to the sapphire plate by using silicone oil. The layers were placed in a liquid-nitrogen dewar and cooled to 85 K.

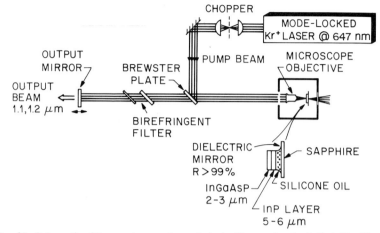

FIG. 27. Schematic of the synchronously mode-locked laser using the InGaAsP epi-layer on an InP substrate and cooled through the sapphire plate. [From Putnam *et al.* (1982).]

The external cavity consisted of the dielectrically coated sapphire plate and the partially transmitting output mirror. The Kr⁺-ion laser was mode locked to produce 100-psec pulses at 0.647 μm. The beam, after passing through the chopper, was injected into the cavity by the Brewster angle plate and focused on the InGaAsP-type layer by a microscope objective. Tuning the cavity length to match the repetition rate of the pump was achieved by translation of the output mirror. Pulses as short as 6.5 psec were obtained by assuming single-sided exponential pulse shapes.

An interesting advantage of the thin-film laser is the ability to vary the output wavelength by several techniques. Tuning over 80 Å was obtained by varying the birefringent filter. Translating the slightly nonuniformly thick crystal caused a shift in the Fabry–Perot resonance and yielded a tuning range of 260 Å. By increasing the sample temperature a tuning of 80 Å to longer wavelength was obtained. The emitting wavelength can, in principle, also be tuned by varying the composition from InP ($\lambda = 0.9$ μm) to InGaAs ($\lambda = 1.6$ μm).

Mode locking has also been observed at 0.4965 μm by using a CdS-type platelet in a similar external cavity arrangement (Roxlo and Salour, 1981a,b). When the CdS platelet was cooled down to 95 K and pumped with 100-psec-long pulses from a 0.476-μm actively mode locked Ar⁺ laser, single-sided exponential pulses as short as 4.7 psec have been observed. However, because of the large signals in the tails of the pulse the actual pulse width is estimated to be ≈ 8 psec.

By using a similar arrangement, mode locking of CdSe and CdSSe has also been achieved, thus extending the wavelength range from 0.500 to 0.700 μm

(Roxlo *et al.,* 1982a,b). By replacing the output mirror with a wavelength-selective tuner consisting of a prism and an output reflector, the tuning characteristics of the CdS-type platelet laser has been demonstrated (Roxlo *et al.,* 1981).

4. PASSIVE MODE LOCKING

a. *Effect of Saturable Absorption*

Passive mode locking requires a small, localized region of saturable absorption in addition to a region of gain. Such lasers by themselves tend to emit self-persistent pulsations, and the mode-locking and self-pulsation phenomena have many features in common. In the case of self-persistent pulsation, the cycle begins at the end of the preceding pulse with the exponential buildup of the depleted carrier density and gain and the subsequent recovery of the saturable absorption (van der Ziel, 1979). The threshold is impeded by the absorption in the saturable loss region. As the stimulated emission begins to increase exponentially, the absorption in the lossy region begins to decrease rapidly resulting in a rapid increase in the net gain. The rapid increase in the net gain results in the emission of a short pulse of light that is limited by the corresponding decrease in carrier density and the associated reduction in the net gain. After the emission of the pulse, the cycle repeats itself. With self-pulsating lasers, the pulse in each cycle begins from the spontaneous emission, and the pulse widths are typically 50 – 100 psec. The pulse-to-pulse repetition rate is determined by the time required to reach the threshold condition and increases with current.

By using an external resonator to store the emitted pulse, the requirement for the pulse to start up again each time from spontaneous emission is eliminated. When the stored pulse is reinjected into the active region, the amplitude increases rapidly in the high-gain region, and the high amplitude bleaches the saturable absorber, resulting in significant pulse narrowing. For successful mode locking, the absorbing region must saturate more easily than the gain decreases. High gain and efficient coupling to the cavity are also required. This can be achieved by antireflective coating of the interior laser – cavity facet. In addition, the mode-locking frequency must be higher than the self-pulsation rate; otherwise, self-pulsation will predominate over passive mode locking. In practice, this is not a severe limitation because the antireflective coating can effectively eliminate self-lasing and the pulsations of the laser. The upper-frequency limit of passive mode locking is given by the requirement that sufficient population inversion to initiate mode locking is obtained during the round-trip time. The frequency range can be increased by operating at higher injection currents.

The first observations of passive mode locking were made by using strip

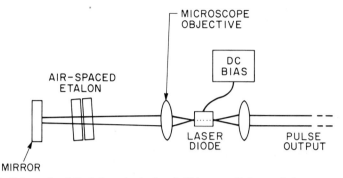

FIG. 28. Schematic of diode laser in the bandwidth-controlled extended resonator. [From Ippen *et al.* (1980a).]

buried heterostructure lasers (Ippen *et al.*, 1980a,b). The experimental arrangement (Fig. 28) consists of a strip-buried heterostructure laser with an antireflective-coated interior facet, a collimating microscope objective, an air-spaced etalon used to select the bandwidth, and a reflecting mirror. The repetition rate was 850 MHz. The shortest pulses had a 5.1-psec FWHM with a sech²-type pulse shape. The relatively smooth structure of the autocorrelation suggests that the pulses are approximately transform limited.

The saturable loss in these lasers results from degraded regions that can occur randomly throughout the active region (Paoli, 1979; Hartman *et al.*, 1979). With continued operation, the size of the degraded regions increases. Considerably shorter pulses have been obtained from such lasers. The optical field intensity is highest near the exit facet and pulses as short as 1.3 psec were observed in which the degradation is localized at the exit facet (Ippen *et al.*, 1980b).

b. Controlled Introduction of Saturable Absorption

Saturable absorption has been introduced in GaAs-type lasers in a controlled manner by the proton bombardment of one of the mirror facets (van der Ziel *et al.*, 1981a). With proton bombardment, the range of penetration and the proton dosage are externally variable. The resulting saturable absorption as evidenced by laser self-pulsations was studied under controlled conditions. The bombardment introduces deep levels that greatly reduce the radiative efficiency and the carrier lifetime of the bandgap emission. The electrical resistance of the damaged region is higher than for the bulk of the active region; hence, the current and carrier densities are reduced. In addition, the optical loss is increased. Both effects can be modified by an appropriate heat treatment that reduces the optical absorption by a low-temperature anneal, while the high electrical resistivity is removed by a longer heat treatment at higher temperatures.

The GaAs-type lasers used to study the effect of proton bombardment had the separate optical confinement buried optical guide geometry that utilizes strong refractive index guiding and is highly stable against self-focusing and self-pulsation (Henry *et al.,* 1981a). Pulsating outputs are frequently observed in gain-guided lasers in which the laser stripe is defined by proton bombardment. The pulsations in these lasers have been attributed to self-focusing of the optical beam (Lang, 1980; van der Ziel, 1981a) and to nonradiative regions that introduce saturable absorption (Paoli, 1979; Dixon and Joyce, 1979; van der Ziel *et al.,* 1979; Henry, 1980).

Several dosage levels were studied. Of the untreated chips all were lased and had approximately an 80-mA threshold, but none exhibited pulsations. Five usable lasers were obtained from the bar with the 10^{15}-cm^{-2} dosage. The threshold of these lasers had increased to approximately 120 mA, but the dosage level was inadequate to cause pulsations. With a 3×10^{15}-cm^{-2} dosage the threshold increased to approximately 145 mA, and four out of five lasers showed pulsation behavior. For the 10^{16}-cm^{-2} dosage, the chips did not lase initially. After annealing for 5 min at 500°C to remove some of the optical loss, the lasers exhibited a pulsating output.

The 300-keV protons have a projected range ≈ 2.4 μm in GaAs. The distribution is highly asymmetrical and is skewed toward the facet surface. Hence, the proton damage is reasonably uniformly distributed over the first 2.4 μm. Due to carrier diffusion, the carrier density will also be significantly reduced for the next few micrometers. Hence, the damaged region extends for 4–5 μm from the facet. A theory of pulsation behavior resulting from a nonradiative region has recently been described (Henry, 1980; Henry *et al.,* 1981b). It closely describes the effects reported here. The minimum length of a nonradiative region required for the onset of pulsation was ≈ 3 μm. The length depends on the detailed values of the parameters that enter the calculation and is consistent with the observations reported here.

The pulsation behavior as a function of pulsed current amplitude is shown in Fig. 29 for a laser with a 10^{16}-cm^{-2} implant and an anneal at 500°C for 5 min. The top curve is a monitor of the approximately 14-nsec-long current pulse and the lower traces depict the optical intensity. At the threshold (≈ 105 mA) the optical emission rises approximately exponentially and weak lasing begins at the end of the pulse. The emission decays after termination of the current pulse. With increasing current, the output exhibits strong pulsations starting at a repetition rate of ≈ 500 MHz and increases with increasing current. The intensity minima between the pulses correspond to the spontaneous emission levels. As the current is further increased, the pulsations

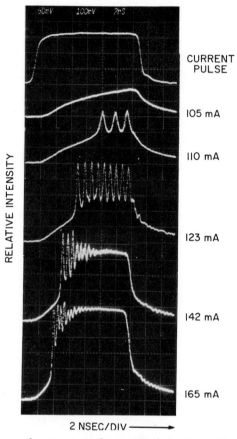

FIG. 29. Pulse response of a separate-confinement buried optical guide laser with 10^{16} cm^2 proton bombardment at 300 keV and annealed at 500°C for 5 min. The top curve shows the current pulse. The lower curves show the optical emission at increasing pulsed currents and illustrate the evolution of the pulsation behavior. [From van der Ziel *et al.* (1981a).]

become quenched. This effect results from the increase in the optical damping of the nonlinearity at high output power.

c. Subpicosecond Pulse Generation

Subpicosecond pulses from passively mode-locked lasers were first observed by using buried optical guide geometry lasers (van der Ziel *et al.*, 1981b). Saturable absorption was introduced at the exit facet by bombardment with 3×10^{15} cm^{-2} at 600 keV (van der Ziel *et al.*, 1981a,b). After bombardment, the lasers were annealed at 500°C for 3 min. The 600-keV protons have a range of ≈ 6 μm in GaAs. Carrier diffusion will reduce the

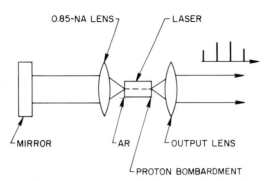

FIG. 30. Schematic of the passive mode-locked buried optical guide laser in an external cavity. [From van der Ziel *et al.* (1981b).]

carrier density for the next several micrometers. Thus, the nonlinear region extends for 8–10 μm from the exit face. The round-trip time through the nonlinear region is 0.23–0.29 psec, which is less than half the measured pulse width. The interior laser facet was AR coated with a $\lambda/4$ layer of SiO.

The experimental arrangement (Fig. 30) was similar to Figs. 12 and 28 except that the bandwidth-limiting etalon was removed and no rf current was applied. The external cavity consisted of a 0.85-NA microscope objective that collimated the light and a mirror. The optical cavity length was ≈ 15 cm, corresponding to a round-trip transit time of $\approx 10^{-9}$ sec. The emission from the AR-coated facet was reinjected into the active region by careful adjustment of the lens and mirror.

The autocorrelation consists of a series of peaks of 1-psec FWHM (Fig. 31). Assuming a sech^2-type pulse intensity, the laser pulses have a 0.65-psec FWHM. The multiple pulses result from the reflection of part of the pulse at the AR-coated facet and its subsequent amplification by the gain medium and pulse shaping by the saturation absorber. This process continues until the gain medium is exhausted and the pulse train has left the active medium. The autocorrelation at zero delay in Fig. 31 is the sum of the self-correlations of the pulses. The additional peaks at multiples of the 11-psec round-trip transit-delay time is the sum of the correlations of pulses that are separated by this time interval. The correlations have identical widths that indicate that the pulse widths are equal. By varying the length of the external cavity, subpicosecond pulses have been obtained with repetition rates between 0.6 and 2.7 GHz. The upper frequency is the short-cavity limit determined by the length of the collimating objective and is not a fundamental limitation of the mode-locking repetition rate.

The time dependence of the emission is the Fourier transform of the optical spectrum. The multiple pulse behavior is explained by the incom-

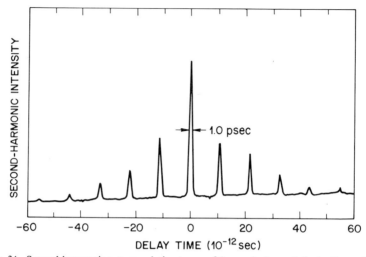

FIG. 31. Second-harmonic autocorrelation traces of the passively mode-locked laser. Assuming sech^2 pulses, the autocorrelation FWHM of 1 psec corresponds to a 0.65-psec FWHM of the pulse intensity. The additional peaks at nonzero delay time are the correlations of pulses separated in time by multiples of the 11-psec round-trip transit time through the laser. $I_0 = 93$ mA, and $f = 1.040$ GHz. [From van der Ziel *et al.* (1981b).]

plete filling of the optical spectrum shown in Fig. 32, which consists of several Fabry–Perot modes of the laser cavity resulting from the residual reflectivity of the AR-coated laser facet. The mode spacing is 2.1 Å, and the corresponding round-trip transit time through the laser is $T_L = \lambda^2/\Delta\lambda c = 11$ psec, which agrees with the separation in Fig. 31. The modes are broadened to approximately 1 Å. This results from a mode shift to longer wavelength during the emission of the pulse, which is due to the increase in the carrier-dependent refractive index as the population is reduced by stimulated emission. The short pulses result from the interference of coherent waves of the mode groups. Measurement of the spectral FWHM corresponding to the autocorrelation of Fig. 31 yields $\Delta\lambda = 12$ Å. Hence $\Delta\nu t_p = 0.34$, which agrees with $\Delta\nu t_p = 0.3148$ for sech^2-type pulses.

From Fig. 31, the FWHM of the pulse intensity envelope is estimated to be 20 psec. The corresponding spectral width of a mode in Fig. 32, after deconvolution of the mode shift and resolution of the spectrometer, is approximately 0.4 Å. This width corresponds to approximately 17 modes of the composite cavity. The bandwidth product is 0.34, which is in good agreement with the assumption of a sech^2-type shaped pulse envelope.

In order to confirm the importance of the wide spectral bandwidth required for subpicosecond pulses, a Fabry–Perot etalon that limited the spectral width to a single laser-mode group was inserted in the cavity. The

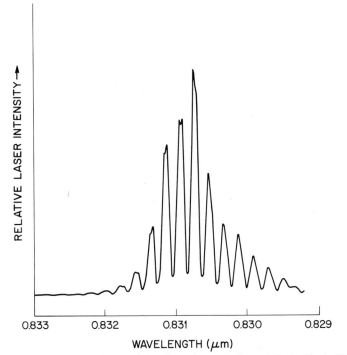

FIG. 32. Optical spectrum of the emission from the passively mode-locked laser. [From van der Ziel *et al.* (1981b).]

mode locking was enhanced by injecting several milliamps of rf current at 1 GHz. The pulse width obtained from the autocorrelation curve was 20 psec. Assuming the same spectral bandwidth, the bandwidth product is 0.35, again indicating transform-limited operation.

These results clearly indicate that the modes over the whole mode spectrum are locked. This differs from previous mode-locking experiments in which only the composite cavity corresponding to a Fabry–Perot mode group of the laser was locked, and the composite cavity modes of different Fabry–Perot mode groups were not locked. It is the coherence between the mode groups that is responsible for the subpicosecond transform-limited pulses. The coherence of these modes is evident from the absence of a broad underlying band in the autocorrelation (Fig. 31) as well as the remarkable enhancement (by a factor ≈ 10) in the second-harmonic intensity.

Subpicosecond pulses have also been generated by using AlGaAs-type channel substrate-planar (CSP) lasers (Yokoyama *et al.,* 1981). The laser diode has been aged until the threshold had increased by about 50% and exhibited self-pulsations suggesting the presence of a substantial absorbing

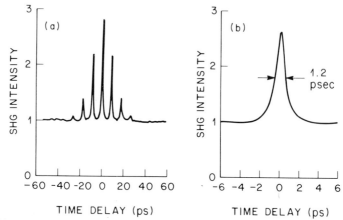

FIG. 33. (a) Second-harmonic-generation autocorrelation measurement of generated subpicosecond coherent optical pulses. (b) Magnification of the central part of trace (a). [From Yokoyama *et al.* (1982).]

region. With a cavity length of 23 mm and $I = 1.5I_{th}$, the autocorrelation traces of Fig. 33 were obtained. The separation between the pulses is 8.8 psec, the round-trip time in the laser cavity, and the time interval between pulses is ≈ 150 psec, corresponding to the round trip through the composite cavity. Assuming a $sech^2$-type shape, the envelope has a 16-psec FWHM and is seen to consist of approximately three relatively intense short pulses. The central part of the autocorrelation trace has a FWHM ≈ 1.2 psec, and assuming a $sech^2$-type pulse shape, a pulse width of 0.78 psec is estimated. The authors assumed a Lorentzian profile and obtained a pulse width of 0.58 psec. The corresponding spectral FWHM is 35 Å, which yields a bandwidth–pulse width product larger than the transform limit. The pulse width increases with increasing cavity length, and passive mode locking could not be observed for external cavity lengths greater than 6 cm.

d. Nonuniform Current Injection

An alternative technique for providing sufficient saturable absorption to induce passive mode locking is to utilize the nonlinearity introduced by nonuniform current injection obtained in a split contact geometry (Harder *et al.*, 1983). This method does not rely on any damage of the crystal and may consequently be more reliable. For a complete description of such an optoelectronic device, the electrical aspects of the driving circuit have to be included. The necessary condition for small-signal instability can be shown to be less restrictive than the condition that the loss saturates more easily than does the gain. Thus a nonuniform pumped laser having a region with a linear gain dependence on carrier density and a region of absorption with a bimolecular recombination rate can either be bistable or pulsating, depending on the biasing condition (Harder *et al.*, 1981, 1982). Switching of such a bistable laser can also be obtained by varying the amount of optical feedback as well as by changing the voltage across the absorber section (Lau *et al.*, 1982).

The lasers had the buried heterostructure geometry and were provided with split contacts consisting of a long gain section and a short absorbing section (Harder *et al.*, 1981, 1982). An SiO antireflective coating was applied to the mirror facet of the gain section in order to increase the coupling to the external optical cavity and to suppress the Fabry–Perot resonances of the laser cavity. The laser was operated by forward biasing the gain section by a current source I above threshold and the absorbing section with a voltage source V with a series resistor R_2 (Fig. 34). The optical cavity length was varied between 10 and 30 cm, corresponding to a pulsation frequency between 500 MHz and 1.5 GHz.

The dependence on the light output of current through the gain section I_1 is shown in Fig. 34. The light versus current $(L–I)$ curve depends on the biasing resistor R_2 that is in series with the absorber section. For large R_2 $(R_2 = 200\ \text{k}\Omega)$, corresponding to current biasing, the pump rate is constant and the device displays bistability with a large hysteresis. For small R_2 $(R_2 = 330\ \Omega)$, corresponding to voltage biasing, the carrier density in the absorber section is clamped to a value determined by the applied voltage. The hysteresis loop of the device is reduced to a single curve with a very high

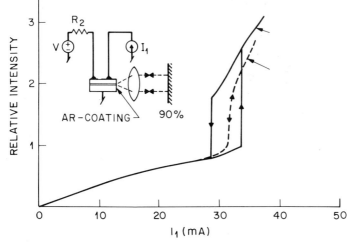

FIG. 34. Light-current characteristic of a double contact laser in an external optical cavity for current biasing $(R_2 = 200\ \text{k}\Omega,\ V = -20\ \text{V};\ \text{———})$ and voltage biasing $(R_2 = 330\ \Omega,\ V = 1.05\ \text{V};\ \text{– – –})$ as measured with the avalanche photodiode (APD). [From Harder *et al.* (1983).]

differential quantum efficiency. Through such a voltage biasing the laser can be forced into operating points not attainable with current biasing (Harder *et al.,* 1982, 1983). When the absorber section is current biased, the light output is stable. Only weak noise peaks can be seen at multiples of the frequency corresponding to the round-trip time. Voltage biasing of the laser causes unstable operation and self-pulsations. The shortest pulses were observed when the laser was biased just above the light jump. For a 30-cm-long external cavity, an autocorrelation width of 53 psec was obtained, corresponding to a 37-psec-FWHM pulse width in the Gaussian pulse-shape approximation.

The three-terminal, split-contact laser operating in a nonuniform current injection regime appears to offer a simple technique for obtaining relatively long pulses. A better understanding of the device characteristics and operation may lead to a significant further reduction in pulse width.

An alternative to the split-contact laser—the cleaved, coupled, cavity laser—has recently been devised (Tsang *et al.,* 1983a,b). In this geometry the laser is cleaved into two sections with the sections separated by an air gap of $1-5$ μm. The separation and accurate alignment of the exit facets are most easily achieved with lasers having a relatively thick $(3-5$-μm) bonding pad that is used as a hinge between the two active-laser sections. The device operates as two electrically isolated lasers coupled by the narrow air gap. By an appropriate adjustment of the air space, a coherent interference of the output is obtained at one wavelength, and the coupled laser emits in a single mode.

Mode-locked-operation of such two-section lasers has been obtained by using the rear section as an electronically controllable saturable absorber and the first section as the gain medium (Tsang *et al.* 1983c). The light from the rear absorber facet was collimated by a lens, and a grating served as a reflecting element. By appropriate adjustment of the saturable absorber section, detector-limited pulses (200 psec) were obtained.

In addition, gated operation of the mode-locked laser was obtained by using a laser split into three sections, with the middle and rear sections operating as before and the front output section acting as an electronically controlled gate (Fig. 35) (Tsang *et al.,* 1983c). By using such a gate, the number of mode-locked pulses in a burst was controlled by pulse modulation of the gate section.

e. *Pulse Broadening Due to Dispersion*

The dispersion of the refractive index of the laser causes a temporal broadening of a short pulse as it passes through the medium. It adds quadratically to the spectral bandwidth-limiting term in determining ω_D in Eq. (32), which in turn enters in the coupled mode locking equations [Eqs. (71)–(73)].

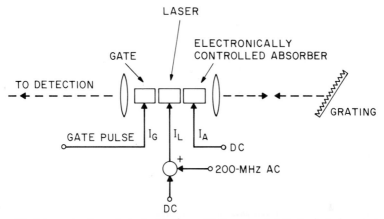

FIG. 35. Schematic of a mode-locked C³ laser with three coupled diodes for lasing, electronic gating, and optical absorption control. [From Tsang *et al.* (1983c).]

The broadening tends to counteract the pulse-narrowing effect of the gain and saturable absorbing region. The dispersive effect has been calculated (Ho, 1979b; van der Ziel and Logan, 1983).

For the passive mode-locking case, the half-pulse energy E_0 is obtained by eliminating $(\omega_D \tau_p)^{-2}$ from Eqs. (71) and (73). Substitution back into Eq. (71) or (73) yields the pulse width τ_p and into Eq. (72) yields the phase shift δ. Assuming E_0 has been determined from Eqs. (71)–(73), we let

$$K = [L_{a0}(E_0/E_a)^2 - G_{i0}(E_0/E_g)^2]^{1/2}, \tag{83}$$

and from Eq. (73), the pulse width is

$$\tau_p = 2/\omega_D K. \tag{84}$$

We now discuss two limits of ω_D given by Eq. (32). For the first, an external bandwidth-limiting element that limits the emission to a narrow band is assumed to exist in the cavity. An example given earlier is the Fabry–Perot etalon, which limits the emission to a single, composite cavity mode of width $\Delta\nu$. In this case, the first term in Eq. (32) dominates, and the pulse width is given by

$$\tau_p = \sqrt{L_s}/(\pi K \, \Delta\nu), \tag{85}$$

where $\omega_s = 2\pi \, \Delta\nu$.

For sech²-type pulse shapes, the transform limit is

$$\tau_p = 0.3148/\Delta\nu, \tag{86}$$

and hence $\sqrt{L_s}/K \approx 1$. For a laser operating at 0.83 μm with a spectral bandwidth of 0.4 Å, and assuming that the transform limit applies, the calculated pulse width is 18 psec, which is in good agreement with measured values.

In the second case, we assume that all spectral bandwidth-limiting elements have been removed. The laser is mode locked over the full spectral range, and ω_D is limited by the dispersion term in Eq. (32). The pulse width is given by

$$\tau_p = 2 \left(\frac{d^2 k}{d\omega^2} \frac{l_\phi}{L_0} \right)^{1/2} \Bigg/ K. \tag{87}$$

The dispersion is given by

$$\frac{d^2 k}{d\omega^2} = \frac{2\pi}{\omega_0^2} \left(\lambda \frac{\partial^2 n}{\partial \lambda^2} \right), \tag{88}$$

where $\lambda \, \partial^2 n / \partial \lambda^2$ is the dispersion $\partial n^* / \partial \lambda$ of the group velocity refractive index

$$n^* = n - \lambda(\partial n/\partial \lambda). \tag{89}$$

The wavelength dependence of n^* for GaAs is shown in Fig. 36 (van der Ziel and Logan, 1983) and was obtained from the measured refractive index data (Casey *et al.,* 1974). It contains a relatively constant term from $n \approx 3.6$ that contains a weak band edge contribution at $\lambda \approx 0.875$ μm. The derivative term in Eq. (89) contributes to the strong dispersion of n^* in Fig. 35. The theory of group refractive index in symmetric slab waveguides has been discussed by Buus and Adams (1979). The dispersion of n^* has been obtained experimentally as a function of wavelength from the measured wavelength-dependent mode spacing $\Delta\lambda$ shown in Fig. 37 (van der Ziel and Logan, 1983). AlGaAs-type buried optical guide lasers that were 137 μm long and contained a 0.12-μm-wide GaAs-type active layer, symmetric 0.5-μm-wide Al$_{0.15}$Ga$_{0.85}$As-type optical waveguide layers, and Al$_{0.3}$Ga$_{0.7}$As-type cladding layers were used. This geometry results in weak confinement to the active layer with most of the light contained in the transparent optical guiding layers. Consequently, the Fabry–Perot fringes are visible over a much wider wavelength range than is usually obtained with lasers having larger confinement factors. The wavelength dependence of n* obtained from

$$n^* = \lambda^2/(2l \, \Delta\lambda) \tag{90}$$

is shown in Fig. 38. The dispersion of n^* contains the dispersion curve due to the active layer that is centered at the bandgap energy and the long-wavelength tail of n^* dispersion from the optical guiding layers. The arrows denote two wavelengths of laser emission that occur on the long-wavelength

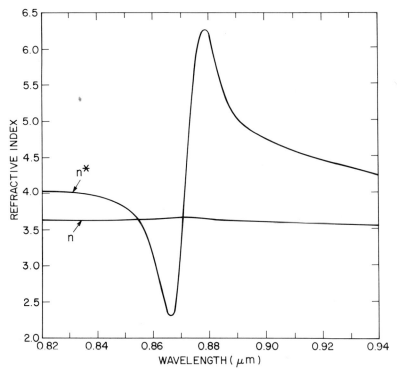

FiG. 36. Plot of the wavelength dependence of the refractive index n for GaAs and the group velocity refractive index n^* as a function of wavelength. [From van der Ziel and Logan (1983). Dispersion of the group velocity refractive index in GaAs double heterostructure lasers. *IEEE J. Quantum Electron.* **QE-19**, 164. © IEEE 1983.]

side of the dispersion curve. The first (a) corresponds to the initial wavelength and has $\partial n^*/\partial\lambda = -21.2\ \mu m^{-1}$. After aging the laser for a sufficient length of time to induce pulsations, the wavelength was found to shift to the longer wavelength (b), and the dispersion was reduced to $-7.9\ \mu m^{-1}$. Using the value at (b), for a 380-μm-long laser, we obtain the round-trip loss of the mode amplitude $L_0 = 0.95$ and $K = 0.5$. With a dispersion of 8 μm^{-1}, the pulse width from Eq. (87) is 0.25 psec, which, while approximately a factor of 2 smaller than the shortest pulse width observed in passive mode locking, is in good agreement with experimental results.

Short-pulse operation in this limit is enhanced by reducing the total dispersion effect. This can be achieved by using short lasers and by modification of the material to reduce $\lambda\ \partial^2 n/\partial\lambda^2$. For the latter, the large optical cavity design reduces the optical confinement to the active region and thus minimizes the strong dispersion associated with the band edge refractive index of

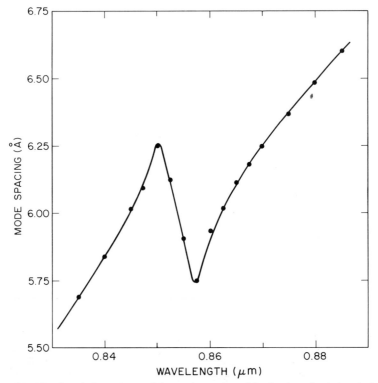

FIG. 37. Wavelength dependence of the mode spacing of the five-layer buried optical guide laser that exhibits a longitudinal-mode spectrum over a wide spectral range. $l_1 = 137 \mu m$. [From van der Ziel and Logan (1983). Dispersion of the group velocity refractive index in GaAs double heterostructure lasers. *IEEE J. Quantum Electron.* **QE-19**, 164. © IEEE.]

this layer. In addition, the damage introduced by the saturable absorption shifts the emission to longer wavelengths, also reducing the dispersion. By optimizing these effects, dispersion-limited pulses of less than 10^{-13} sec appear possible from semiconductor lasers. Such short pulse widths correspond to a transform-limited bandwidth of 72 Å for a 0.83-μm AlGaAs-type laser. This is also the approximate bandwidth of the spectral gain, a fact which suggests that it will be difficult to achieve pulses much shorter than 10^{-13} sec from mode-locked semiconductor lasers.

5. Injection Locking

The mode locking of a single laser in an external cavity results in the repetitive amplification of the reinjected component of the previous pulses. Somewhat different effects can be obtained when the modes of two separate

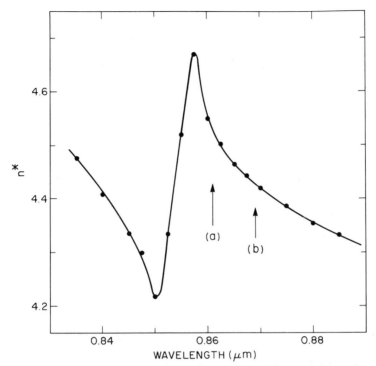

FIG. 38. Wavelength dependence of n^* for the five-layer laser. The arrows denote the peak of the emission (a) before and (b) after degradation. $l_1 = 137 \ \mu$m. [From van der Ziel and Logan (1983). Dispersion of the group velocity refractive index in GaAs double heterostructure lasers. *IEEE J. Quantum Electron.* **QE-19,** 164. © IEEE.]

laser oscillators are coupled. Typically, one is interested in using a highly single mode master oscillator and coupling the constant output of this oscillator through appropriate Faraday rotator isolators to a current-modulated slave oscillator. This requires the lasers to be reasonably well separated in space, which means that lenses are required to couple the light between the lasers. A number of investigators have studied the effect of injection coupling of longitudinal modes of two such laser oscillators (Kobayashi and Kimura, 1980, 1981; Kobayashi *et al.,* 1980; Otsuka and Tarucha, 1981; Lang, 1982).

By removing the optical isolation, each laser will inject a fraction of its emitted radiation into the other laser, thus coupling both lasers. The slight differences in the mode spacing of the two lasers produce an interference between the laser mode amplitudes. For an appropriate choice of drive currents, this interference is constructive only at a single wavelength, thus resulting in highly single mode emission even at high modulation speeds (Tsang *et al.,* 1983c).

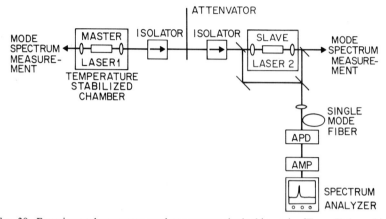

FIG. 39. Experimental apparatus used to measure the locking gain. [From Kobayashi and Kimura (1980). Injection locking characteristics of an AlGaAs semiconductor laser. *IEEE J. Quantum Electron.* **QE-16**, 915. © IEEE 1980.]

The locking of osillators has been described by Adler (1946). For a given injection and output power level P_{in} and P_{osc}, respectively, as measured inside the slave laser, half of the locking bandwidth is given by

$$\Delta f = (4\pi\tau_{ph})^{-1}(P_{in}/P_{osc})^{1/2}, \tag{91}$$

where $\Delta f_c = (4\pi\tau_{ph})^{-1}$ is the half-power bandwidth of the laser cavity.

The effect of the locking gain on the locking bandwidth has been studied by using two nearly identical single-mode AlGaAs-type lasers operating at 0.842 μm with the arrangement shown in Fig. 39 (Kobayashi and Kimura, 1980). Radiation from the master oscillator (laser 1) is collimated by a microlens, passed through two optical isolators and an optical attenuator, and focused on the slave laser (laser 2). The frequency differences between the lasers were adjusted by controlling the temperature of the laser heat sinks and by adjusting the drive current of the master oscillator.

Figure 40 shows the measured locking half-bandwidth as a function of the locked-laser to injected-laser power ratio over a range of injected currents and incident power levels. Good agreement with Eq. (91) was obtained. The maximum locking bandwidth was 1.5 GHz and was obtained with a 23-dB injection-locking gain. The maximum injection-locking gain was 40 dB where the locking bandwidth was 500 MHz and the minimum injected power was −29 dB m.

An advantage of the injection-locking scheme is that the master oscillator can be adjusted to emit in a highly single longitudinal mode. The radiation injected in the modulated slave laser causes the modulated emission from this laser also to be on the same single longitudinal mode. Single lasers that are modulated at gigahertz rates typically operate in several longitudinal

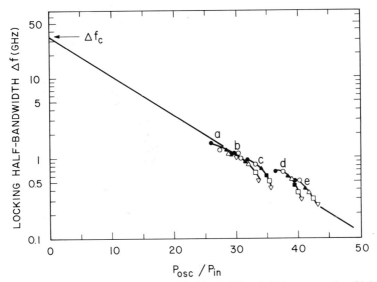

FIG. 40. Experimental measurements of locking half-bandwidth versus ratio of injection-locked laser power P_{osc} to injected power P_{in} in laser 2. ——represents theoretical values obtained from Eq. (91). Values of P_{in} are (a) 16 dB m, (b) 19 dB m, (c) 21 dB m, (d) 26 dB m, and (e) 29 dB m. Values of I/I_{th} are ●, 1.16; ○, 1.20; ▲, 1.22; △, 1.26; ■, 1.30; □, 1.34; ▽, 1.37. [From Kobayashi and Kimura (1980).]

modes, even though essentially single longitudinal mode operation may be obtained for dc operation. The multimode operation results from the transient owing to the current passing through the threshold. Two extremely deleterious effects result when such lasers are used in a single-mode fiber communications system containing dispersion. Multimode pulses propagating through a dispersive fiber are broadened over time by the dispersive effect of the fiber (Gloge, 1971; Gloge et al., 1979). At 1.5 μm, the typical dispersion is ≈ 1.5 psec/km Å (Cohen et al., 1981; Lazay and Pearson, 1982). The mode spacing of a 250-μm-long, 1.5-μm laser is ≈ 11 Å. Assuming that four dominant modes are present, the pulse delay is ≈ 45 psec/km and transmission through a 10-km span results in a pulse broadening of 0.45 nsec. This significantly limits the bit rate at which short pulses can be transmitted over a long fiber link.

The second deleterious effect is known as mode partition noise (Ogawa, 1982; Ogawa and Vodhanel, 1982). During the initial transient of pulsed operation, the emission may occur randomly in several modes, even though at later times the emission may stabilize to a single mode. When transmitted over a long dispersive fiber, the random-mode partition noise is transformed into pulse arrival-time fluctuations that lead to large error rates in high-speed data transmission. The requirements of extreme spectral purity even during

J. P. VAN DER ZIEL

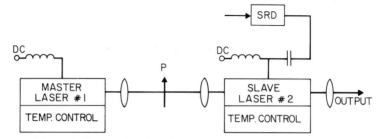

FIG. 41. Schematic diagram of the injection-locking arrangement used to obtain single-mode 58-psec FWHM pulses at 1 GHz. [From Andersson *et al.* (1982).]

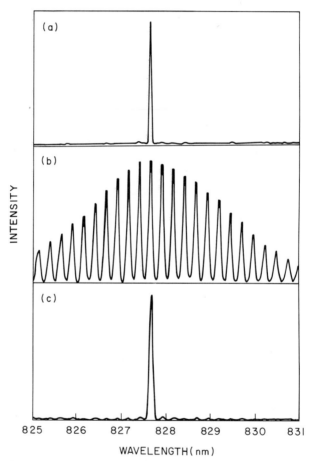

FIG. 42. Optical spectrum of (a) the single-mode master laser, (b) the free-running slave laser, and (c) the injection-locked slave laser. [From Andersson *et al.* (1982).]

the transient turn-on time of the laser have been met by using injection locking. Other single-mode schemes include distributed feedback (Akiba *et al.*, 1983), distributed Bragg reflectors (Utaka *et al.*, 1980; Sakakibara *et al.*, 1980; Koyama *et al.*, 1983), short external cavities (Duguay *et al.*, 1980; Damen *et al.*, 1981), as well as the cleaved coupled cavity (Tsang *et al.*, 1983a,b).

An injection-locking scheme resulting in short pulses at a single wavelength is shown in Fig. 41 (Andersson *et al.*, 1982). The master laser operates stably in a single longitudinal mode (Fig. 42a). The emission from this laser is collimated and injected into a second laser that is biased by both a direct current and 60-psec-long current pulses from a step recovery diode. When operated by itself, the emission from the slave laser is highly multimode (Fig. 42b). With the slave laser injection locked to the master laser, the emission has essentially the same spectral characteristic of the master laser (Fig. 42c). Pulses as short as 58-psec FWHM at 1-GHz repetition rates were obtained with single-mode operation.

By using a similar injection-locking scheme, the output from a single AlGaAs-type cw laser was injected into a second, highly modulated multimode laser. (Otsuka *et al.*, 1982). At low repetition rates (200–700 MHz), multiple relaxation oscillations were observed. At higher frequencies (933 MHz), there is time for only a single relaxation oscillation, and the emission occurs as a single pulse of approximately 47-psec duration. The emission wavelength is locked to the single mode of the master oscillator. This was achieved by using an estimated injected light power of 0.1% of the modulated laser average power. Thus such a weak cw external light source can be very effective in suppressing the multilongitudinal modes of the modulated laser. Single-mode, picosecond pulses have been obtained at up to 2-GHz repetition rates. The pulse width decreases with increasing dc bias applied to the modulated laser. Similar results have been obtained by using InGaAsP/InP-type lasers operating at 1.28 μm.

REFERENCES

Adler, R. (1946). *Proc. IRE* **34**, 351.
Akiba, S., Sakai, K., Matsushima, Y., and Yamamoto, T. (1979). *Electron. Lett.* **15**, 606.
Akiba, S., Williams, G. E., and Haus, H. A. (1981). *Electron. Lett.* **17**, 527.
Akiba, S., Utaka, K., Sakai, K., and Matsushima, Y. (1983). *IEEE J. Quantum Electron.* **QE-19**, 1052.
Andersson, T., Lundqvist, S., and Eng, S. T. (1982). *Appl. Phys. Lett.* **41**, 14.
Aspin, G. J., and Carroll, J. E. (1979). *IEE J. Solid-State Electron Devices* **3**, 220.
Auschnitt, C. P., and Jain, R. K. (1978). *Appl. Phys. Lett.* **32**, 727.
Auschnitt, C. P., Jain, R. K., and Heritage, J. P. (1979). *IEEE J. Quantum Electron.* **QE-15**, 912.

Au Yeung, J. C. (1981). *IEEE J. Quantum Electron.* **QE-17**, 398.
Au Yeung, J. C., and Johnston, A. R. (1981). *SPIE Proc.* **269**, 84.
Au Yeung, J. C., and Johnston, A. R. (1982a). *SPIE Proc.* **322**, 25.
Au Yeung, J. C., and Johnston, A. R. (1982b). *Appl. Phys. Lett.* **40**, 112.
Au Yeung, J. C., Bergman, L. A., and Johnston, A. R. (1982). *Appl. Phys. Lett.* **41**, 124.
Bradley, D. J., Holbrook, M. B., and Sleat, W. E. (1981). *IEEE J. Quantum Electron.* **QE-17**, 658.
Broom, R. F. (1969). *Electron. Lett.* **5**, 571.
Broom, R. F., Mohn, E., Risch, C., and Salathe, R. (1970). *IEEE J. Quantum Electron.* **QE-6**, 328.
Buus, J., and Adams, M. J. (1979). *IEE J. Solid-State Electron Devices* **3**, 189.
Cao, W. L., Vaucher, A. M., and Lee, C. H. (1981a). *Appl. Phys. Lett.* **38**, 653.
Cao, W. L., Vaucher, A. M., Ling, J. D., and Lee, C. H. (1981b). *IEEE J. Quantum Electron.* **QE-17**, 68.
Cao, W. L., Vaucher, A. M., and Lee, C. H. (1981c). *Appl. Phys. Lett.* **38**, 306.
Casey, H. C., Jr., Sell, D. D., and Panish, M. B. (1974). *J. Appl. Phys.* **24**, 63.
Chen, J., Sibbett, W., and Vukusic, J. I. (1982). *Electron. Lett.* **18**, 426.
Cohen, L. G., Mammel, W. M., Stone, J., and Pearson, A. D. (1981). *Bell Syst. Tech. J.* **60**, 1713.
Damen, T. C., Duguay, M. A., Shah, J., Stone, J., Wiesenfeld, J. M., and Logan, R. A. (1981). *Appl. Phys. Lett.* **39**, 142.
Dixon, R. W., and Joyce, W. B. (1979). *IEEE J. Quantum Electron.* **QE-15**, 470.
Duguay, M. A., Damen, T. C., Stone, J., Wiesenfeld, J. M., and Burrus, C. A. (1980). *Appl. Phys. Lett.* **37**, 369.
Figueroa, L. (1981). *IEEE J. Quantum Electron.* **QE-17**, 1074.
Figueroa, L., Lau, K. Y., and Yariv, A. (1980a). *Appl. Phys. Lett.* **36**, 248.
Figueroa, L., Lau, K. Y., Yen, H. W., and Yariv, A. (1980b). *J. Appl. Phys.* **51**, 3062.
Figueroa, L., Lau, K. Y., and Yariv, A. (1981). *SPIE Proc.* **269**, 92.
Fork, R. L., Greene, B. I., and Shank, C. V. (1981). *Appl. Phys. Lett.* **38**, 671.
Glasser, L. A. (1978). *Electron. Lett.* **14**, 725.
Glasser, L. A. (1980). *IEEE J. Quantum Electron.* **QE-16**, 525.
Gloge, D. (1971). *Appl. Opt.* **10**, 2442.
Gloge, D., Marcatili, E. A. J., Marcuse, D., and Personick, S. D. (1979). *In* "Optical Fiber and Telecommunications" (S. E. Miller and A. G. Chynoweth, eds), Chap. 4. Academic Press, New York.
Goodwin, J. C., and Garside, B. K. (1983). *IEEE J. Quantum Electron.* **QE-19**, 1068.
Harder, C., Lau, K. Y., and Yariv, A. (1981). *Appl. Phys. Lett.* **39**, 382.
Harder, C., Lau, K. Y., and Yariv, A. (1982). *IEEE J. Quantum Electron.* **QE-18**, 1351.
Harder, C., Lau, K. Y., and Yariv, A. (1983). *Appl. Phys. Lett.* **42**, 772.
Harris, E. P. (1971). *J. Appl. Phys.* **42**, 892.
Hartman, R. L., Logan, R. A., Koszi, L. A., and Tsang, W. T. (1979). *J. Appl. Phys.* **50**, 4616.
Haus, H. A. (1975a). *IEEE J. Quantum Electron.* **QE-11**, 736.
Haus, H. A. (1975b). *J. Appl. Phys.* **46**, 3049.
Haus, H. A. (1975c). *IEEE J. Quantum Electron.* **QE-11**, 323.
Haus, H. A. (1976). *IEEE J. Quantum Electron.* **QE-12**, 169.
Haus, H. A. (1980a). *J. Appl. Phys.* **51**, 4042.
Haus, H. A. (1980b). *IEE Proc. Part I Solid-State Electron. Devices* **127**, 323.
Haus, H. A. (1981). *Jpn. J. Appl. Phys.* **20**, 1007.
Haus, H. A., and Ho, P. T. (1979). *IEEE J. Quantum Electron.* **QE-15**, 1258.
Haus, H. A., Shank, C. V., and Ippen, E. P. (1975). *Opt. Commun.* **15**, 29.

Henry, C. H. (1980). *J. Appl. Phys.* **51**, 3051.

Henry, C. H., Logan, R. A., and Merritt, F. R. (1981a). *IEEE J. Quantum Electron.* **17**, 2196.

Henry, C. H., Logan, R. A., and Merritt, F. R. (1981b). *J. Appl. Phys.* **52**, 1560.

Hirota, O., and Suematsu, Y. (1979). *IEEE J. Quantum Electron.* **QE-15**, 142.

Ho, P. T. (1979a). *Electron. Lett.* **15**, 526.

Ho, P. T. (1979b). *IEE J. Solid-State Electron Devices* **3**, 246.

Ho, P. T., Glasser, L. A., Ippen, E. P., and Haus, H. A. (1978a). *Appl. Phys. Lett.* **33**, 241.

Ho, P. T., Glasser, L. A., Ippen, E. P., and Haus, H. A. (1978b). *In* "Picosecond phenomena" (C. V. Shank, E. P. Ippen, and S. L. Shapiro, eds.), pp. 114–116. Springer-Verlag, Berlin and New York.

Holbrook, M. B., Bradley, D. J., and Kirkby, P. A. (1980a). *Appl. Phys. Lett.* **36**, 349.

Holbrook, M. B., Sleat, W. E., and Bradley, D. J. (1980b). *Appl. Phys. Lett.* **37**, 59.

Holbrook, M. B., Sleat, W. E., and Bradley, O. S. (1980c). *In* "Picosecond Phenomena II" (R. M. Hochstrasser, W. Kaiser, and C. V. Shank, eds.), Topics in Chemical Physics, Vol. 14, pp. 26–29. Springer-Verlag, Berlin and New York.

Ikushima, I., and Maeda, Y. (1978). *IEEE J. Quantum Electron.* **QE-14**, 331.

Inaba, H. (1983). *In* "Optical Devices and Fibers" (Y. Suematsu, ed.), Japan Annual Reviews in Electronics Computers and Telecommunications, pp. 60–81. North-Holland Publ., Amsterdam.

Ippen, E. P., and Shank, C. V. (1977). *In* "Ultrashort Light Pulses" vol. 18 (S. L. Shapiro, ed.), p. 83. Springer-Verlag, Berlin and New York.

Ippen, E. P., Eilenberger, D. J., and Dixon, R. W. (1980a). *Appl. Phys. Lett.* **37**, 267.

Ippen, E. P., Eilenberger, D. J., and Dixon, R. W. (1980b). *In* "Picosecond Phenomena II" (R. M. Hochstrasser, W. Kaiser, and C. V. Shank, eds.), p. 21. Springer-Verlag, Berlin and New York.

Ito, H., Yokoyama, Y., and Inaba, H. (1980). *Electron. Lett.* **16**, 620.

Johnson, L. F. (1980). *J. Appl. Phys.* **51**, 6413.

Kaminow, I. P., Stulz, L. W., Ko, J. S., Dentai, A. G., Nahory, R. F., DeWinter, J. C., and Hartman, R. L. (1983). *IEEE J. Quantum Electron.* **QE-19**, 1312.

Kobayashi, S., and Kimura, T. (1980). *IEEE J. Quantum Electron.* **QE-16**, 915.

Kobayashi, S., and Kimura, T. (1981). *IEEE J. Quantum Electron.* **QE-17**, 681.

Kobayashi, S., Yamada, J., Machida, S., and Kimura, T. (1980). *Electron. Lett.* **16**, 746.

Koyama, F., Suematsu, Y., Arai, S., and Tawee, T. (1983). *IEEE J. Quantum Electron.* **QE-19**, 1042.

Lang, R. (1980). *Jpn. J. Appl. Phys.* **10**, L93.

Lang, R. (1982). *IEEE J. Quantum Electron.* **QE-18**, 976.

Lau, K. Y., Figueroa, L., and Yariv, A. (1980). *IEEE J. Quantum Electron.* **QE-16**, 1329.

Lau, K. Y., Harder, C., and Yariv, A (1982). *Appl. Phys. Lett.* **40**, 369.

Lazay, P. D., and Pearson, A. D. (1982). *IEEE J. Quantum Electron.* **QE-18**, 504.

Mohn, E. (1969a). *Electron. Lett.* **5**, 261.

Mohn, E. (1969b). *Conf. Ser.—Int. Symp. on GaAs, and Inst. Phys.* No. 7, p. 101.

Morikawa, T., Mitsuhashi, Y., and Shimada, J. (1976). *Electron. Lett.* **12**, 435.

Morozov, V. N., Nikitin, V. V., and Sheronov, A. A. (1968). *JETP Lett. (Engl. Transl.)* **7**, 256.

New, G. H. C. (1972). *Opt. Commun.* **6**, 188.

New, G. H. C. (1974). *IEEE J. Quantum Electron.* **QE-10**, 115.

Ogawa, K. (1982). *IEEE J. Quantum Electron.* **QE-18**, 849.

Ogawa, K., and Vodhanel, R. S. (1982). *IEEE J. Quantum Electron.* **QE-18**, 1090.

Olsson, A., and Tang, C. L. (1981). *IEEE J. Quantum Electron.* **QE-17**, 1977.

Olsson, A., and Tang, C. L. (1982). *IEEE J. Quantum Electron.* **QE-18**, 1982.

Otsuka, K., and Tarucha, S. (1981). *IEEE J. Quantum Electron.* **QE-17**, 1515.

Otsuka, K., Tarucha, S., Kubodera, K., and Noda, J. (1982). *SPIE Proc. Tech. Symp.* **322,** 172.
Paoli, T. L. (1979). *Appl. Phys. Lett.* **34,** 652.
Paoli, T. L., and Ripper, J. (1970). *IEEE J. Quantum Electron.* **QE-6,** 335.
Putnam, R. S., Roxlo, C. B., Salour, M. M., Groves, S. H., and Plonko, M. C. (1982). *Appl. Phys. Lett.* **40,** 660.
Risch, C., and Voumard, C. (1977). *J. Appl. Phys.* **48,** 2083.
Risch, C., Voumard, C., Reinhart, F. K., and Salathe, R. (1977). *IEEE J. Quantum Electron.* **QE-13,** 692.
Roxlo, C. B., and Salour, M. M. (1981a). *Appl. Phys. Lett.* **38,** 738.
Roxlo, C. B., and Salour, M. M. (1981b). *IEEE J. Quantum Electron.* **QE-17,** 68.
Roxlo, C. B., Bebelaar, D., and Salour, M. M. (1981). *Appl. Phys. Lett.* **38,** 507.
Roxlo, C. B., Putnam, R. J., and Salour, M. M. (1982a). *IEEE J. Quantum Electron.* **QE-18,** 338.
Roxlo, C. B., Putnam, R. S., and Salour, M. M. (1982b). *Proc. Tech. Symp. SPIE* **322,** 31.
Sakakibara, Y., Furuya, K., and Utaku, K., and Suematsu, Y. (1980). *Electron. Lett.* **16,** 456.
Sala, K. L., Kenney-Wallace, G. A., and Hall, G. E. (1980). *IEEE J. Quantum Electron.* **QE-16,** 990.
Salathe, R. P. (1979). *Appl. Phys. Lett.* **20,** 1.
Shank, C. V., Fork, R. L., Yen, R., Stolen, R. H., and Tomlinson, W. J. (1982). *Appl. Phys. Lett.* **40,** 761.
Suzuki, S., and Saito, T. (1982). *Electron. Lett.* **18,** 821.
Tsang, W. T., and Logan, R. A. (1979). *IEEE J. Quantum Electron.* **QE-15,** 451.
Tsang, W. T., Olsson, N. A., and Logan, R. A. (1983a). *Appl. Phys. Lett.* **42,** 650.
Tsang, W. T., Olsson, N. A., and Logan, R. A. (1983b). *Electron. Lett.* **19,** 488.
Tsang, W. T., Olsson, N. A., and Logan, R. A. (1983c). *Appl. Phys. Lett.* **43,** 339.
Tucker, R. S., Eisenstein, G., and Kaminow, I. P. (1983). *Electron. Lett.* **19,** 552.
Utaka, K., Kobayashi, K., Kishina, K., and Suematsu, Y. (1980). *Electron. Lett.* **16,** 455.
van der Ziel, J. P. (1979). *Appl. Phys. Lett.* **35,** 116.
van der Ziel, J. P. (1981a). *IEEE J. Quantum Electron.* **QE-17,** 60.
van der Ziel, J. P. (1981b). *J. Appl. Phys.* **52,** 4435.
van der Ziel, J. P. (1983). *SPIE Proc. Tech. Symp.* **439,** 49.
van der Ziel, J. P., and Logan, R. A. (1982). *IEEE J. Quantum Electron.* **QE-18,** 1340.
van der Ziel, J. P., and Logan, R. A. (1983). *IEEE J. Quantum Electron.* **QE-19,** 164.
van der Ziel, J. P., and Mikulyak, R. M. (1980). *J. Appl. Phys.* **51,** 3033.
van der Ziel, J. P., Merz, J. L., and Paoli, T. P. (1979). *J. Appl. Phys.* **50,** 4620.
van der Ziel, J. P., Tsang, W. T., Logan, R. A., and Augustyniak, W. M. (1981a). *Appl. Phys. Lett.* **39,** 376.
van der Ziel, J. P., Tsang, W. T., Logan, R. A., Mikulyak, R. M., and Augustyniak, W. M. (1981b). *Appl. Phys. Lett.* **39,** 525.
van der Ziel, J. P., Logan, R. A., and Mikulyak, R. M. (1981c). *Appl. Phys. Lett.* **39,** 867.
Van Stryland, E. W. (1979). *Opt. Commun.* **31,** 93.
Vaucher, A. M., Cao, W. L., Ling, J. D., and Lee, C. H. (1982). *IEEE J. Quantum Electron.* **QE-18,** 187.
Voumard, C. (1977). *Opt. Lett.* **1,** 61.
Yokoyama, H., Ito, H., and Inaba, H. (1982). *Appl. Phys. Lett.* **40,** 105.

SEMICONDUCTORS AND SEMIMETALS, VOL. 22, PART B

CHAPTER 2

High-Frequency Current Modulation of Semiconductor Injection Lasers

Kam Y. Lau

ORTEL CORPORATION
ALHAMBRA, CALIFORNIA

and

Amnon Yariv

CALIFORNIA INSTITUTE OF TECHNOLOGY
PASADENA, CALIFORNIA

I. Introduction

Semiconductor lasers are potentially devices of great importance for optical-communications systems due to their small size, high efficiency, and
high speed for direct modulation. They emit radiation of sufficient direc-
tivity and beam quality to make coupling into optical fibers (especially
single-mode fibers) very efficient. Furthermore, their narrow spectral width
reduces the effect of the intrinsic chromatic dispersion of the silica fibers.
Semiconductor lasers can thus be used in high data-rate systems that are
composed of either short lengths of multimode fibers or long lengths of
single-mode fibers.

The practical direct-modulation bandwidth of semiconductor lasers is
commonly accepted to be in the lower gigahertz ($< 1 – 2$-GHz) range. Within
this frequency range, however, the modulation characteristics vary from one
laser structure to another. Although most III – V semiconductor lasers can
achieve a modulation bandwidth of $1 – 2$ GHz regardless of their specific
laser structures, only the ones appropriately designed can attain higher modulation bandwidths. Such high-modulation speeds are presently demanded
in some specialized systems such as the laser – optical-fiber system for
synchronization of airborne radar arrays (Yen and Figueroa, 1979) or similar ground systems (Lutes, 1980), and they are certainly required as the data
communications network expands to greet the arrival of the information era.

The object of this chapter is to provide the basic background on direct
modulation characteristics of semiconductor injection lasers of various
structures, to explain their various modes of behavior in physical terms, and

to explore the ultimate limitation in attaining ever higher modulation bandwidths. Various modulation techniques will be discussed and compared. The emphasis of previous experimental and theoretical work was placed more on investigating modulation characteristics of existing laser structures than on developing ideal laser structures for high-speed modulation. This chapter places its emphasis on the latter. A systematic investigation of how various device parameters affect high-frequency characteristics yields general guidelines for constructing a laser geared toward high-speed modulation. Although we will concentrate on GaAlAs/GaAs-type ternary lasers, the GaInAsP/InP-type quaternary lasers are expected to have substantially similar characteristics. Cases for which differences do occur will be pointed out. It will be assumed that the reader is familiar with the basic physics and properties of semiconductor lasers.

II. Laser Kinetics, Rate Equations, and Their Range of Applicability

Laser dynamics are commonly described by a pair of rate equations governing the photon and carrier densities inside the laser medium (Statz and deMars, 1960; Kleinman, 1964) as

$$\frac{dN}{dt} = \frac{J}{ed} - \frac{N}{\tau_s} - \alpha NP, \tag{1a}$$

$$\frac{dP}{dt} = \alpha NP - \frac{P}{\tau_p} + \beta \frac{N}{\tau_s}, \tag{1b}$$

where N is the carrier density, P the photon density in a mode of the laser cavity, J the pump current density, d the thickness of the active layer, τ_s the spontaneous recombination lifetime of the carriers, τ_p the photon lifetime of the cavity, α the optical gain coefficient, β the fraction of spontaneous emission entering the lasing mode, and e the electronic charge. Equation (1a) states that the rate of increase in carrier density is equal to the rate of current injection J/ed less the rate of carrier loss due to spontaneous recombination $-N/\tau_s$ less the loss of carriers due to stimulated recombination $-\alpha NP$. Equation (1b) states that the rate of increase in photon density is equal to the rate of photon generation by stimulated emission αNP less the loss rate of photons due to cavity dissipation $-P/\tau_s$ plus the rate of spontaneous emission into the photon mode $\beta N/\tau_s$. Equations (1) are simply a bookkeeping of the supply, annihilation, and creation of carriers and photons inside the laser cavity and describe laser dynamics in a most basic manner. More detailed explanations of laser modulation behaviors can be obtained by addition and/or modification of the terms in Eqs. (1). Although extensively used, Eqs.

(1) are obviously only approximate equations that do not take into account the fact that light actually propagates and bounces back and forth inside the laser cavity. The actual averaging process leading to Eqs. (1) and its range of validity are seldom discussed. It seems appropriate, therefore, to summarize the results of Moreno (1977) regarding the conditions under which the rate equations are applicable.

1. THE TRAVELING-WAVE RATE EQUATIONS

The starting point for the analysis of laser kinetics involves the coupled rate equations, which are basically local photon and injected-carrier conservation equations (Icsevgi and Lamb, 1969; Lamb, 1969)

$$\frac{\partial X^+}{\partial t} + c\,\frac{\partial X^+}{\partial z} = c\kappa N X^+ + \beta\frac{N}{\tau_s}, \tag{2a}$$

$$\frac{\partial X^-}{\partial t} - c\,\frac{\partial X^-}{\partial z} = c\kappa N X^- + \beta\frac{N}{\tau_s}, \tag{2b}$$

$$\frac{dN}{dt} = \frac{J}{ed} - \frac{N}{\tau_s} - \kappa c N(X^+ + X^-), \tag{2c}$$

where X^+ and X^- are the forward and backward propagating photon densities (which are proportional to the light intensities), c the group velocity of the waveguide mode, κ the gain constant in reciprocal centimeters (unit carrier density), and z the distance along the active medium ($z = 0$ is the center of the laser). The following simplifying assumptions were made in writing Eqs. (2).

(1)   The quantities $X^\pm$ describe the photon densities at a given position in the laser cavity at time t integrated over the lasing linewidth of the mode of oscillation, which is assumed to be much narrower than the homogeneously broadened laser transition-line profile.

(2)   The optical gain is a linear function of the injected carrier density N.

(3)   Variations of the carrier and photon densities in the lateral dimension are not significant.

(4)   Diffusion of carriers can be ignored.

Here (1) and (2) are very reasonable assumptions that can be deduced from detailed analysis (Stern, 1973; Casperson, 1975; Henry et al., 1980). The representation of the semiconductor laser as a homogeneously broadened system can also be derived from basic considerations (Pantell and Puthoff, 1969) and has been confirmed experimentally (Brosson et al., 1981). Transverse modal and carrier diffusion effects, ignored in assumptions (3) and (4), can lead to modifications of the dynamic behavior of lasers (Chinone et al., 1978a; Wilt et al., 1981). This will be discussed in Part III.

Equations (2) are to be solved subject to the boundary conditions

$$X^-(\tfrac{1}{2}L) = RX^+(\tfrac{1}{2}L),\tag{3a}$$

$$X^+(-\tfrac{1}{2}L) = RX^-(-\tfrac{1}{2}L),\tag{3b}$$

where L is the length of the laser and R the reflectivity of the end mirrors. The steady-state solution of Eqs. (2) gives the static photon and electron distributions inside the laser medium and has been solved by Casperson (1975, 1977). The solution is summarized as

$$X_0^+(z) = (ae^{u(z)} - \beta)/\kappa c\tau_s,\tag{4a}$$

$$X_0^-(z) = (ae^{-u(z)} - \beta)/\kappa c\tau_s,\tag{4b}$$

where the zero subscript denotes steady state quantities and a is a quantity given by the transcendental equation

$$(1 - 2\beta)\zeta + 2a \sinh \zeta = gL/2,\tag{5}$$

where

$$e^\zeta = \frac{1}{2}\left(\sqrt{\frac{(R-1)^2\beta^2}{(Ra)^2} + \frac{4}{R}} + (R-1)\frac{\beta}{Ra}\right),\tag{6}$$

$g = \kappa J_0\tau_s/ed$ is the unsaturated gain, and $u(z)$ is given transcendentally by

$$(1 - 2\beta)u(z) + 2a \sinh u(z) = gz.\tag{7}$$

The electron density $N_0(z)$ is given by

$$\kappa c N_0(z) = g/[1 + 2a \cosh u(z) - 2\beta].\tag{8}$$

Figure 1 shows plots of $X_0^+(z)$, $X_0^-(z)$, and $g_0(z) = \kappa c N_0(z)$ for a 300-μm laser with three values of end-mirror reflectivities. The high nonuniformity in the distributions becomes apparent at low reflectivities.

2. SPATIALLY AVERAGED RATE EQUATIONS AND THEIR RANGE OF VALIDITY

Equations (2) constitute a set of three coupled, nonlinear differential equations in two variables and do not lend themselves to easy solution. Considerable simplification can be made if the spatial variable is integrated over the length of the laser. Such simplification is valid only when the end-mirror reflectivity is sufficiently large. A more precise definition of the range of validity is given in what follows.

We start by integrating Eqs. (2a) and (2b) in the z variable and taking their sum:

$$\frac{dP^*}{dt} + \frac{2c(1 - R)P(L/2)}{L(1 + R)} = c\kappa(NP)^* + 2\beta\frac{N^*}{\tau_s},\tag{9}$$

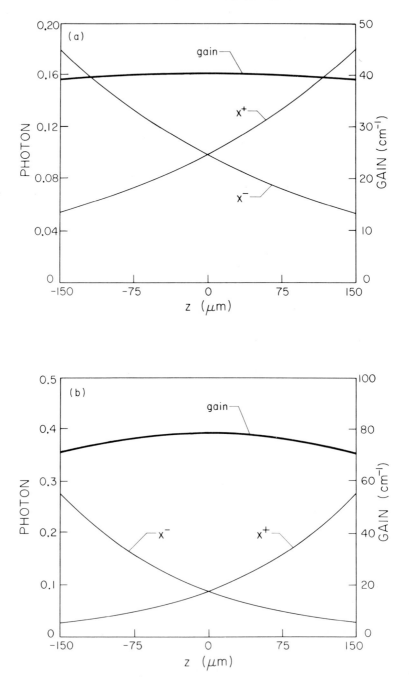

FIG. 1. Static photon and electron density distributions inside laser diodes with mirror reflectivities of (a) 0.3, (b) 0.1, and (c) 0.9.

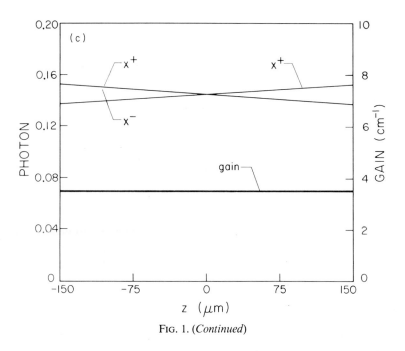

FIG. 1. (*Continued*)

where the asterisk denotes the spatial average $\int_{-L/2}^{L/2} dz/L$, $P = X^+ + X^-$ is the total local photon density, and the boundary conditions [Eqs. (3)] have been used. Equation (2c) integrates straightforwardly to

$$\frac{dN^*}{dt} = \frac{J}{ed} - \frac{N^*}{\tau_s} - \kappa c(NP)^*, \tag{10}$$

where we have assumed a uniform pump current density; J = const in z.
   Introducing factors f_1 and f_2,

$$f_1 = (NP)^*/N^*P^*, \tag{11}$$

$$f_2 = P(\tfrac{1}{2}L)/P^*(1 + R), \tag{12}$$

one can write the spatially averaged rate equations [Eqs. (9) and (10)] in the form

$$dP^*/dt = c\kappa f_1 N^*P^* - 2c(1 - R)f_2(P^*/L) + 2\beta(N^*/\tau_s), \tag{13}$$

$$dN^*/dt = (J/ed) - (N^*/\tau_s) - \kappa c f_1 N^*P^*, \tag{14}$$

which are recognized as the commonly used rate equations if the conditions

$$f_1 = 1, \tag{15}$$

$$f_2 = -\tfrac{1}{2} \ln R/(1 - R) \tag{16}$$

are satisfied. The first of these conditions requires that for the quantities N and P, the spatial average of the product equals the product of the spatial averages. This condition is not satisfied in general but will be true if the electron density N is uniform, as in the case when R approachs unity, which is apparent from Fig. 1c. The second condition requires that the photon loss rate [Eq. (13)] be inversely proportional to the conventional photon lifetime. It will also be satisfied if R is very close to unity, since both Eqs. (12) and (16) converge to $\frac{1}{2}$ in this limit.

A more precise delineation of the range of validity for Eqs. (15) and (16) is obtained by calculating f_1 and f_2 from the exact, steady-state solution [Eqs. (4)–(8)] and comparing them with Eqs. (15) and (16). Figure 2 shows plots of f_1 and $1/f_2$ as a function of end-mirror reflectivity R. The calculation was done with the laser biased above the threshold. The dotted lines are the ideal values of f_1 and f_2 given by Eqs. (15) and (16). The figure indicates that the usual rate equations will hold for R larger than approximately 0.2. Figure 3 shows plots of $1/f_2$ as a function of the pumping level expressed in terms of the total unsaturated gain gL. It shows that the usual rate equations are not valid below the lasing threshold as well as for low reflectivities or a low spontaneous emission factor β.

These results lead to the conclusion that the simple rate equations, expressed in the form (where the N and P now denote averaged quantities)

$$\frac{dN}{dt} = \frac{J}{ed} - \frac{N}{\tau_s} - \alpha NP, \qquad (17a)$$

$$\frac{dP}{dt} = \alpha NP - \frac{P}{\tau_p} + \beta \frac{N}{\tau_s}, \qquad (17b)$$

where $1/\tau_p = (c/L) \ln(1/R)$ is the classical photon lifetime and $\alpha = \kappa c$, will hold if the end-mirror reflectivity is greater than 0.2 and the laser is above the threshold. The spontaneous emission factor β in Eq. (17b) is a factor of two higher than that defined in Eqs. (2), due to the inclusion of photons propagating in both directions. Common GaAs-type lasers with mirrors formed by the cleaved crystal facets have a reflectivity of 0.31 and are thus well within the scope of Eqs. (17).

Another factor that can render the spatially uniform assumption invalid is evident when fast phenomena, i.e., occurring on the time scale of the cavity transit time) are considered. It is obvious that the cavity lifetime and the concept of cavity modes appearing in Eq. (17b) are no longer applicable on that time scale. In common semiconductor lasers, in which the cavity length is approximately 300 μm, the cavity transit time is about 3.5 psec. The usual rate equations are therefore not applicable in describing phenomena shorter than about 5 psec or at modulation frequencies higher than 60 GHz.

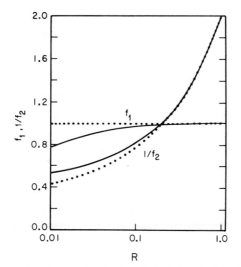

FIG. 2. Variations of f_1 and $1/f_2$ with R when $\beta \leq 10^{-3}$ and $gL > 10$; ——, exact; $\cdots$, averaged. [From Moreno (1977).]

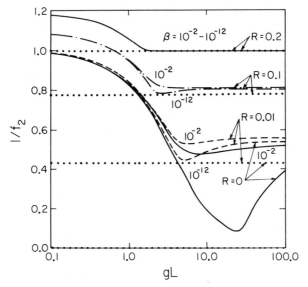

FIG. 3. Dependence of $1/f_2$ on the parameters gL, β, and R; $\cdots$, averaged; ——, - - -, and − · −, exact. [From Moreno (1977).]

In the following sections, we shall make heavy use of Eqs. (17), which will serve as the basis for most of the analysis of the modulation characteristics of lasers.

3. THE OPTICAL CONFINEMENT FACTOR

Thus far we have assumed that the entire optical mode lies within the active region so that the rate of stimulated emission is αNP. In reality, the cross section of the optical mode extends beyond the active region so that only part of the mode lies within the gain region. For simplicity, we assume that the part of the optical mode lying outside the gain region does not experience any loss. One is easily tempted to modify the rate equations [Eqs. (17)] for the photon and electron density by replacing the stimulated gain term αNP by $\Gamma \alpha NP$, where Γ is the fraction of the optical mode lying inside the active region. This procedure is *incorrect,* although intuitively it appears otherwise. The correct rate equations for the photon and electron *densities* should be

$$\frac{dN}{dt} = \frac{J}{ed} - \frac{N}{\tau_s} - \alpha NP, \tag{18a}$$

$$\frac{dP}{dt} = \Gamma \alpha NP - P_{\tau_s} + \Gamma \beta \frac{N}{\tau_s}. \tag{18b}$$

Notice in Eqs. (18) that Γ appears only in the photon rate equation. To see that Eqs. (18) are correct, we assume that the cross-sectional area of the active region is A. The area of the optical mode is therefore A/Γ, and the length of the laser is L. Let us convert Eqs. (18) into rate equations for the *total* number of electrons and photons. This is done by multiplying Eq. (18a) by AL and Eq. (18b) by AL/Γ, resulting in

$$\frac{dN^{tot}}{dt} = \frac{I}{e} - \frac{N^{tot}}{\tau_s} - \alpha \left(\frac{\Gamma}{AL}\right) N^{tot} P^{tot}, \tag{19}$$

$$\frac{dP^{tot}}{dt} = \Gamma \alpha \frac{N^{tot}}{AL} P^{tot} - \frac{P^{tot}}{\tau_p} + \beta \frac{N^{tot}}{\tau_s}, \tag{20}$$

where N^{tot} and P^{tot} denote the total number of electrons and photons, respectively, and I is the total current. Since the stimulated recombination of every electron must give birth to a photon, the rate of loss of the *total number* of electrons due to stimulated recombination must be equal to the rate of increase in the *total number* of photons by stimulated emission. We can see that Eq. (20) [and therefore Eq. (18)] is indeed correct by noting that the stimulated terms in the rate equations for the total number of electrons and photons are equal in magnitude.

III. Transient Response and Small-Signal Modulation Characteristics of Semiconductor Injection Lasers

Most types of lasers emit a series of sharp optical spikes in response to a step increase in the excitation level and continue to ring for some time before the output settles to a steady-state value. This phenomenon, called relaxation oscillation, also occurs in some semiconductor lasers. Due to the small size of the lasers and the short spontaneous lifetime (a few nanoseconds), the frequency of this ringing is higher (in the gigahertz range) than those in most other types of lasers. This presents a problem in high-speed digital modulation of semiconductor lasers in optical communications, since the highly irregular spiking can interfere with the bit pattern. This temporal instability is manifested as a sharp resonance in the modulation frequency response of the laser at the same frequency as the ringing. This puts an upper frequency limit on the analog modulation of the laser diode. However, injection lasers of different structures exhibit relaxation oscillations of different strengths, while some do not exhibit any at all. This can be accounted for by two features that are unique to injection lasers: Carriers can diffuse laterally within the active region, and injection lasers have an unusually high amount of spontaneous emission entering the lasing mode owing to the waveguiding structure. Both of these factors vary with laser structures and can be controlled to some extent by pertinent laser design. In this section, we shall examine and compare some basic small-signal modulation responses in lasers of different structures.

4. STEADY-STATE CHARACTERISTICS OF INJECTION LASERS

Relaxation oscillation in semiconductor lasers and its frequency dependence on the pump current can be accurately predicted from the simple, spatially uniform rate equations described in Sections 2 and 3 [see Eqs. (17)]. In this section, we first examine the steady-state solution of the rate equations that serves to illustrate the basic light-versus-pump-current characteristics of an ideal injection laser. The rate equations are [Eqs. (18)]

$$\frac{dN}{dt} = J - N - (N - N_{om})P, \tag{21a}$$

$$\frac{dP}{dt} = \gamma[\Gamma(N - N_{om})P - P + \Gamma\beta N], \tag{21b}$$

where, for convenience, the variables have been normalized as follows: The electron density N and the electron density for transparency N_{om} have been normalized by $1/\alpha\tau_p$; the photon density P has been normalized by $1/\alpha\tau_s$; the pump current density J has been normalized by $(ed/\alpha\tau_s\tau_p)$; the time t has been normalized by τ_s; and $\gamma = \tau_s/\tau_p$. An additional parameter N_{om} has been added to the stimulated emission terms in Eqs. (21). This is a more accurate description of stimulated gain in GaAs, in which the electron density must exceed a certain level N_{om} for the medium to exhibit positive gain. The inclusion of N_{om} in the rate equations would introduce an offset in the variable N but would not affect the physics of the system in a significant way (except for a corresponding offset in the threshold current). The inclusion of the optical confinement factor Γ in the photon rate equation [Eq. (21b)] has been explained in Section 3; its inclusion will also not affect gravely the physics of the system. Therefore, in some later sections when these factors are not essential to the results and when simplicity is desired, N_{om} and Γ will be set without loss of generality to zero and one, respectively.

Typical values of these parameters are as follows (Kressel and Butler, 1977): $\tau_s = 4$ nsec, $\tau_p = 2$ psec, $d = 0.2$ μm, $\alpha = 2.8 \times 10^{-6}$ cm^3 sec^{-1}, $N_{om} = 7.5 \times 10^{17}$ cm^{-3}, $10^{-5} < \beta < 10^{-3}$, depending on the laser structure and guiding mechanism. The confinement factor Γ depends strongly on the optical guiding structure. With these numbers, $\gamma = 2000$ and $N_{om} = 2.5$. The steady-state solution of Eqs. (21), in the limit as $\beta \to 0$, assumes the simple form (where the zero subscript denotes steady-state quantities)

$$P_0 = 0, \qquad\qquad N_0 = J, \qquad\qquad J < N_{om} + (1/\Gamma), \tag{22a}$$

$$P_0 = \Gamma J - (\Gamma N_{om} + 1), \qquad N_0 = N_{om} + (1/\Gamma), \qquad J > N_{om} + (1/\Gamma). \tag{22b}$$

The optical output (proportional to the photon density) remains at zero up to the threshold pumping level and increases linearly with further increase in the pump current. The electron density is clamped to the value $N_{om} + (1/\Gamma)$ when pumped above a current density equal to $N_{om} + (1/\Gamma)$, which is defined to be the threshold current density J_{th}. When the laser is operating above the threshold, additional electrons injected into the active region recombine to emit photons. The solutions are slightly different when we take into account spontaneous emission, whose major effect is to smooth out the threshold point. The steady-state solutions in which $\beta \neq 0$ are expressed to the lowest order in β: above the threshold, $J \gg N_{om} + (1/\Gamma)$,

$$N_0 = \left(N_{om} + \frac{1}{\Gamma}\right) - \beta\left[\frac{N_{om} + (1/\Gamma)}{J - N_{om} + (1/\Gamma)}\right], \tag{23a}$$

$$P_0 = \Gamma J - (\Gamma N_{om} + 1) + \beta\left[\frac{(\Gamma N_{om} + 1)(J - N_{om})}{J - N_{om} + (1/\Gamma)}\right]; \tag{23b}$$

at the threshold, $J = N_{om} + (1/\Gamma)$,

$$N_0 = N_{om} + \frac{1}{\Gamma} - \frac{1}{\Gamma}\sqrt{N_{om} + 1)\beta}, \tag{23c}$$

$$P_0 = \sqrt{(N_{om} + 1)\beta}, \tag{23d}$$

and below the threshold, $J \ll N_{om} + (1/\Gamma)$

$$N_0 = J\left[1 + \beta\frac{J - N_{om}}{J - N_{om} - (1/\Gamma)}\right], \tag{23e}$$

$$P_0 = \frac{\beta J}{N_{om} + (1/\Gamma) - J}. \tag{23f}$$

Figure 4 shows plots of the electron density and photon density as a function of bias current for the cases $\beta = 0$ and $\beta \neq 0$. The approximate formulas [Eqs. (23)] agree with the exact results extremely well except near the threshold.

5. RELAXATION OSCILLATION AND EFFECTS OF SPONTANEOUS EMISSION

Relaxation oscillation results from the interplay between the optical field and the population inversion, as governed by the rate equations [Eqs. (21)]. The relative temporal instability results from the following mechanism: An increase in the optical intensity causes a reduction in the inversion due to the increased rate of stimulated transitions, which in turn causes a reduction in the gain that tends to decrease the field intensity. Figure 5 shows the result of a numerical integration of Eqs. (21) in which $\beta = 10^{-4}$ and $N_{om} = 2.5$, so that the threshold current $J_{th} = 3.5$. The laser was biased at $J = 3$ at $t < 0$ and was step excited to $J = 3.55$ at $t = 0$. We can notice a time delay between the onset of the current step and the emission of the first optical pulse. During this time delay the electron density builds up from the initial value to above the threshold inversion level.

Much information about the oscillation behavior can be gained by small-signal analysis of the rate equations, which is a procedure that linearizes Eqs. (21). Writing

$$N = N_0 + ne^{i\omega t}, \tag{24a}$$

$$P = P_0 + pe^{i\omega t}, \tag{24b}$$

$$J = J_0 + je^{i\omega t}, \tag{24c}$$

we separate the variables into the steady-state part and a small sinusoidally varying part. Upon substitution into Eqs. (21) and ignoring the products of

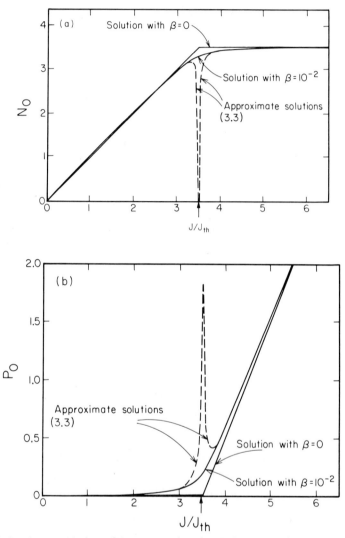

FIG. 4. Steady-state solutions of the rate equations for (a) electron density N_0 and (b) photon density P_0 versus pump current density J.

the small terms, we obtain

$$i\omega n = j - n - nP_0 - (N_0 - N_{om})p, \tag{25a}$$

$$i\omega p = \Gamma\gamma(nP_0 + (N_0 - N_{om})p + \beta n). \tag{25b}$$

This represents a conjugate pole–pair type of frequency response, and the

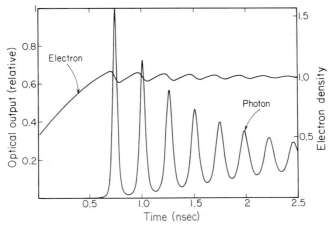

FIG. 5. Relaxation oscillation resulting from a step change in excitation current. $N_{om} = 2.5$; $t < 0$, $J = 3.0$; $t > 0$, $J = 3.55$; $\beta = 3 \times 10^{-4}$.

system is overdamped, critically damped, or underdamped depending on whether F^2 is larger than, equal to, or smaller than $4G$, where

$$F = \gamma[N_{om} + (1/\Gamma) - N_0] + P_0 + (1/\Gamma), \tag{26a}$$

$$G = \gamma[(1 + P_0/\Gamma) - (1 - \beta)(N_0 - N_{om})]. \tag{26b}$$

The underdamped case corresponds to relaxation oscillation, and the critically damped case gives a flat frequency response with maximum bandwidth.

We can see easily with the help of Fig. 4 that F increases and G decreases as β is increased. This suggests that with a sufficiently high spontaneous emission factor, the system will be critically damped and relaxation oscillation will eventually cease to occur. This was first pointed out by Boers and Vlaadingerbroek (1975). One can calculate from Eqs. (23), (25), and (26) a condition for β that produces critical damping. Expressed to the lowest order in $1/\gamma$

$$\beta_{min} = 2\frac{[J - N_{om} - (1/\Gamma)]^{3/2}}{(1/\Gamma) + N_{om}}\gamma^{-1/2}. \tag{27}$$

A plot of β_{min} versus pump current J is shown in Fig. 6 with $\gamma = 2000$, $\Gamma = 1$, and $N_{om} = 2.5$. Typical values of β_{min} range between 10^{-3} and 10^{-2}. Figure 7 shows results of numerical calculations of the rate equations with the same parameters for N_{om} and γ as in Fig. 5. The system is biased at $J = 3.6$ at $t < 0$ and excited by a step current to $J = 5.6$ at $t = 0$. The successive increment in

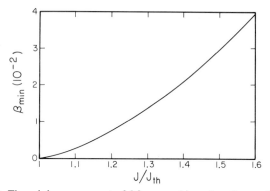

FIG. 6. The minimum amount of β for quenching relaxation oscillation.

β from 2.8×10^{-4} to 1.4×10^{-2} clearly illustrates the damping effect in the response.

Lang and Kobayashi (1976) first suggested and experimentally verified that injecting external, coherent photons into the laser cavity can have the same effect as having a large, spontaneous emission factor, thereby suppressing the relaxation oscillation. Otsuka (1977) gave a detailed analysis of the effect, and an integrated version of the device was recently fabricated (Fekete *et al.*, 1980). The analysis was done in a straightforward manner by adding an external injection term P_i into the right-hand side of the photon rate equation [Eq. (21b)]. Small-signal analysis indicates that quenching of the relaxation oscillation occurs for $P_i > 10^{-2}$, i.e., injecting 10^{-2} of the photons emitted from a similar laser is sufficient to quench the relaxation oscillation.

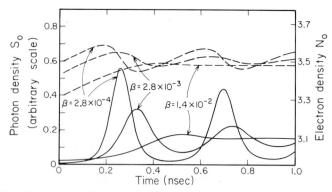

FIG. 7. Transient response of lasers with different β. $N_{om} = 2.5$; $t < 0$, $J = 3.6$; $t > 0$, $J = 5.6$.

6. COMPARISON OF TRANSIENT RESPONSES OF LASERS WITH VARIOUS STRUCTURES

Figure 8a shows the response of a proton-bombarded stripe laser that belongs to the general class of stripe geometry lasers. This type of laser has neither carrier nor optical confinement in the transverse direction (along the junction plane) and has a typical stripe width of 10 μm. The transient response of this type of laser shows strong relaxation oscillation. Figure 8b shows the transient response of a channeled substrate planar (CSP) laser (Aiki *et al.*, 1978) that belongs to a second class of lasers with transverse optical guiding but no transverse carrier confinement. The response shows little relaxation

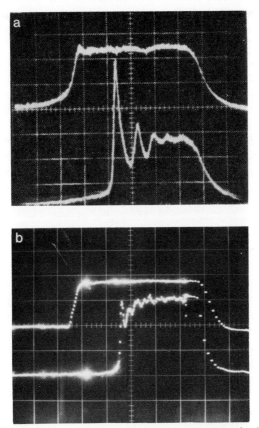

FIG. 8. Observed transient responses of (a) a proton-bombarded stripe laser and (b) a CSP laser. Top traces represent current, and bottom traces represent light; horizontal scale: 1 nsec per division.

oscillation and is suitable for medium-bit-rate (< 1 Gbit sec^{-1}) optical transmitters. A third important class of lasers has transient responses lying somewhere between the first two classes; the spiking is more pronounced than the second type but less drastic than the first type. This class of lasers, exemplified by the buried heterostructure (BH) (Tsukada, 1974) and embedded lasers (Katz et al., 1980), have both carrier confinement and optical guiding in the transverse direction. Schematic diagrams of the cross sections of these three classes of lasers are shown in Fig. 9. Another laser structure that exhibits excellent characteristics (in terms of low threshold and single stable transverse and longitudinal modes) is the transverse junction stripe (TJS) laser, fabricated on a semi-insulating substrate (Lee et al., 1987; Nita et al., 1979) and shown in Fig. 9e. The TJS laser is different from most lasers in that the carriers are injected transversely across a $p-n$ homojunction instead of across the GaAs/GaAlAs heterojunctions. The transient response of the TJS laser resembles that of the CSP laser and is, thus, suitable for transmitters.

The different transient behaviors of various lasers can only be partly explained by the calculations in Section 5. The spontaneous emission factor β varies among structures and directly influences the strength of the relaxation oscillation. The calculation of β is done by first evaluating the radiation efficiency of an infinitesimal dipole radiator into the specific optical mode, which is assumed to be known for the laser structure. The total spontaneous emission factor is obtained by integrating the dipole radiation efficiency over the entire carrier distribution (Suematsu and Furuya, 1977; Peterman, 1979). The result is given by

$$\beta = \lambda^4 K / 4\pi^2 n n_e n'_e V_{\text{eff}} \, \Delta\lambda, \tag{28}$$

where λ is the optical wavelength, K the astigmatism factor (which is an indication of the degree of gain guiding in a laser structure), n the refractive index of the active layer material, n_e the effective refractive index of the active layer, n'_e the differential effective refractive index of the active layer, V_{eff} the mode volume of the optical field, and $\Delta\lambda$ the half-width of the Lorentzian optical gain spectrum of the material. We can see immediately that the spontaneous emission factor β is inversely proportional to the optical mode volume. Therefore, lasers with a tight optical confinement in the transverse direction, such as the BH and the embedded lasers, have a high value of β and hence show relatively weak relaxation oscillation. However, values of β calculated from Eq. (28) for BH lasers lie around 10^{-4} and, according to results of Section 5, are not sufficient to quench the relaxation oscillation to the degree observed experimentally. This can be explained by the fact that during transients more than one longitudinal mode can be excited and should be described by a more complicated set of rate equations

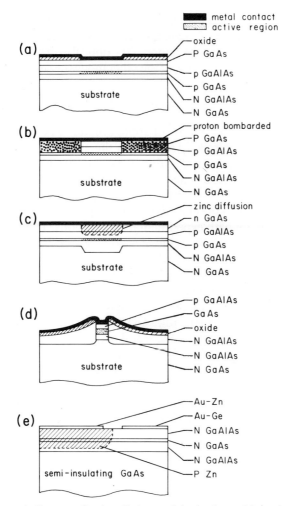

FIG. 9. Schematic diagrams of various GaAs-type injection lasers: (a) simple stripe geometry, (b) proton-defined stripe, (c) channeled substrate planar, (d) buried heterostructure, and (e) transverse junction stripe.

that includes more than one mode (Peterman, 1978):

$$\frac{dN}{dt} = J - N - N \sum_i g_i P_i, \qquad (29a)$$

$$\frac{dP_i}{dt} = N g_i P_i - P_i + \beta_i N, \qquad (29b)$$

for $i = -\infty \to \infty$. Here P_i denotes the photon density in the ith mode. The factor g_i is the gain factor of mode i, where mode 0 is taken to be the mode whose wavelength resides at the center of the optical gain spectrum. When g_0 is taken to be 1, the gain factors g_i of the other modes (assuming a Lorenztian gain lineshape) are given by

$$g_i = 1/(1 + bi^2), \tag{30}$$

where b typically assumes a value of approximately 5×10^{-4} for a 300-μm device (Renner and Carroll, 1978). When the entire set of rate equations [Eqs. (29)] is taken into consideration, we can derive a solution similar to those of Eqs. (21) but with an effective value of β given approximately by the original β multiplied by the number of modes participating in the transient process. This increase in the effective value of β puts the calculated results in reasonable agreement with experimental observations. There are, however, exceptional cases in which the value of β alone cannot explain what is being observed. It does not explain, for example, why a CSP laser (which has a lower value of β) and a TJS laser (which has a value of β comparable to that of the BH laser) show weaker relaxation oscillations than a BH laser. There is probably more than one factor that contributes to these discrepancies. Likely candidates are transverse carrier diffusion effects, which have been neglected in the simple rate equation approach, and parasitic capacitances associated with the laser structure, which will be considered in a later section. The separate roles played by spontaneous emission and lateral carrier diffusion in the transient response were first clarified by Chinone et al. (1978b), by numerically solving the full rate equations, including the diffusion term, and comparing them to real lasers. The results will be summarized in the next section.

7. EFFECT OF LATERAL CARRIER DIFFUSION

Except for those lasers that have built-in structures for confining carriers in the lateral direction, lateral carrier diffusion within the active layer plays a role (sometimes very significant and sometimes less so) in dampening the transient spiking. The mechanism can be visualized through the following heuristic consideration: Assume, for simplicity, that the injection current and the optical mode are in the form of a δ function in the transverse direction along the junction plane (i.e., in the limit of zero stripe width and infinite optical confinement). The diffusion of carriers in the transverse direction, neglecting the optical field, is described by

$$\frac{\partial N}{\partial t} = D \frac{\partial^2 N}{\partial x^2} + J(x, t) - \frac{N}{\tau_s}, \tag{31}$$

where $J(x, t) = J(t) \delta(x)$, x is the transverse coordinate, and D is the diffusion constant. The optical mode, having a transverse distribution $\delta(x)$, sees a gain proportional to $N(x = 0, t)$. Separating the variables into steady-state and sinusoidally varying parts

$$N = N_0(x) + n(x)e^{i\omega t}, \tag{32a}$$

$$J = J_0(x) + j(x)e^{i\omega t}, \tag{32b}$$

the small-signal diffusion equation reads

$$D\,(\partial^2 n/\partial x^2) - n[(1/\tau_s) + i\omega] = j\,\delta(x), \tag{33}$$

which has the solution

$$n(x) = (jl/2D)e^{-|x|/l}, \tag{34}$$

where $l = \sqrt{D\tau_s}$ is the small-signal diffusion length and $1/\tau_s' = (1/\tau_s) + i\omega$. The amplitude of the sinusoidal modulation in the optical mode gain is proportional to $n(0)$, given by

$$n(0) = (j/2D)\sqrt{D\tau_s}/(1 + i\omega\tau_s). \tag{35}$$

Hence, diffusion effectively introduces a *one-half pole* in the modulation response and acts to dampen the resonance. A self-consistent analysis including the optical field has been worked out by Wilt *et al.* (1981) in this limit of zero stripe width.

Chinone *et al.* (1978a) classified injection lasers into three main categories: type I, those with neither carrier nor optical confinement in the lateral direction; type II, those with lateral optical confinement but with no lateral carrier confinement; and type III, those with both lateral carrier and optical confinement. The analysis was based on the following local rate equations that include spatial dependence in the lateral direction

$$\frac{\partial N}{\partial t} = D\frac{\partial^2 N}{\partial x^2} + \frac{J(x, t)}{ed} - \frac{N}{\tau} - g_t P(x), \tag{36a}$$

$$\frac{d}{dt}\int_{-\infty}^{\infty} P(x)\,dx = \int_{-\infty}^{\infty}\left(\Gamma g_t - \frac{1}{\tau_p}\right)P(x)\,dx + \beta\int_{-\infty}^{\infty}\frac{N}{\tau}\,dx, \tag{36b}$$

where $P(x)$ is the optical mode profile, Γ the confinement factor, g_t the local gain coefficient. The optical mode profile $P(x)$ is solved independently in the case when a built-in waveguide exists and must be solved self-consistently with the carrier distribution $N(x)$ when guiding is provided by the gain profile. Equations (36) are excited with a step increase in the pump current, and the strengths of the resulting relaxation oscillations (defined by the peak to valley ratio P_{max}/P_{min} of the first spike of the relaxation oscillation) are shown in Fig. 10 for various types of lasers. Figure 10a shows the relaxation oscillation strength versus pump current for type-II lasers (e.g., CSP) with a

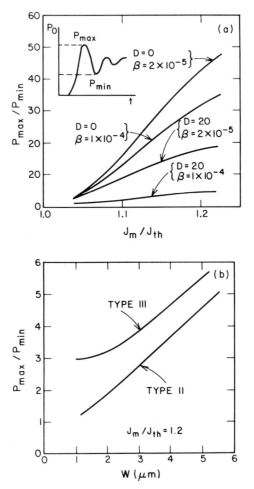

FIG. 10. Comparison of the strength of relaxation oscillations (a) with various β and diffusion coefficients D for type-II lasers ($W = 5$ μm), and (b) between type-II and type-III lasers ($J_m/J_{th} = 1.2$). [From N. Chinone, K. Aiki, M. Nakamura, and R. Ito (1978). *IEEE Journal of Quantum Electronics,* **QE-14,** 625. Copyright © 1978 IEEE.]

5-μm stripe width. It can be seen that both carrier diffusion and spontaneous emission have substantial effects on the strength of the relaxation oscillation. However, it is also observed that carrier diffusion is more effective in this respect. In type-III lasers (BH, embedded) no carrier diffusion occurs and spontaneous emission is the primary factor for reducing the spiking. Figure 10b shows a comparison between type-II and type-III lasers of various stripe widths. Narrow-stripe lasers always show larger damping since for type-III lasers the spontaneous emission factor is larger and for type-II lasers the effect of diffusion is more substantial. For type-I lasers with stripe widths larger than the carrier diffusion length, neither carrier diffusion nor spontaneous emission is significant enough to reduce spiking, and P_{max}/P_{min} ratios of over 100 can be observed. In addition, stripe geometry lasers can have mode instabilities during transients that significantly enhance the spiking behavior. This mode instability can lead to self-pulsation of the optical radiation in a manner described by Lang (1980).

In summary, in lasers whose stripe width is narrower than ~ 10 μm, carrier diffusion plays a heavier role than spontaneous emission in suppressing the relaxation oscillation, especially for lasers without a lateral carrier confinement structure. On the other hand, the fraction of spontaneous emission going into the lasing mode is primarily responsible for suppressing the spiking in lasers with a lateral carrier confinement structure.

8. SMALL-SIGNAL ANALOG MODULATION RESPONSE OF INJECTION LASERS

Starting with the simplest form of the rate equations [Eqs. (21)], the small-signal analysis of Section 5 leads to a second-order, low-pass filter type of frequency response. Equations (25) lead to the following transfer function for the small-signal photon density

$$p = \frac{\Gamma j\gamma(P_0 + \beta)}{(i\omega + 1 + P_0)[i\omega + \gamma(1 + \Gamma N_{om} - \Gamma N_0)] + \gamma(\Gamma N_0 - \Gamma N_{om})(P_0 + \beta)}. \tag{37}$$

The corner frequency of this response function occurs at

$$f_r = (1/2\pi)\sqrt{\gamma P_0} + \text{higher-order terms in } \beta, \tag{38}$$

where P_0 is the normalized photon density [see definitions following Eqs. (21)], $\gamma = \tau_s/\tau_p$, and the dimensionless f_r is normalized by $1/\tau_s$. The spontaneous emission factor β determines the Q of the resonance but has no significant effect on the corner frequency itself. In the case of a high Q, the phase of the modulated output undertakes an abrupt transition from 0 to $-180°$ at the corner frequency but exhibits a soft transition in the case of a low Q. At low-bias currents, a prominent resonance peak seldom exists, since a very low β is sufficient to produce a very low Q. All of these features follow directly from the typical characteristics of a second-order transfer function.

The kinds of amplitude and phase responses described earlier are actually observed in various types of injection lasers, and the corner frequency is found to follow the square-root dependence of Eq. (38) extremely well. Figure 11 shows the results of measurements on a proton stripe laser. The relationship between the phase and the Q of the amplitude response is

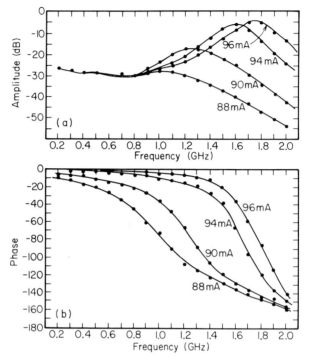

FIG. 11. Measured (a) amplitude and (b) phase responses of a proton stripe laser. Threshold $\simeq 80$ mA.

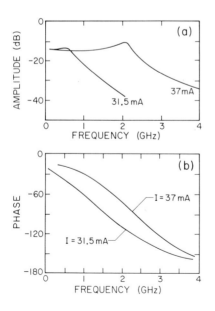

FIG. 12. Measured (a) amplitude and (b) phase response of a TJS laser. $I_{th} \simeq 30$ mA.

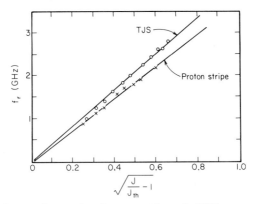

FIG. 13. Measured corner frequencies of a proton stripe and a TJS laser versus pump current.

evident. Figure 12 shows the amplitude and phase response of a TJS laser (Figueroa *et al.*, 1982). The frequency response is relatively flat, as would be expected from the absence of the relaxation oscillation as described in Section 6. Figure 13 shows plots of the corner frequency versus $\sqrt{(J/J_{\text{th}}) - 1}$ (which is proportional to $\sqrt{P_0}$) for the two lasers. The modulation bandwidths of the two lasers are very similar.

IV. Ultimate Frequency Response of Semiconductor Injection Lasers

The ability of GaAs-type semiconductor lasers to be modulated in the gigahertz frequency range opens up some interesting applications in the transmission of microwave signals through optical fibers (Yen and Barnoski, 1978). From the analysis of previous sections, it is clear that the useful small-signal modulation frequency range is limited to that below the relaxation oscillation resonance, which occurs at around 2–3 GHz, depending on the structure and internal parameters of the laser. Questions are being raised as to whether it is fundamentally possible to push the modulation frequency into the X-band region (~ 10 GHz) or even higher, how we should design such a laser, and what price should we have to pay in order to obtain such a wide bandwidth. In Section 9, we shall first explore the theoretical ultimate limit of the high-frequency modulation response of semiconductor lasers. These results are utilized in obtaining design guidelines for very high-frequency lasers. Since these results are based on small-signal modulation theory, they can be applied to analog modulation as well as to digital pulse code modulation (PCM) formats with sufficiently small modulation amplitudes. What should be regarded as small requires further delineation and is the subject of Section 13. A rough conclusion that can be drawn from Section 13 is that laser modulation behavior can be fairly satisfactorily explained by the small-signal model when the optical modulation depth (defined here as the ratio of the modulation amplitude to the peak of the modulated waveform) is less than ~ 70%. Within this small-signal regime, an analog modulation bandwidth of x GHz would correspond roughly to a maximum digital bit rate of x Gbit sec^{-1} before significant degradation in bit error rate results. Depending on whether the digital scheme involves a return-to-zero or a non-return-to-zero format, the digital bit rate can be higher than the analog rate by a factor approximately equal to two. The correspondence is not as clear in the large-signal regime, since the digital bit error rate depends not only on the signal-to-noise level but also on the quality of the pulses. Nevertheless, an estimate of the digital modulation bandwidth can be obtained from an analog large-signal analysis. The ultimate limit in the frequency response and the guidelines for designing high-frequency lasers to be described in Section 9 will still apply in the large-signal regime, since results obtained from the large-signal analysis do not constitute a drastic departure from that of the small-signal analysis and involve merely a moderate modification in specific details. Experimental results of high-speed modulation of semiconductor lasers will be described in Sections 12 and 13. The effect of parasitic elements on the modulation bandwidth will be discussed in Part V.

9. ULTIMATE LIMITS OF HIGH-SPEED SEMICONDUCTOR LASERS

This section is intended to resolve and clarify various factors that influence the modulation bandwidth of injection lasers. Do lasers with shorter carrier lifetimes have higher modulation bandwidths, as in the case of light-emitting diodes (Lindstrom *et al.*, 1983)? Are shorter lasers intrinsically faster? To answer these questions we start out with the expression for the laser relaxation resonance frequency given by Eq. (38). In the unnormalized form it reads

$$f_r = (1/2\pi)\sqrt{\alpha(P_0/\tau_p)}. \tag{39}$$

The modulation bandwidth of the laser is widely accepted to be equal to f_r. Equation (39) can be re-expressed as

$$f_r = (1/2\pi)\sqrt{\Gamma\alpha[(J - J_{th})/ed]}, \tag{40a}$$

where J_{th} is the threshold current density given in Section 4. Equation (40a) can be converted into a more familiar form

$$f_r = \frac{1}{2\pi}\sqrt{\Gamma\alpha\frac{J_{th}}{ed}\left(\frac{J}{J_{th}} - 1\right)} = \frac{1}{2\pi}\sqrt{\frac{\Gamma N_{om}\alpha\tau_p + 1}{\tau_s\tau_p}\left(\frac{J}{J_{th}} - 1\right)}. \tag{40b}$$

If we set N_{om} to zero, we obtain the well-known relation (Ikegami and

Suematsu, 1968; Paoli and Ripper, 1970)

$$f_r = (1/2\pi) \sqrt{(1/\tau_s\tau_p)[(J/J_{th}) - 1]}.$$ (41)

However, in GaAs- and GaInAsP-type laser materials, the factor $\Gamma N_{om}\alpha\tau_p$ in Eq. (40b) is several times larger than 1, and, therefore, Eq. (41) is not a very accurate description of the relaxation oscillation frequency. The failure of Eq. (41) has been noticed by numerous workers (Gonda and Mukai, 1975; Angerstein and Siemsen, 1976) who, in trying to fit experimental data to Eq. (41), obtained inconsistent values for the spontaneous lifetime τ_s or photon lifetime τ_p, both of which can be independently measured or calculated. Paoli (1981) pointed out that a better fit to experimental data can be obtained if we use in Eq. (41) the photon lifetime of the unpumped cavity, which includes the intrinsic loss of the semiconductor material (in additional to the mirror loss, scattering loss, and free-carrier absorption loss). This is exactly what is given in Eqs. (40), which follows directly from the rate equations including a nonzero N_{om}.

Although widely in use, Eq. (41) is a nonilluminating (and, in fact, somewhat misleading) form of expressing the laser modulation bandwidth. It suggests, for instance, that the modulation bandwidth increases with decreasing spontaneous and photon lifetimes. However, if we try to shorten the internal laser parameters such as τ_s and τ_p to obtain a higher f_r, we simultaneously increase the threshold current density J_{th}, and it is not clear that f_r can be increased in this manner. We can see clearly from Eqs. (39) and (40) that f_r depends on the photon density P_0 and hence on the *absolute* current density $J - J_{th}$, not the ratio J/J_{th}. The $(J/J_{th}) - 1$ dependence of Eq. (41) is, therefore, somewhat artificial. One can also obtain a wider bandwidth simply by raising the pump current density J. Operating at a high current density is not compatible with reliability, and the corresponding increase in optical power may cause catastrophic facet damage (Lau *et al.,* 1981). The pertinent questions are these: given certain limits in the operating current density[†] and optical power density, what is the highest achievable f_r, and how can it be achieved?

To answer these questions, we shall resort to the more direct expression for the relaxation oscillation frequency [Eq. (39)]. It suggests three obvious ways to increase the relaxation frequency: by increasing the optical gain coefficient, by increasing the photon density, and by decreasing the photon lifetime. The gain coefficient α can be increased roughly by a factor of five by cooling the laser from room temperature to 77 K (Stern, 1976). Biasing the

---

[†] Actually, the limit to be imposed should be the injected carrier per unit volume, but we shall assume a uniform active layer thickness of 0.2 μm and express the limiting quantity in terms of the more familiar current density.

laser at higher currents would increase the photon density in the active region, which simultaneously increases the optical output power density I_{out} according to

$$I_{out} = \frac{1}{2} \hbar\omega \frac{P_0 \times \text{mode volume}}{\tau'_p \times \text{cross-sectional area}} \qquad (42)$$

or

$$I_{out} = \frac{1}{2} \hbar\omega \frac{P_0 L}{\tau'_p}, \qquad (43)$$

where P_0 is the static photon density, ω_0 the optical frequency, and τ'_p the photon lifetime associated with mirror loss only

$$\tau_p = \frac{L}{v \, ln(1/R)}. \qquad (44)$$

Experimental evidence has shown (Casey and Panish, 1978; Kressel and Butler, 1977) that at least in the case of pulse operation, critical catastrophic optical damage is related to the electric field E_t at the inside of a mirror facet. This electric field is related to the output optical power density I_{out} by

$$I_{out} = n\sqrt{\varepsilon_0/\mu_0}[(1 - R)E_t^2/2(1 + \sqrt{R})^2]. \qquad (45)$$

Catastrophic mirror damage occurs at a critical electric field E_t^{max}, which in the case for a laser with a mirror reflectivity of $R = 0.3$, corresponds to an output power density of $I_{out}^{max} \simeq 1$ MW cm^{-2} (Wakao et al., 1982). This sets an upper limit on the maximum permissible photon density and, hence, the maximum modulation bandwidth. This limit can be increased very considerably by using a window structure such as the crank TJS laser (Takamiya et al., 1980), the window stripe laser (Yonezu et al., 1979), or the window buried laser (Blauvelt et al., 1982).

The third way to increase the modulation bandwidth is to reduce the photon lifetime by decreasing the length of the laser cavity. Such a laser must be driven at higher current densities or else thermal effects due to excessive heating will limit the maximum attainable modulation bandwidth. To illustrate these points the relaxation frequency as a function of the cavity length and pump current density is plotted in Fig. 14, using Eq. (39) and the static solutions of the rate equations at 300 and 77 K. The output power density is also plotted in Fig. 14 by using Eq. (43). As an example, a common laser with a cavity length of 300 μm operating at an output optical power density of 0.8 MW cm^{-2} possesses a bandwidth of 5.5 GHz, and the corresponding pump current density is 3 kA cm^{-2}. Operating at an identical power density, the bandwidth is 8 GHz for a shorter laser with a cavity length of 100 μm, but

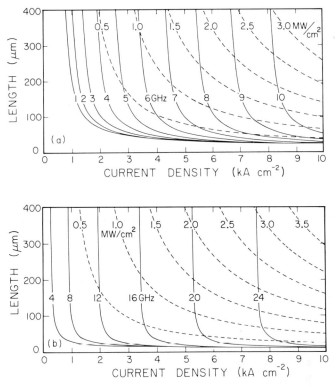

FIG. 14. (a) Relaxation frequency v_{rel} (——) and optical power density outside the mirrors (---) as a function of the cavity length and pump current density at $T = 300$ K. Active layer thickness = 0.15 μm, $R = 0.34$, $v = 8 \times 10^9$ cm sec^{-1}, $\alpha = 2.56 \times 10^{-6}$ cm^3 sec^{-1}, $\Gamma = 0.5$, $N_{om} = 1 \times 10^{18}$ cm^{-3}, $\hbar\omega = 1.5$ eV. A bimolecular spontaneous recombination is assumed in this calculation. (b) Same as (a) but at $T = 77$ K. The same parameters as in (a) are used except $\alpha = 1.45 \times 10^{-5}$ cm^3 sec^{-1}, $N_{om} = 0.6 \times 10^{17}$ cm^{-3}. [From Lau et al. (1983a).]

the corresponding current density is 6 kA cm^{-2}. Figure 14b shows corresponding plots for a laser operating at liquid-nitrogen temperature. The increase in bandwidth is a direct result of the increase in the gain constant α. It can be seen that a modulation bandwidth beyond 20 GHz can be achieved; however, incorporation of a short optical cavity and/or a window structure is imperative under these operating conditions.

10. OPTIMAL LASER DESIGN FOR HIGH-FREQUENCY MODULATION

It is clear from Fig. 14 that for conventional 300-μm lasers, we usually run into the optical power limit before the current limit as we increase the pump current. This can be avoided by shortening the laser cavity so that we can bias the laser at a higher current density without running into the optical power constraint. There are, however, no solid data on the maximum current density that a laser can tolerate before heating and reduced reliability result.

The degradation of lasers under high-current operation may not be due to the passage of current alone. The simultaneous presence of intense optical radiation at the facet can play a part as well. For example, when we compare two lasers of nearly identical construction except that one of them incorporates a window structure [transverse junction stripe (TJS) and the crank TJS (Takamiya *et al.,* 1980; Takahashi *et al.,* 1978)], the window laser can operate at twice the current density of its conventional counterpart without suffering loss in reliability. A schematic diagram of the crank TJS laser is shown in Fig. 15a. The purpose of the window is twofold: It reduces mirror degradation by making the window section transparent to the optical radiation, and it broadens the optical mode near the end facet. In this particular window laser structure, the maximum current density for reliable operation exceeds 10 kA cm^{-2}.

It should be noted that the optical power limit considered so far refers to the optical power limit at the mirror facet of the laser. It is known that the optical power that can be tolerated at the mirror facet is considerably lower than that inside the bulk of the crystal. (The reason, it is believed, is due to existence of surface states and absorption centers at the crystal surface, in addition to the fact that such exposed surfaces are vulnerable to contamination during the fabrication process.) It must be emphasized that the modula-

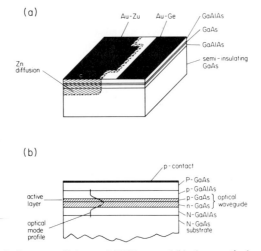

FIG. 15. Schematic diagrams of (a) a crank TJS laser and (b) a large optical cavity (LOC) laser.

tion bandwidth is proportional to the square root of the optical power density [Eq. (39)], so that a high optical power density at the active region is needed for high-frequency operation. The optical damage that might be inflicted on the laser mirror must be averted by some means *without lowering the optical power density at the active region.* The window structure illustrated earlier satisfies this requirement. The optical power in that structure is concentrated along the whole length of the laser except the ends. Other structures designed for high-power operations, notably the large-optical-cavity (LOC) structure (Lockwood *et al.,* 1970; Chinone *et al.,* 1979), *do not* satisfy this requirement. An LOC structure is shown in Fig. 15b. The $Ga_{1-x}Al_xAs$ layer or layers adjacent to the GaAs active layer act as the optical waveguide, which results in a significantly broader optical mode profile in the vertical direction. The optical power *density* is thus reduced at the mirror facet (which permits the device to operate at a higher total optical power) and also along the length of the laser. No improvement in the modulation bandwidth of LOC lasers is thus expected.

In summary, analysis indicates that high-frequency operation of semiconductor lasers is limited first by the maximum optical power density and second by the maximum current density at which the laser can reliably operate. A laser with high gain would have a higher-frequency response as a direct consequence of Eq. (39). The optical limit can be overcome by keeping the laser length short and incorporating window structures. All of the structural designs aiming toward a low-threshold laser, such as reduced current leakage and increased optical confinement factor, would also be helpful in the quest for higher-frequency response. An improvement in the modulation speed is not expected from LOC lasers despite its capability for higher output power. The basic properties of the GaAs-material system, however, impose a fundamental limit on the frequency response as depicted in Fig. 14.

Given the basic problems involved in high-frequency operations, the optimal high-speed laser should have the following characteristics: It should be a short-cavity laser (as short as permitted by the current density limit), preferably with a window structure, having a high quantum efficiency (to ensure no current leakage or other internal losses that might cause local heating) and a low lasing threshold (so that it can operate at a high optical power without running into the current limit). For example, if a window structure can raise the optical power limit by a factor of five to 5 MW cm^{-2} (as reported in Takamiya *et al.,* 1980) and the maximum operating current density is set at 5 kA cm^{-2},[†] then from Fig. 14a the modulation bandwidth at 300 K can reach 7–8 GHz for a laser with a length of 250 μm. Experimental results confirming these predictions will be described in next section. Further in-

[†] Data from Mitsubishi Corporation.

creases in the modulation bandwidth can be accomplished by cooling the laser.

Although the treatment in this section has been aiming toward GaAs/GaAlAs-type lasers, the fundamental physics involved apply to GaInAsP/InP-type lasers as well. The general conclusions are therefore valid for the quaternary system, with some modifications in the numerical details. The major differences between quaternary lasers and GaAs-type lasers, as far as high-frequency calculations are concerned, are in the optical gain constant and in the value for optical power limitation. It is known that quaternary lasers show a higher tolerance to facet damage by intense optical power than do GaAs-type lasers. Quaternary lasers are therefore potentially more promising in high-frequency operation. On the other hand, it is not certain whether short-cavity GaInAsP-type lasers will behave in the same way as their GaAs counterparts. Auger recombination (Thompson, 1981) in quaternary laser materials is believed to be a major cause for the temperature sensitivity of these lasers. The effect of Auger recombination is especially significant at high carrier concentrations, which occur in short-cavity lasers. A major reduction in the optical power output and in the quantum efficiency could then result, which would defeat the purpose for having a short cavity in the high-frequency laser design. These are questions that should form interesting areas of investigation.

11. PREVIOUS EXPERIMENTS ON HIGH-FREQUENCY, SMALL-SIGNAL MODULATION

Experimental work on probing the high-frequency modulation capability of semiconductor lasers began even before the first cw operation of GaAlAs-type lasers at room temperature was reported (Ikegami and Suematsu, 1968, 1970; Paoli and Ripper, 1970). The modulation responses of most high-quality lasers constructed to date have been examined, including the BH laser (Maeda et al., 1978), the CSP laser (Furuya et al., 1978), stripe geometry lasers of various stripe widths (Kobayashi et al., 1977), the V-groove laser (Marshall et al., 1979), and the TJS laser (Nagano and Kasahara, 1977). The emphasis of the experiments was placed more on improving the modulation characteristics (usually in terms of reducing the magnitude of the relaxation oscillation resonance) than on pushing the resonance to higher frequencies. The recent work by Figueroa et al. (1982) presented a comprehensive comparison of the high-frequency characteristics of a number of laser structures and attempted to push the relaxation oscillation of existing lasers to higher frequencies. A highest frequency of around 5 GHz was attained by several lasers, including the BH, CSP, and TJS lasers, while operating at twice the threshold current. At that bias condition, the optical power output of these commercial lasers (with standard lengths of about 250–300 μm) reaches a level such that the operation lifetime of the device would be drastically reduced by a factor of more than 10 (according to manufacturer's data). This can be avoided, according to the results of Section 9, by reducing the length of the laser as well as incorporating a window structure into the laser.

12. SHORT-CAVITY LASERS

Experiments have been performed to determine the modulation bandwidth achievable in a short-cavity laser. The lasers used were buried heterostructure lasers fabricated on a semi-insulating substrate (BH on SI) (Bar-Chaim *et al.*, 1981), a schematic diagram of which is shown in Fig. 16. In addition to a low lasing threshold (typically ≤ 15 mA), which is necessary to avoid excessive heating when operated at high above threshold, these lasers possess very low parasitic capacitance (Ury *et al.*, 1982), which otherwise would obscure modulation effects at high frequencies (≳ 5 GHz). The standard measurement system, shown in Fig. 17, consists of a sweep oscillator (Hp8350) used in conjunction with a network analyser (Hp8410 series) and a microwave *s*-parameter test set (Hp8746B). The photodiode used was a high speed GaAs-type pin diode fabricated on a semi-insulating substrate. The photodiode was an improved version of the one reported previously (Bar-Chaim *et al.*, 1983); its frequency response was calibrated from dc to 15 GHz using a mode-locked dye laser and a microwave spectrum analyser. The − 3-dB point of the photodiode response is at 7 GHz and the − 5-dB point at 12 GHz. The observed modulation response of the laser is normalized by the photodiode response at each frequency. The modulation responses of a

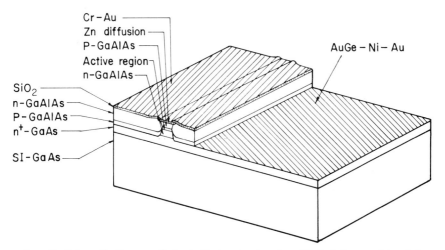

FIG. 16. Schematic diagram of a buried heterostructure laser fabricated on semi-insulating GaAs substrate.

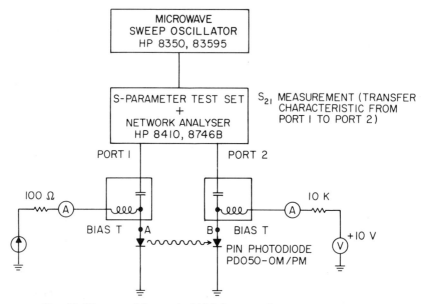

FIG. 17. Diagram of the standard high-frequency laser measurement setup.

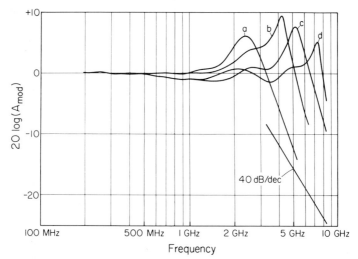

FIG. 18. Modulation characteristics of short-cavity (120-μm) BH-on-SI laser at bias power levels of (a) 1 mW, (b) 2 mW, (c) 2.7 mW, and (d) 5 mW.

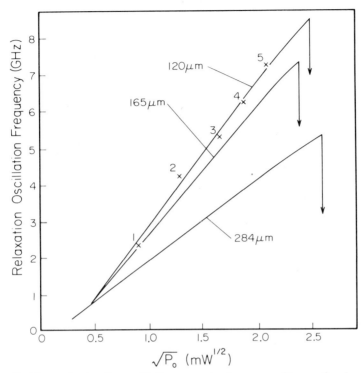

FIG. 19. Measured relaxation oscillation resonance frequency of lasers of various cavity lengths, as a function of $\sqrt{P_0}$, where P_0 is the cw output optical power. The points of catastrophic damage are indicated by downward pointing arrows. [From Lau *et al.* (1983a).]

120-μm-long BH on SI laser at various bias levels are shown in Fig. 18 (Lau *et al.*, 1983a). The modulation bandwidth can be pushed to beyond 8 GHz as the catastrophic damage point is approached. Figure 19 shows the relaxation oscillation frequency of this laser as a function of $\sqrt{P_0}$ where P_0 is the output optical power, together with that of similar lasers with longer cavity lengths. The advantage of a short-cavity laser in high frequency modulation is evident.

It must be emphasized that a very low-threshold laser is required for practical realization of these short-cavity lasers so that catastrophic damage (not heating) is the basic limitation in order to take full advantage of the reduced photon lifetime, and the increased current density has no significant effect on device reliability. As we shorten the laser cavity, the current density will continue to rise until a point is reached beyond which long term reliability would be compromised. Life-test data of commercial BH-on-SI lasers

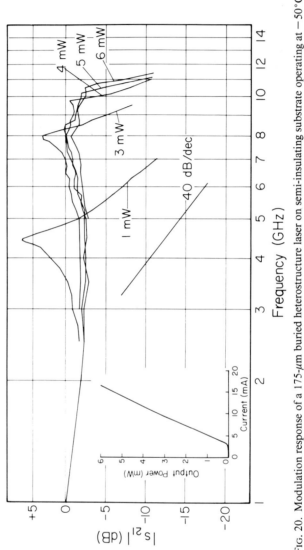

Fig. 20. Modulation response of a 175-μm buried heterostructure laser on semi-insulating substrate operating at $-50°$C.

have shown that, at least for the laser structure considered above, that point occurs at L ≅ 150 μm.

13. Low-Temperature Operation

In this section, we describe experimental results on direct amplitude modulation of low threshold BH on SI lasers operating at below room temperature. This was the first demonstration of a direct modulation bandwidth of beyond 10 GHz in a semiconductor laser.

The lasing threshold current of a 175-μm-long BH on SI laser at room temperature is 6 mA, dropping to ~ 2 mA at − 70°C. The $I-V$ curves reveal a drastic increase in the series resistance of the laser below ~ 60°C. This is believed to be due to carrier freeze-out at low temperatures, since the dopants being used, Sn (n-type) and Ge (p-type), in GaAlAs have relatively large ionization energies. Modulation of the laser diode becomes very inefficient as soon as freeze-out occurs because of a reduction in the amplitude of the modulation current due to a higher series resistance.

The frequency response of the lasers was measured using the standard arrangement. Figure 20 shows the response of a 175-μm-long laser at − 50°C at various bias levels. The relaxation resonance is quite prominent at low optical power levels. As the optical power is increased, the resonance gradually subsides, giving way to a flat overall response. The modulation bandwidth, taken to be corner frequency of the response (the frequency at the relaxation resonance peak or at the − 3-dB point in cases when it is absent), is plotted against the square root of the emitted optical power ($\sqrt{P_o}$) in Fig. 21 at room temperature and at − 50 and − 70°C. The relative slopes of the plots in Fig. 21 yield values for the relative change in A as the temperature is varied. The ratio of the slope at 22°C to that at − 50°C is 1.34 according to Fig. 21. This factor is fairly consistent (between 1.3 and 1.4) among all the lasers tested, even including those from different wafers. According to these measurements we deduce that the intrinsic differential optical gain increases by a factor of ~ 1.8 by cooling from 22 to − 50°C. This is consistent with theoretical gain calculations (Casey and Panish, 1978).

14. High Photon-Density Devices

Although the results of the previous sections demonstrated that a direct modulation bandwidth beyond 10 GHz can be accomplished by lowering the operating temperature of the laser to − 50°C, it is more desirable to attain the bandwidth under room temperature operation. This can be accomplished by tackling the third parameter in Eq. (39), the optical power density P_0. As we pointed out before, increasing the optical power can bring about undue degradation or even catastrophic damage to the laser unless the structure of the laser is suitably designed. One common means of raising the

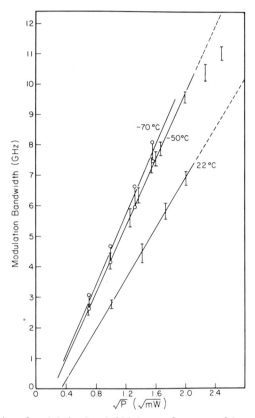

FIG. 21. Variation of modulation bandwidth (corner frequency of the modulation response) with the square root of the emitted optical power $\sqrt{P}$.

ceiling of the reliable operating power of semiconductor lasers is by means of a large optical cavity (Kressel and Butler, 1977). The mechanism responsible for a higher catastrophic damage power in these devices is a lowering of the optical power density at the active layer, since such damage commonly originates from the active layer near the crystal facet. This maneuver, however, serves little to increase the modulation bandwidth since the quantity of concern here, the photon density *within* the active region (P_0 in Eq. (39)), remains unchanged. A laser suitable for high-speed operation should, therefore, be one with a tight optical confinement in the active region along the entire length of the laser with a transparent window at the end regions that is able to withstand much larger photon densities without catastrophic damage. The use of a transparent window structure to increase the catastrophic damage level has already been demonstrated before (Takahashi *et al.,* 1978;

Blauvelt *et al.*, 1982). Using a window buried heterostructure laser fabricated on a semi-insulating substrate, a direct modulation bandwidth of beyond 10 GHz for a semiconductor laser operating at room temperature was demonstrated.

The laser used in this experiment is shown in Fig. 22. The device is structurally similar to the BH-on-SI laser (Fig. 16) except that here the end regions near the facets are covered by a layer of unpumped GaAlAs that forms a transparent window. The laser was fabricated using a two step liquid phase epitaxial (LPE) growth process, followed by mesa stripe etching on which part of this stripe was deleted in order to grow the GaAlAs windows. In the second LPE growth, two blocking GaAlAs layers were grown (Lau *et al.*, 1984d). Individual devices were fabricated using a precise cleaving technique in the window regions within several micrometers from the edge of the double heterostructure. The optical wave propagates freely in the end-window region. As a result of diffraction, only a small amount of light reflected from the crystal facet couples back into the active region. The theoretical value of the effective reflectivity, assuming a fundamental gaussian beam profile, is reduced to 5% for $L = 5$ μm. The actual values of L for the devices fabricated lie around this value.

The threshold current of window BH-on-SI lasers ranges from 14 to 25 mA. The catastrophic damage threshold in these devices is beyond 120 mW under pulse operation. Under cw operation, the maximum operating power is limited by heating to 50 mW. The normalized modulation response of a window laser at various bias optical power levels is shown in Fig. 23. The conspicuous absence of the relaxation oscillation peak should be contrasted with the responses of similar devices that are capable of being modulated to comparably high frequencies (~ 10 GHz) as described pre-

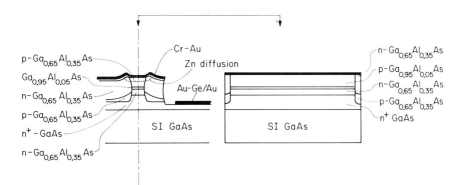

FIG. 22. Schematic diagram of the window buried heterostructure laser on semi-insulating substrate.

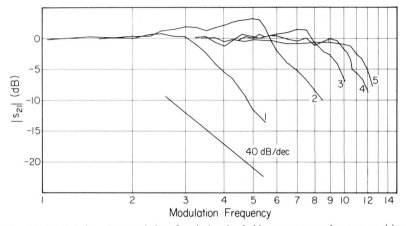

FIG. 23. Modulation characteristics of a window buried heterostructure laser on semi-insulating substrate at various bias optical power levels at room temperature. The curves correspond to bias optical powers of (1) 1.7 mW, (2) 3.6 mW, (3) 6.7 mW, (4) 8.4 mW, and (5) 16 mW.

viously. This is due to superluminescent damping effect (Lau and Yariv, 1982a; Lau et al., 1983b) as a result of the reduced facet reflectivity due to presence of the window. A plot of the -3-dB modulation bandwidth of the window BH-on-SI laser against the square root of the bias optical power is shown in Fig. 24. Contributions from parasitic elements are believed to be at least partly responsible for the departure of the observed data from a linear relationship at high frequencies. Also shown in Fig. 24 are the bandwidth characteristics of a large-optical-cavity, BH-on-SI laser. The substantially lower modulation speed at comparable output power levels bears witness to the fact that the optical power density at the active region of these devices is lower than that in lasers of conventional construction.

The results described in this section, together with those of the last two sections, complete the verification of the modulation bandwidth dependence on three fundamental laser parameters as given in Eq. (39). With these results in hand, it is quite conceivable that the direct modulation bandwidth of semiconductor lasers can be extended to the 20-GHz range by optimizing all of the three parameters simultaneously according to the theoretical results in Eq. (39). However, as the results in this and the last section show, one has still to reckon with the problem of laser parasitic elements before such bandwidth can be realized. The subject of parasitic elements will be discussed in the next section.

The modulation curves shown earlier have been carefully normalized by the photodiode response and the lasers are sufficiently well matched to 50 Ω so that electrical reflections are insignificant. (This is ascertained by micro-

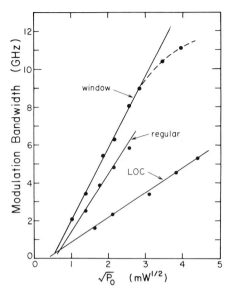

FIG. 24. −3-dB modulation bandwidth versus square root of the emitted optical power.

wave s parameter measurements on the operating laser. See Section 16 for a discussion of this subject.) A dip in the frequency response of the *photodiode* before rolloff can sometimes be observed owing either to the mounts or, more often the case in avalanche photodiodes, to carrier diffusion that can be observed as long tails in their responses to short optical pulses. For example, a commercial avalanche photodiode (Telefunken SP171) has a typical rise time of ∼ 120 psec but a fall time > 200 psec. Figure 25 shows the measurement data before and after normalization by the photodiode response. Calibration of the photodiode was done best by using a microwave spectrum analyser to observe the response of the photodiode to a short optical pulse whose width is much shorter than the response time of the photodiode. A common light source used in this measurement is a picosecond mode-locked dye laser (Bradley, 1977). A more conveniently obtained source is a mode-locked GaAs-type laser (Ho, 1978; van der Ziel, 1981; Auyeung, 1981; Auyeung and Johnston, 1982); pulsing a GaAs-type laser with short intense electrical pulses is another simple source (Liu *et al.*, 1980; Lau and Yariv, 1980). The use of a GaAs-type laser as an optical source is preferred not only because of simplicity, but also because the calibration can be done at the same wavelength range as the lasers to be tested. Failure to correctly calibrate the photodiode can lead to stray features in the modulation characteristics that can be erroneously taken to be intrinsic to the laser.

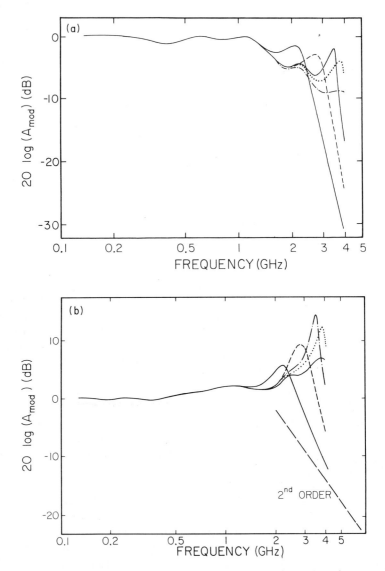

FIG. 25. Measured frequency response of a BH-on-SI laser (a) before and (b) after normalization by the photodiode response.

15. LARGE-SIGNAL EFFECTS AND PULSE CODE MODULATION

In the preceding sections we were mainly concerned with small-signal modulation characteristics of semiconductor lasers. The predictions on the ultimate frequency response and the means of achieving it are based on the small-signal theory with the underlying assumption that the amplitude of the modulation current (and hence the optical modulation depth) is small. What should be regarded as small can be ascertained from a comparison between the small-signal theory and the results obtained from a direct numerical solution of the rate equations (Ikegami and Suematsu, 1970). The results will be illustrated in this section and the small-signal regime can be defined to be all cases in which the optical modulation depth is $\leq 70\%$. (The optical modulation depth is defined as the ratio of the modulation amplitude to the peak of the modulated waveform.) For reasons of noise reduction and reliable digital threshold detection in optical receivers, the optical background preferably should be minimized (and hence the optical modulation depth be maximized). Most experiments in the digital modulation of laser diodes were performed with the laser biased just slightly above threshold and modulated with positive current pulses of relatively large amplitude. Suppose, for the moment, that the laser is biased with a direct current I_b and superimposed on it are positive modulation current pulses of amplitude I_p. Then, provided that the PCM signal is sufficiently pseudorandom such that long strings of zeros and ones do not occur,[†] the laser can be regarded as biased at an average current of $(I_b + I_p)/2$ and modulated with a microwave signal of amplitude $I_p/2$. The modulation depth in this case is close to 100%, and the modulation characteristics can be described by a large-signal theory.

Ikegami and Suematsu (1970) studied modifications to the results of the small-signal analysis due to large-signal effects by direct numerical integration of the rate equations. They found that under a large sinusoidal modulation current, the optical response becomes pulselike, and the relaxation oscillation resonance occurs at a frequency lower than that predicted by the small-signal theory. Thus, under large-signal modulation, the maximum modulation bandwidth of the laser would be reduced. Figure 26 shows the frequency of the large-signal relaxation oscillation resonance, normalized by f_r versus the optical modulation depth. One can observe from Fig. 22 that the relaxation oscillation frequency is reduced by a factor of 0.7 when the optical modulation depth is 70% and drops to 0.6 times the small-signal resonance frequency when the modulation depth approaches 100%. It should be noted that the results shown in Fig. 22 are obtained from the rate equations [Eqs. (21)] with the spontaneous emission factor β set to zero. Since the shift of the resonance to lower frequencies results from the highly spiked optical response near the resonance, taming the spiky behavior by using a finite β should result in smaller shifts. The results shown in Fig. 22 can be regarded as a worst-case calculation.

[†] A similar problem exists in electronic digital communication systems in which a long run of the same digit or some other unfortunate patterns can cause malfunctions in digital circuits intended to extract the pulse rate. A scrambler to randomize the digits is commonly interposed between a binary source and the data transmission system (see, e.g., Pierce and Posner, 1980).

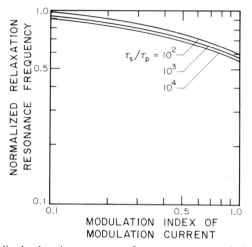

FIG. 26. Normalized relaxation resonance frequency versus optical modulation depth $(I_b = 2 J_{th})$. The reduction factor is defined as the ratio of the relaxation oscillation frequency obtained under large-signal conditions to that under small-signal conditions. [From Ikegami and Suematsu (1970).]

Large-signal effects in the modulation of semiconductor lasers have also been considered analytically by Harth and Siemsen (1974; Harth, 1973). An expression was obtained for the time variation in the photon density under a modulation current of $j \cos \omega t$

$$p(t) = [P_0/I_0(a)] \exp[a \cos(\omega t + \vartheta)], \tag{46}$$

where $I_0(a)$ is the modified Bessel function of the first kind of order zero, and a is an amplitude factor shown in Fig. 27. Equation (46) reduces to the usual

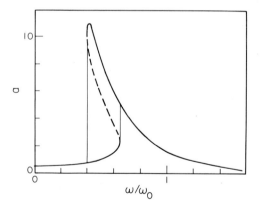

FIG. 27. Calculated large-signal frequency response of injection lasers based on a model by Harth. [From Harth (1973).]

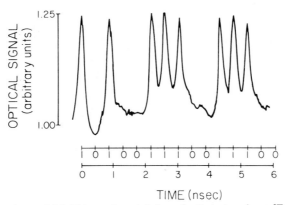

FIG. 28. Experimental 2.3-Gbit sec⁻¹ modulation of a GaAs-type laser. [From Russer and Schultz (1973).]

sinusoidal response for small a (when the amplitude of modulation is small) but gives a periodic pulselike behavior for large a. The hysteretic behavior of the modulation amplitude versus frequency characteristic has not been confirmed experimentally, although discontinuities have been observed in the large-signal modulation characteristics (Ikegami and Suematsu, 1970).

There were numerous reports on high-speed PCM modulation of semiconductor lasers. Russer and Schultz (1973) reported a modulation bit rate of 2.3 Gbit sec⁻¹ as early as 1973, even though the modulation depth was only 20% and the bit pattern was less than perfect (Fig. 28). Biasing the laser slightly above the threshold, Abbott *et al.* (1978) obtained a modulation depth of 80% with good pulse quality at 1.1-Gbit sec⁻¹ pseudorandom PCM by using both GaAs- and GaInAsP-type lasers. Yamada *et al.* (1979) and Hagimoto *et al.* (1982) (Fig. 29) achieved modulaton bit rates of 2 and 4 Gbit sec⁻¹, respectively, by using a 1.3-μm GaInAsP-type laser. Tell and Eng (1980) modulated a TJS laser at 8 Gbit sec⁻¹ in a non-return-to-zero format (Fig. 30), showing discernible optical bit patterns even though substantial reduction in the pattern effect is still required for practical applications.

The predictions concerning the ultimate frequency response of semiconductor lasers and the guidelines for laser designs should be applicable to the large-signal case, provided that one includes the appropriate frequency reduction factor. The correspondence between the maximum modulation bit rate and the cutoff frequency under sinusoidal, large-signal modulation is not as straightforward as the small-signal case, because the bit error rate depends not only on the pulse amplitude but also on the pulse shape. Nevertheless, for the sake of comparison between different laser structures, the PCM bit rate and the analog bandwidth can be taken as equal, adding a factor of two to the PCM bit rate when a non-return-to-zero format is used.

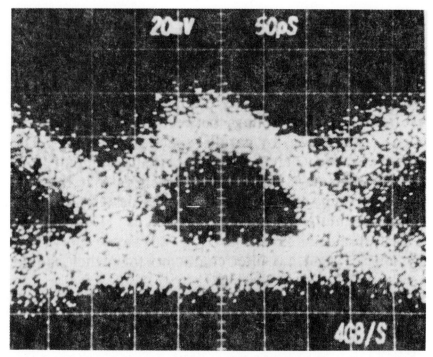

FIG. 29. Experimental 4-Gbit sec^{-1} modulation of a GaInAsP-type laser. What is shown here is known as an *eye diagram,* which is a continuous display of the detected optical signal on an oscilloscope when the laser is modulated by a train of pseudorandom digital pulses. The degree to which the eyes are open indicates the ability to distinguish the high and low digital levels and, hence, the signal quality. [From Hagimoto *et al.* (1982).]

16. PULSE CODE MODULATION UTILIZING NONLINEAR TRANSIENT EFFECTS

In attempts to overcome the basic limitations in the maximum modulation bit rate of semiconductor lasers as predicted by the small-signal or large-signal theory, nonlinear transient effects were considered as potentially useful. In the modulation schemes discussed so far in which transient effects are to be avoided, lasers with a highly suppressed relaxation oscillation resonance are the candidates of choice. The opposite is true when we actually try to use the transients as signal pulses.

Both the numerical simulation and the experimental evidence of Sections 5 and 6 indicate that the first spike of the relaxation oscillation is extremely short. Common experimental observations indicate detector-limited pulse widths of approximately 70 psec. Ultrafast streak camera measurements (Kobayashi *et al.,* 1980) and optical autocorrelation measurements (Liu *et*

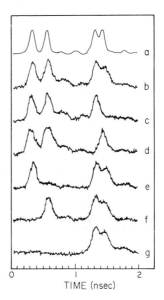

FIG. 30. 8-Gbit sec⁻¹ non-return-to-zero modulation of a TJS laser: (a) Output electrical pulses from word generator and (b)–(g) output pulses from detector with different ones removed. [From Tell and Eng (1980).]

TIME (nsec)

al., 1980) show pulse widths on the order of 15–30 psec generated in some lasers under very high-current pulse excitation. Thus it appears that by employing the first spike of the relaxation oscillation, modulation pulses can be spaced approximately 30–40 psec apart, corresponding to bit rates of 25–30 Gbits sec⁻¹! Two factors render the preceding proposition improbable. First, there is a finite time delay between the onset of the current pulse and the activation of the laser equal to the time required for the electrons to fill up to above the threshold level, as illustrated in Fig. 5. This turn-on delay can be minimized by prebiasing the laser at or slightly above the lasing threshold, although this will introduce a small optical background. A more serious problem is the intersymbol interference (the pattern effect). This arises from the relatively long spontaneous carrier lifetime of 1–3 nsec. After emission of an optical pulse, the electron population inside the laser is usually different from the prepulse level; i.e., the system will take a length of time (on the order of a spontaneous lifetime) to relax back to the equilibrium prepulse level. When a second modulation pulse is applied during this interim period, the resulting optical pulse will not be identical to the preceding one because of a different starting condition. Danielsen (1976) suggested that by carefully adjusting the area of the modulating current pulse, it should be possible to make the electron density at the end of the current pulse return to the prepulse level. In this way, a second modulation pulse immediately following the first one will see the same starting conditions as the first pulse and hence will produce an identical optical pulse. Figure 31 shows a numeri-

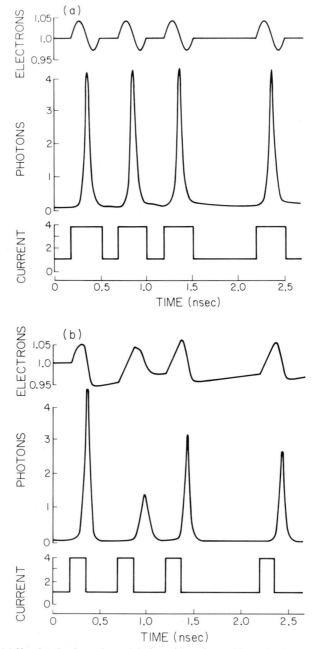

FIG. 31. (a) Simulated pulse code modulation with optimum bias and pulse area. The upper plot shows the electron density, the middle plot the photon density, and the lower plot the drive current. The current pulses assume a bit pattern of 11101. (b) The unoptimized case.

cal simulation of a laser modulated by a string of current pulses with different pulse widths. This was verified experimentally by Torphammar *et al.* (1979), whose result was a 1.1-Gbit sec^{-1} pattern-effect-free modulation capability with clean, background-free optical pulses (Fig. 32). As predicted theoretically, the bias current and the drive current pulse area had to be carefully controlled to within 1 and 10%, respectively, to attain the pattern-effect-free condition. Lee and Derosier (1974) have suggested that the pattern effect can be reduced by adding a backward swing to the drive current pulse (bipolar pulsing), which removes the excess carriers left in the active region after the optical pulse. This also requires a carefully controlled current pulse to ensure identical before-and-after conditions and has only been demonstrated at 500 Mbits sec^{-1} (Fig. 33). However, such careful control of current parameters (which can be difficult at high speed) is not required if no dc bias is applied to the laser, since in this case the carrier density before the current pulse would be zero and the carrier density after the current pulse would also be zero if the backward swing were sufficiently large that all the excess carriers were withdrawn.

The highest modulation bit rate that can be achieved with this scheme is apparently *not* limited by the response of the laser diode itself. Provided that the drive current amplitude is sufficiently large to drive the carrier density

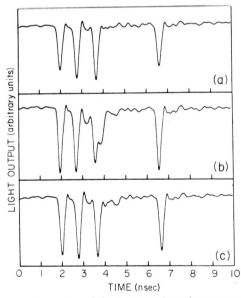

FIG. 32. Experimental laser output for a short pseudorandom sequence 0111001000 at 1.1 Gbit sec^{-1}. (a) and (b) Response when the drive current pulses are too small and too large, respectively; (c) optimum case. [From P. Torphammar, R. Tell, H. Eklund, and A. Johnston, (1979). *IEEE Journal of Quantum Electronics,* **QE-15,** 1271. Copyright © 1979 IEEE.]

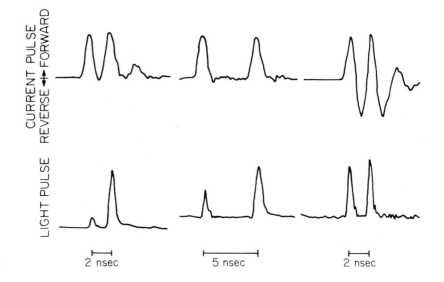

FIG. 33. Experimental demonstration of reducing pattern effect by bipolar pulsing. [From T. P. Lee and R. M. Derosier, (1974). Charge storage in injection lasers and its effect on high-speed pulse modulation of laser diodes, *Proceedings of IEEE*, **62**, 1176. Copyright © 1974 IEEE.]

above the threshold level, the laser responds immediately with a single optical pulse that has been shown to be as short as 15–30 psec. The response of the laser can be made arbitrarily fast, provided that the drive current pulse can be made arbitrarily short and its amplitude arbitrarily large *simultaneously*. For a very short drive current pulse width τ, its minimum amplitude I_{min} required for lasing to occur without any dc bias is given approximately by

$$I_{min} = I_{th}(\tau_s/\tau), \tag{47}$$

where I_{th} is the cw threshold current of the laser. With these parameters, the minimum pulse amplitude is about 30 times that of the cw threshold current. The highest bit rate that can be attained is thus limited by the ability to generate short current pulses of sufficient amplitude.

An experiment has been performed to demonstrate the scheme just outlined by using combinations of step-recovery-diode impulse generators and high-pass filters for generating picosecond bipolar pulses (Lau *et al.*, 1980). The pulses have a width of 70 psec and an amplitude of 12 V (into 50 Ω) in each swing. Figure 34 shows two such consecutive current pulses separated by 300 psec; the amplitudes of the pulses were 12 V (into 50 Ω). The laser

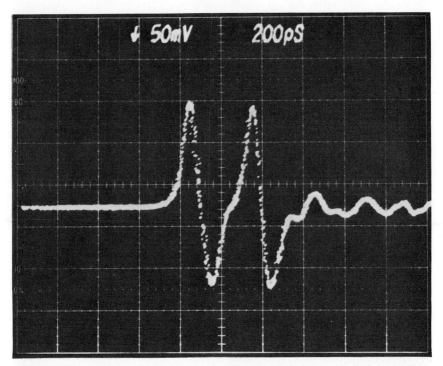

FIG. 34. Electrical pulses generated with a step-recovery diode followed by a high-pass filter. The pulse amplitude was 12 V (into 50 Ω). Horizontal scale is 200 psec per division. [From Lau and Yariv (1980).]

used was a BH laser with a 9-mA threshold. Such a low-threshold laser was needed in order that the lasing condition [Eq. (47)] be satisfied with a reasonable pulse amplitude. The response of the laser to the drive pulse pattern of Fig. 34 is shown in Fig. 35a. The output was detected with a fast avalanche photodiode (APD) (Telefunken BPW28) with about a 150-psec rise time. No significant pattern effect was seen for pulse separations as small as 250 psec. This was not the case when there was no backward swing in the drive current pulse or when a dc bias current was applied, which would have been required had the lasing threshold been higher. The pulses are not completely return-to-zero, which is mostly due to the detector-limited response. At a lower modulation rate of 1.9 Gbit sec^{-1}, the pulses are completely return-to-zero, as shown in Fig. 35b.

Although PCM modulation of injection lasers using nonlinear transient effects requires more complicated electronic drive circuitries, the lasers do potentially offer better performances in optical-fiber transmission as com-

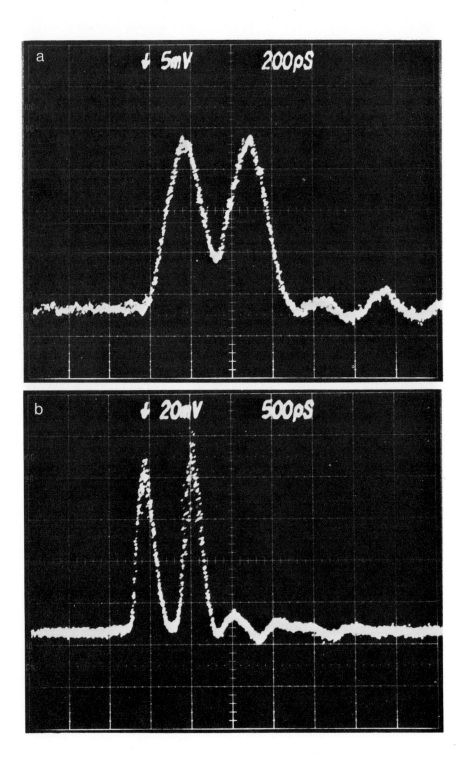

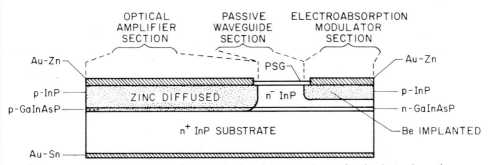

FIG. 36. Schematic diagram of a Q-switched semiconductor laser using the electroabsorption effect. The amplifier and modulator sections range in length from 150 to 200 and 50 to 75 μm, respectively. [From Tsang *et al.* (1981).]

pared to PCM modulation in the small-signal or large-signal regimes. Even at similar bit rates, the narrow, intense optical pulses obtained in the nonlinear regime can tolerate a much larger amount of dispersion in optical fibers (and hence can increase the transmission length) and offer better bit error rates. The trade-off is, as usual, between performance and complexity.

A different approach to digital modulation has been demonstrated by Tsang *et al.* (1981; Tsang and Walpole, 1982), who fabricated an actively Q-switched quaternary laser. The absorber is produced by the Franz–Keldysh electroabsorption effect in one section of the laser cavity. A schematic diagram of this device is shown in Fig. 36. An applied reverse-biasing voltage on this section produces a shift in the band edge, making that section opaque to the optical radiation. With the absorber in effect, the gain section of this laser can be biased to a very high level without causing the device to lase. A sudden reduction of the absorber voltage switches the Q of the cavity to a high value, lasing commences, and a sharp optical pulse is emitted. After the optical pulse is delivered, the electron density in the gain section can rapidly return to the prepulse condition, since this section is continuously pumped at a very high level. The time constant for returning to equilibrium is still the spontaneous lifetime, and therefore the pattern effect discussed will *not* be alleviated in this device. Nevertheless, by using a special pulse-position modulation scheme, Tsang *et al.* (1981) showed that a bit rate of 10 Gbit sec^{-1} is theoretically possible. Actual devices fabricated achieved Q-

FIG. 35. (a) Response of a low-threshold (9-mA) BH laser to the drive pulse in Fig. 34 at 3.3-Gbit sec^{-1} return-to-zero format. (b) The received pulses totally resolved at a lower bit rate of 1.9 Gbit sec^{-1}. [From Lau and Yariv (1980).]

switching by using a continuous microwave drive signal at the absorber section that produced a continuous train of optical pulses with full on–off modulation up to a frequency of 3 GHz.

V. Parasitic Elements and Circuit Considerations in High-Frequency Modulation

A common source of concern when dealing with high-frequency electronics is the adverse effects that parasitic elements might have on device performance. Modulation of semiconductor laser diodes at high frequencies is no exception. Are the observed modulation characteristics due to the laser alone or are they due to parasitic elements? Will bond wires limit the frequency response of laser diodes? Does one iaser structure possess a larger parasitic capacitance than others? To answer these questions we must treat the laser as an electrical element and establish an equivalent circuit for it, together with the parasitic elements. Characterization of networks at high frequencies is most commonly done with s parameters (s stands for scattering). The ease with which these parameters are measured makes them especially suited for describing microwave devices. Scattering parameters can be easily converted to other more familiar descriptions of the network. When a Smith chart is laid over a polar display of s_{11} or s_{22}, for example, the input or output impedance can be read directly. If a swept-frequency source is used, the display becomes a graph of input or output impedance versus frequency.

17. THE INTRINSIC EQUIVALENT CIRCUIT OF A SEMICONDUCTOR LASER

Parasitic elements aside, let us first consider the equivalent circuit of the intrinsic laser diode. The elements of the laser-equivalent circuit are derived from the rate equations [Eqs. (21)], which describe the interplay between the optical intensity (or photon density) and the injected carriers. These equations are augmented by the voltage–carrier density characteristics of the injection junction. Morishita $et\ al.$ (1979) have used this approach to show that the laser diode can be modeled by a parallel R-L-C circuit (see Fig. 37a) whose elements are

$$R_i = R_p[(I_{th}/I^0) - 1], \tag{48a}$$

$$L_i = R_p\tau_p/(I^0/I_{th}) - 1], \tag{48b}$$

$$C_i = \tau_s/R_d, \tag{48c}$$

where

$$R_p = (2kT/e)(1/I_d) \tag{49}$$

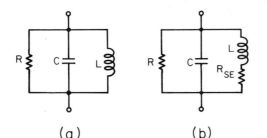

(a)   (b)

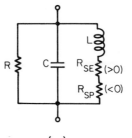

(c)

FIG. 37. Intrinsic equivalent circuit of a laser diode: (a) intrinsic equivalent circuit without spontaneous emission, (b) with spontaneous emission included, and (c) including saturable absorption and/or superlinear gain. [From J. Katz, S. Margalit, C. Harder, D. P. Wilt, and A. Yariv (1981). *IEEE Journal of Quantum Electronics*, **QE-17**, 4. Copyright © 1981 IEEE.]

is the usual expression of the differential resistance of a diode, I^0 the bias current of the laser diode, and I_d a normalized current given by

$$I_d = (N_s^0/\tau_s)/a_d ed, \tag{50}$$

where N_s^0 is the dc value of the injected electron density (in a p-type material), a_d the diode area, d the thickness of the active region, and e the electron charge. Here I_d is approximately equal to the current in the diode below the threshold and to the threshold current for currents above the threshold. In AlGaAs-type lasers, typical values of R_i are about or less than 1 Ω; values of L_i are about 1 pH, and values of C_i are about 3 nF. Katz *et al.* (1981) and Harder *et al.* (1982c) augment this model with the inclusion of spontaneous emission and saturable absorber and/or superlinear gain.[†] The resulting equivalent circuit is shown in Fig. 37b–c, in which the extra resistor R_{se} due

---

[†] A saturable absorber or a superlinear gain in a laser can lead to self-pulsations (Lau *et al.*, 1980). For a more detailed description of self-pulsations in injection lasers, see Section 22.

to the presence of spontaneous emission is given as

$$R_{se} = \beta R_p \frac{n_e}{n_p^0(n_p^0 + \beta)(n_e^0 - n_{om})} \approx \beta R_p \frac{n_e^0}{(n_p^0)^2}, \tag{51}$$

where n_p^0 is the steady-state photon density and n_e^0 is the steady-state electron density. The value of R_{se} is of the order of β and is consequently quite small (a few milliohms). It is interesting to see that the equivalent circuit is that of a parallel $R-L-C$ circuit with a resistor R_{se} in series with the inductance. The actual light output is proportional to the current flowing through the inductor L_i. One should note that the capacitance C_i in Fig. 33a corresponds to the junction (diffusion) capacitance of the diode. The fact that the value of C_i is *huge* when the laser is above the threshold (~ 3000 pF) should *not* be interpreted as a cause for reduced modulation bandwidth or for dips in the modulation characteristic, since it is in parallel with an inductor L_i, whose value is very small (on the order of picohenries above threshold). The L_i-C_i combination produces the relaxation oscillation resonance. Dumant *et al.* (1980) included the C_i but omitted the L_i in the equivalent circuit of the intrinsic diode (which then constitutes an equivalent circuit of a light-emitting diode but not of a laser) and might explain the discrepancy of their results from others.

The magnitude of the impedance of the entire equivalent circuit $|Z(\omega)|$ is plotted in Fig. 38 as a function of frequency ω. The impedance is essentially zero at all frequencies except near the relaxation oscillation resonance frequency, at which its value does not exceed ~ 1 Ω. This reflects the fact that gain clamping at these frequencies is manifested in the equivalent circuit as an ac short so that no ac voltage, and hence Fermi-level fluctuations, can develop. Consequently, in comparison to relatively large external elements, the intrinsic laser diode can be regarded as a short circuit at all frequencies as

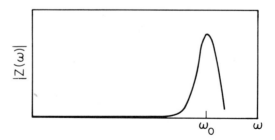

FIG. 38. Impedance of a laser diode $|Z(\omega)|$. [From J. Katz, S. Margalit, C. Harder, D. P. Wilt, and A. Yariv (1981). *IEEE Journal of Quantum Electronics*, **QE-17**, 4. Copyright © 1981 IEEE.]

long as the laser is biased above threshold. Having established this fact, we proceed to consider the parasitic elements.

18. EFFECT OF PARASITICS ON THE MODULATION RESPONSE

Figure 39 shows the general geometry of a laser fabricated on a conductive substrate and a simple equivalent circuit of this laser. It includes the major parasitic elements involved. We can also include a submount capacitance by connecting a shunt capacitance to the entire circuit in Fig. 39. For a well-constructed mount, the value of this capacitance is usually $\lesssim 0.2$ pF and can then be neglected. The value of the bond wire inductance L depends not only on the length of the wire, but also on its diameter and its proximity to the ground plane (see, e.g., Jackson, 1975). A good bond wire should not constitute an inductance much larger than 1 nH. The parasitic capacitance involved is the capacitance between the top and bottom contact in the area outside the lasing region and obviously depends on the material between the contacts and its area. The value of this capacitance, as we shall see later, varies widely from laser to laser. As it turns out, this capacitance is the major element of concern at high frequencies ($\lesssim 5$ GHz).

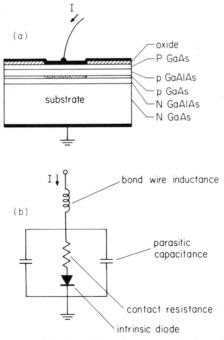

FIG. 39. General geometry of (a) a laser fabricated on a conductive substrate and (b) a simple equivalent circuit.

Before discussing how the values of the parasitic elements depend on laser structure, we shall first look at how the parasitics affect the modulation response of injection lasers. We must now take into account how the laser is being driven. In typical experiments in which a laser is mounted at the end of a 50-Ω transmission line and no provision is made for impedance matching (it is not possible to match the impedance of the entire laser-equivalent circuit with a pure resistor anyway), reflections of the microwave drive signal from the laser must be taken into account. This is equivalent to driving the laser with a pure voltage signal source having a source impedance of 50 Ω (Fig. 40b). In an experiment in which a network analyzer is used to measure the response of the laser, the ac signal from the photodiode is compared to the microwave signal feeding the laser. This microwave feed signal is represented by the pure voltage source in Fig. 40b. (Measurement techniques will be discussed in Section 17.) The overall modulation response is then obtained by multiplying the intrinsic response of the laser by the ratio

$$\eta = \frac{\text{current flowing into the intrinsic diode}}{\text{voltage of the signal source}}. \tag{52a}$$

An analysis of the equivalent circuit in Fig. 40b (under a 50-Ω drive) gives the

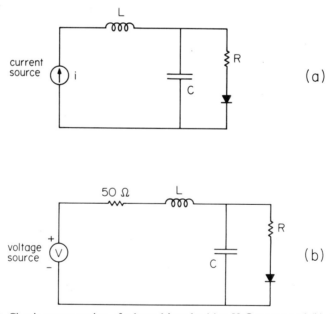

FIG. 40. Circuit representation of a laser driven by (a) a 50-Ω source and (b) an infinite resistance pure current source.

following for η:

$$\eta \sim \left[\frac{(i\omega)^2}{\omega_0^2} + \frac{i\omega}{\omega_0 Q} + 1 \right]^{-1}, \tag{52b}$$

where

$$\omega_0 = \sqrt{\frac{50 + R}{LRC}}, \tag{52c}$$

$$Q = \sqrt{\frac{LRC(50 + R)}{L + 50RC}}, \tag{52d}$$

and L, R, and C are the parasitic components as shown in Fig. 40b. The response of Eq. (52a) is in form of a second-order, low-pass filter type. For reasonable values of L, R, and C, the Q of this response is close to or less than $\frac{1}{2}$, which means that no resonance occurs in the injection ratio η at $\omega = \omega_0$. It is intuitively obvious, by observing the equivalent circuit of Fig. 40b, that if the inductance L is very small, then the principal effect of the parasitic elements comes from the RC low-pass combination, where the presence of the capacitor bypasses an increasing amount of drive current at high frequencies. We shall illustrate by the following examples the range of value of L that are to be considered very small. First, we take the following values for the parasitic elements: $L = 1$ nH, $C = 4$ pF, and $R = 4$ Ω [which, as we shall illustrate later on, are typical values for a BH-on-SI laser with a fairly long (~ 1.5 mm) bond wire]. The value of ω_0 and Q are [Eqs. (52c) and (52d)] 9.25 GHz and 0.516, respectively. A plot of the injection ratio versus frequency is shown in Fig. 41 (labelled case A in figure). Next, we take $C = 4$ pF and $R = 4$ Ω as before, but let $L = 0.2$ nH (which corresponds to a short bond wire ~ 0.5 mm in length). The values of ω_0 and Q in this case are 20.7 GHz and 0.416, respectively. The injection ratio in this case is plotted in Fig. 41 (case B). There are now two distinct corner frequencies at 11.1 and 38.7 GHz. This is what is commonly refered to as an underdamped, second-order filter response and arises from a splitting of the poles on the negative real axis. Physically, this is just the case we have mentioned earlier, that L is sufficiently small so that the RC combination dominates (which produces the roll-off at the first corner frequency at 11.1 GHz). As mentioned before, a good wire bond in a well-designed high-frequency package should not contribute more than 1 nH of inductance, with typical values of 0.2–0.3 nH, and therefore the inductance is really of no concern. The RC combination is, in practice, the principal limiting factor in high-speed modulation at frequencies up to at least 15 GHz. In some lasers where the parasitic capacitance was measured to be as large as 40 pF, a dip in the modulation response in the lower gigahertz range is expected and has been observed by numerous

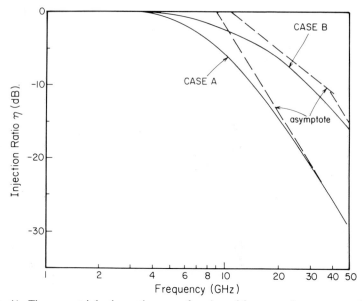

FIG. 41. The current injection ratio η as a function of frequency for two cases. Case A: $R = 4\ \Omega$, $C = 4$ pF, and $L = 1$ nH; case B: $R = 4\ \Omega$, $C = 4$ pF, and $L = 0.2$ nH.

researchers including Harth (1975), Aiki *et al.* (1978), Furuya *et al.* (1978), Maeda *et al.* (1978) and Figueroa *et al.* (1980).

In some experiments, lasers are driven by a pure current source instead of being driven by a 50-Ω microwave source (Fig. 40a). Examples of these arise when a laser is driven with a field-effect transistor or a bipolar transistor (Ury *et al.*, 1982; Katz *et al.*, 1980; Bar-Chaim *et al.*, 1982; Ostoich *et al.*, 1975) in close proximity to the laser, either in hybrid or integrated form. In this case, it is obvious that the parasitic inductance L need not be considered. The bypass of current into the parasitic capacitance C occurs above the corner frequency of $\omega = 1/(CR)$. Calculations show that in this case as long as C is below ~ 3 pF, the parasitic elements do nothing more than lower the magnitude of the resonance peak, which, incidentally, is not unwelcomed.

19. PARASITIC CAPACITANCE OF INJECTION LASERS

Experimentally, the value of the parasitic elements in an equivalent circuit of a semiconductor laser (Fig. 39) can be determined by measuring the microwave s parameter s_{11} (reflection coefficient) at various frequencies and then fitting the data to the calculated reflection coefficient of the assumed network. A satisfactory fit cannot usually be obtained at all frequencies unless the assumed circuit is reasonably similar to the actual circuit (i.e., it contains all the major elements in the right topology). Extensive measurements reveal that the equivalent circuit in Fig. 39 produces reasonable fits to most lasers fabricated on a conductive substrate. Two representatives of this class of lasers, the CSP laser and the BH laser, have been measured by Maeda *et al.* (1978) and Figueroa *et al.* (1982). The magnitude and phase of s_{11} are usually plotted in polar form with frequency as the varying parameter. The plot is superimposed on a Smith chart, from which the total impedance of the network under test can be read off at any frequency from the value of s_{11} at that frequency.

Most lasers fabricated to date on a conductive substrate are of the type shown in Fig. 39. Examples of these are the CSP, proton-bombarded stripe, V-groove, and constricted double heterostructure (CDH) lasers, hereafter collectively referred to as CSP-type lasers. The equivalent circuit of this class of lasers is shown again in Fig. 42a. The BH laser has an equivalent circuit basically similar to the CSP-types; the only difference arises from the existence of a *p*-doped blocking laser in the BH structure. The reverse-biased junction across the blocking layer drastically reduces the parasitic capacitance between the top and bottom contacts. Measurements from different sources (Maeda *et al.*, 1978; Figueroa *et al.*, 1982) have confirmed that the value of C is around 10 pF in BH lasers, while in CSP lasers it is more than 40 pF. This does not necessarily mean that BH lasers are less affected by parasitics, since the contact resistance R_s of the BH laser, due to a narrow stripe width (1 – 2 μm), is about five times that of CSP-type lasers (typically

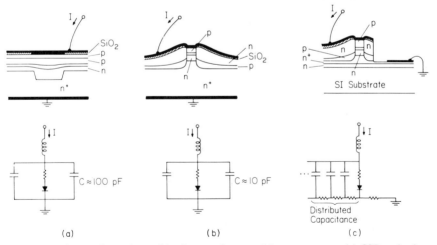

FIG. 42. Comparison of parasitic elements in several laser structures: (a) CSP and other stripe-geometry-type lasers on conductive substrate, (b) BH laser on a conductive substrate, and (c) BH laser on a semi-insulating substrate.

$R_s \sim 10\text{--}15\ \Omega$ in BH lasers and $\sim 2\ \Omega$ in CSP-type lasers). The RC time constants in both cases turn out to be comparable. Both of these lasers, according to the calculations of the preceding section, will display a significant dip at high frequencies, as observed in numerous experiments.

A separate class of lasers can be defined that is based on semi-insulating substrates instead of conventional n^+-doped substrates (Lee *et al.*, 1978; Nita *et al.*, 1979; Bar-Chaim *et al.*, 1981; Yu *et al.*, 1981). A BH-on-SI laser is shown in Fig. 42c, together with its equivalent circuit. The major difference between the equivalent circuit of this laser and other lasers is that the parasitic capacitance in the BH-on-SI laser is in form of a distributed network, due to the particular geometry of this device. The resistance of the bottom n^+ contact layer provides an effective shielding of the capacitances far away from the junction, especially at high frequencies. Assuming that the resistance per unit length of the bottom n^+ layer is R_{dist} and that the parasitic capacitance per unit length between the top contact and the bottom n^+ layer is C_{dist}, the effective impedance Z of the entire distributed $R\text{--}C$ ladder network is given by analogy to transmission line theory as

$$Z = Z_0(1 + e^{-2kW})/(1 - e^{-2kW}), \tag{53}$$

where W is the width of the top contact, Z_0 the characteristic impedance of the ladder network

$$Z_0 = \sqrt{R_{\text{dist}}/j\omega C_{\text{dist}}}, \tag{54a}$$

and k the propagation constant

$$k = \sqrt{j\omega R_{\text{dist}} C_{\text{dist}}}. \tag{54b}$$

At very high frequencies, the real part of k is large, with the physical significance that the electric field does not penetrate too far beyond the lasing junction. Capacitance elements existing far away from the junction, therefore, make no contribution to the parasitics. The total impedance of the distributed network is Z_0 in this limit.

The effect of distributed capacitance can be understood in a more intuitive manner by considering the limit of small R_{dist} and expanding Eq. (53) in this limit:

$$Z = Z_0 \coth(kW)$$
$$= (1/i\omega C_{\text{eff}}) + R_{\text{eff}}, \tag{55}$$

where

$$C_{\text{eff}} = C/(1 + \omega^2 R^2 C^2/45), \tag{56}$$
$$R_{\text{eff}} = (R/3) - \tfrac{2}{945}\omega^2 R^3 C^2, \tag{57}$$

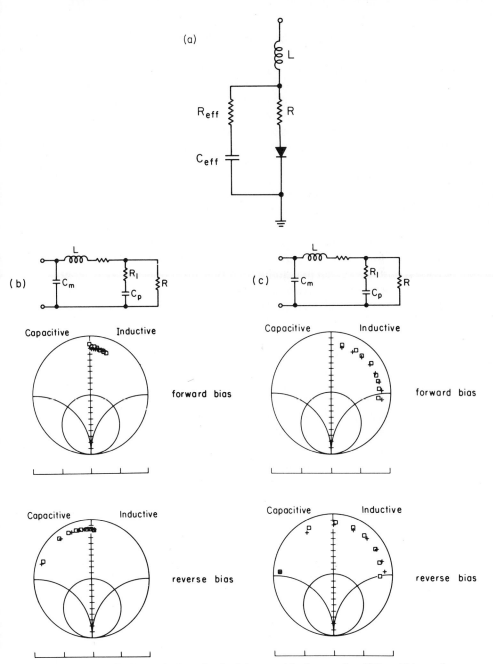

FIG. 43. (a) Effective equivalent circuit of the parasitic elements in a BH-on-SI laser; $L = 0.18$ nH, $C_m = 0.07$ pF, $R = 2\ \Omega$ (forward bias) or ∞ (reverse bias), $C_p = 4.1$ pF, and $R_1 = 0.1\ \Omega$. s_{11} parameter of BH-on-SI lasers and theoretical fit to the equivalent circuit shown on the top of the figures: (b) 250-μm laser with a short (<0.3 mm) bond wire; (c) 175-μm laser with a relatively long (1.0 mm) bond wire; frequency: 1 GHz–8 GHz in steps of 1 GHz; s_{11} parameter plots: $\square$, theoretical and $+$, experimental.

and R and C are, respectively, $R_{dist}W$ and $C_{dist}W$, the total parasitic capacitance and lateral resistance of the bottom conductive n-layer, and only the first four terms in the taylor expansion of Eq. (53) are included. It follows that the equivalent circuit of a laser on SI substrate is that shown in Fig. 43a. The effective capacitance C_{eff} thus decreases at high frequencies and the existence of R_{eff} further reduces the amount of current diverted into the parasitic capacitance at high frequencies. Actual s_{11} measurements of BH-on-SI lasers indeed confirm the equivalent circuit of the configuration shown in Fig. 43a The magnitude and phase of the s_{11} parameter is usually plotted in polar form, with the frequency as the varying parameter. The plot is superimposed on a Smith Chart, from which the total impedance of the network under test can be read off at any frequency from the value of s_{11} at that frequency. Figures 43b and c show such plots of the experimentally measured data (crosses) and theoretical fit (square) of a BH-on-SI laser of cavity lengths of 175 and 250 μm, respectively, at forward bias and reverse bias conditions. The values of the circuit elements used to produce the theoretical fits are also shown in Fig. 43. The values of the effective capacitance and resistances are exceptionally low in these devices, with an equivalent RC time constant of < 10 psec, which accounts for the total absence of the dip in the modulation response of these devices even at frequencies approaching 10 GHz, as shown in the results described in the last few sections. The -3-dB cutoff frequency in the drive current into the laser due to the parasitic elements with values shown in Fig. 43 is ~ 12 GHz. This can account for the fact that the observed bandwidth of the lasers cannot be pushed much beyond 12 GHz despite various techniques are used to enhance its bandwidth. It is therefore necessary to reduce the parasitic elements by suitable electrode designs before further improvements in bandwidth, of which several possible schemes are described in the next section, can be realized.

In conclusion, it appears that parasitic elements are not detrimental to high-speed modulation of semiconductor lasers for frequencies up to $\sim 3-4$ GHz. Lasers that are not specifically designed for high-frequency operation cannot operate above this frequency range anyway owing to limitations described in Sections 9–10. These limitations can be overcome by using suitable laser structures, as experimental evidence described in Section 12 showed, and only then do parasitic elements warrant attention.

VI. Effect of Superluminescence and Reduced Mirror Reflectivity on the Modulation Response

20. THE FREQUENCY RESPONSE OF SUPERLUMINESCENT DIODES

The superluminescent diode is one alternative to the well-established injection lasers and LEDs as light sources for optical-fiber communications. A superluminescent diode is a laser diode without mirrors. The first investigation of the superluminescent diode was carried out by Kurbatov *et al.* (1971), and its static properties were evaluated in detail by Lee *et al.* (1973) and Amann *et al.* (1978; Amann and Boeck, 1979a). Superluminescent diodes have also been integrated monolithically with detectors for optical memory readout, their fabrication being simpler than that involved in laser–detector integration since a mirror facet is not required within the integrated device Amann *et al.* (1980). Because of its relatively wide spectral width, superluminescent diodes are suggested as an ideal source in optical-fiber gyroscope systems in which excessive noise can be created by the coherence of the optical source (see, for example, Giallorenzi *et al.*, 1982). It has also been observed that the optical modulation bandwidth increases substantially as LEDs enter the superluminescent regime. This regime is characterized by a rapid increase in the optical power output and a narrowing of the emission spectrum. An increase in the modulation capability also occurs in some edge-emitting LEDs, though they were not purposely operated in the superluminescent regime (Olsen *et al.*, 1980). This increase in modulation speed was attributed to the shortening of the carrier lifetime due to the stimulated emission of photons.

The discussion in Section 2 clearly indicates that due to the nonuniformity in the longitudinal distribution of the photon and carrier densities in the active region, the modulation response of superluminescent lasers cannot be described by the usual spatially uniform rate equations. Rather, the local rate equations [Eqs. (2)] of Section 2 should be used in their original form. Numerical calculations of the small-signal modulation response based on those equations have been performed by Boeck *et al.* (1980). The case in which the reflectivities from both mirrors are absolutely zero was considered. The small-signal analysis employed the expansion of the photon $X^{\pm}$ and electron N density variables

$$X^{\pm}(z, t) = X_0^{\pm}(z) + x^{\pm}(z)e^{i\omega t}, \tag{58a}$$

$$N(z, t) = N_0(z) + n(z)e^{i\omega t}, \tag{58b}$$

where $x^{\pm}$ and n are small variations about the steady states. This assumes that the electron and photon densities throughout the length of the diode vary in unison. This is true when propagation effects are not important, i.e., when modulation frequencies are small compared with the inverse of the photon transit time. This would amount to over 15 GHz, even for very long diodes (0.25 cm).

Upon substitution of Eqs. (58) into the local rate equations [Eqs. (2)], a set of linearized equations can be obtained for $x^{\pm}$:

$$dx^+/dz = Ax^+ + Bx^- + C, \tag{59a}$$

$$dx^-/dz = Dx^+ + Ex^- + F, \tag{59b}$$

where $A, B, C, D, E,$ and F are given by

$$A = g_0 - \frac{i\omega}{c\tau_s} - \frac{(X_0^+ + \beta)g_0}{1 + i\omega + (X_0^+ + X_0^-)}, \tag{60a}$$

$$B = \frac{-(X_0^+ + \beta)g_0}{1 + i\omega + (X_0^+ + X_0^-)}, \tag{60b}$$

$$C = \frac{g_m(X_0^+ + \beta)}{1 + i\omega + (X_0^+ + X_0^-)}, \tag{60c}$$

$$D = \frac{(X_0^- + \beta)g_0}{1 + i\omega + (X_0^+ + X_0^-)}, \tag{60d}$$

$$E = -\left[g_0 - \frac{i\omega}{c\tau_s} - \frac{(X_0^- + \beta)g_0}{1 + i\omega + (X_0^+ + X_0^-)}\right], \tag{60e}$$

$$F = \frac{-g_m(X_0^- + \beta)}{1 + i\omega + (X_0^+ + X_0^-)}, \tag{60f}$$

where $g_0(z) = \alpha N_0(z)$ is the small signal gain distribution, $g_m = \alpha j \tau_s/(ed)$ is the small-signal gain due to the rf pump current, and ω has been normalized by the inverse of the spontaneous lifetime.

The boundary conditions for solving Eqs. (59) are the same as those used for solving the steady-state case, namely,

$$x^+(0) = x^-(0), \tag{61a}$$

$$x^-(L/2) = 0 = x^+(-L/2). \tag{61b}$$

The frequency response curve is obtained by numerically solving Eqs. (59) for each ω. One set of results is shown in Fig. 44. In addition to predicting the observed increase in modulation bandwidth as the pump current increases, it also predicts a flat modulation response at all pump levels. This is in sharp contrast to the existence of a resonance peak in injection lasers. Additional calculations showed that the modulation bandwidth is a strong function of the spontaneous emission factor (Lau, 1981), which is also in contrast to injection lasers.

21. EFFECT OF ANTIREFLECTION COATINGS ON THE MODULATION RESPONSE OF LASERS

The results in Fig. 42 seem to indicate that a flat-frequency response up to the multigigahertz range can be obtained if superluminescent diodes are operated at high current levels (corresponding to unsaturated optical gains of $400-500$ cm^{-1}). In practice, it is not clear whether such high gains can be attained reliably. An additional problem arises when the mirror reflectivity

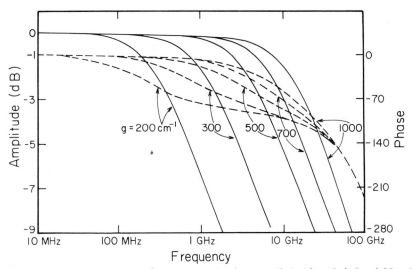

FIG. 44. Frequency response of a pure, superluminescent diode ($\beta = 10^{-4}$, $L = 0.05$ cm). The end mirror reflectivities are assumed to be absolute zero, which is unattainable in practice.

of real devices takes on a finite value. Even when the reflectivity is as small as 10^{-6}, the conventional lasing threshold gain (neglecting internal absorption loss) of a 500-μm device is $(1/L)\ln(1/R) \approx 276$ cm^{-1}. The minimum reflectivity achievable in practice (Lee *et al.*, 1973) is of the order of 10^{-4}. Thus at high pump levels, superluminescent diodes can rarely be operated as what they are named for, but rather as a laser. One would then naturally ask, what would the modulation response of a laser with a very low reflectivity be like? Undoubtedly, superluminescence inside the cavity of such a laser will lead to highly nonuniform carrier and photon density distributions even above the lasing threshold. The local rate equations [Eqs. (2)] must again be used.

To facilitate a fair comparison between lasers with various reflectivities, the length of each laser under consideration is adjusted so that they all possess the same photon lifetime. The calculated characteristics of these five laser diodes, based on the simple rate equations, is exactly identical. Using the full set of local rate equations would yield quite different results. Figure 45 shows the amplitude and phase responses of five devices obtained from a numerical calculation based on the local rate equations. One device corresponds to a common laser with a mirror feedback of 0.3 from both facets and a length of 300 μm. In each of the other diodes, one of the facets has a reflectivity of 0.3 and the other facet takes on various reflectivity values. In each of these four cases, the diode length is adjusted so that they possess a photon lifetime identical to that of the first laser diode. The values of ρ (the internal distributed loss), β, and τ_s are taken to be 60 cm^{-1}, 10^{-4}, and 3 nsec,

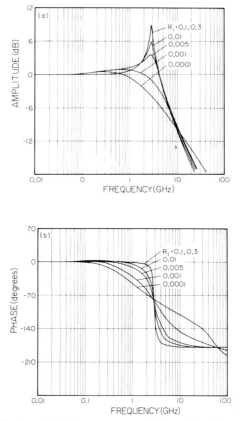

FIG. 45. (a) Amplitude and (b) phase responses of diodes with different mirror reflectivities and different lengths (all possess the same photon lifetime of 1 psec). They are pumped to an unsaturated gain of 200 cm^{-1}, which is above the classical threshold of all of these lasers. $R_2 = 0.3$, $\beta = 2 \times 10^{-4}$; and $g_h = 200$ cm^{-1}. [From Lau and Yariv (1982a).]

respectively. These devices possess the same threshold gain of 100 cm^{-1}. The devices are pumped to an identical level corresponding to an unsaturated gain g_h of 200 cm^{-1}, which translates into a J/J_{th} ratio of 1.25 when κN_{om} assumes the measured value of ~ 300 cm^{-1} (Hakki and Paoli, 1975; Bakker and Acket, 1979; Henry *et al.*, 1980). The commonly employed rate-equation analysis predicts that these five diodes should have identical frequency responses, which is in obvious disagreement with the data shown in Fig. 43. Notice that for the two cases in which $R_1 = 0.3$ and 0.1, the amplitude and phase curves are almost identical to those calculated from the simple rate equations and display a resonance at 2.9 GHz. The resonance is suppressed in devices with a smaller feedback from a mirror, and for $0.005 < R_1 < 0.01$,

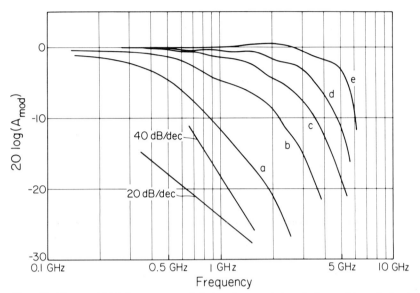

FIG. 46. Measured high-frequency modulation characteristics of a laser with antireflection coatings on both facets. The reflectivities of both facets are estimated to be below 2% as determined by the resultant rise in the lasing threshold. The curves a–e are obtained under increasing bias currents. [From Lau *et al.* (1983b).]

the response is maximally flat. A further decrease in R_1 reduces the modulation bandwidth. These results show that the simple rate equations fail to give the correct frequency response characteristic when the reflectivity of one of the mirrors is reduced to below 0.1. The simple approach *does,* however, give the correct *corner frequency* in the small-signal response even when the reflectivity of one of the mirrors is as low as 10^{-3}, as is evident from Fig. 45. These conclusions are not restricted to this particular case but are true in general over a wide range of pump currents and photon lifetimes.

These predictions have indeed been observed experimentally by Lau *et al.* (1983b). Figure 46a shows the modulation characteristics of a BH-on-SI laser with AR coating on both mirror facets, as compared to an uncoated laser of identical construction. The contrast in the modulation characteristics is evident.

The results presented have important implications in the design of high-frequency semiconductor lasers. First, the prescription for a high-frequency laser described in Sections 9–11, which is based on the spatially uniform rate equations, will remain valid over a wide range of mirror reflectivities down to 10^{-3}. Second, it can provide a means of dampening the relaxation oscillation resonance in lasers of any structure, even at very high frequencies.

VII. Other Modulation Formats and Effects

22. COHERENT OPTICAL DETECTION AND FREQUENCY/PHASE MODULATION OF SEMICONDUCTOR LASERS

Direct-intensity modulation of semiconductor lasers is perhaps the most straightforward way of imprinting a signal on an optical carrier. One of the most important properties of the laser, that it is an essentially single-frequency oscillator, is not utilized in the direct-modulation scheme. In fact, in terms of modulation methods, the direct-intensity modulation of semiconductor lasers is no more sophisticated than signaling by electrical noise bursts before the days of Marconi. Tremendous advances in electronic modulation–demodulation methods since the early days of wireless communication have inspired workers in the field of optical communications to push for similar advances in their field. The book by Gagliardi and Karp (1976) provides an extensive discussion on various coherent and incoherent optical-communication schemes. Recent interests in coherent optical communication using semiconductor lasers were revived by advances in the quality of these lasers. Stable, single-mode semiconductor lasers can now be routinely fabricated and are required in coherent optical communication both as the source and the local oscillator.

Various coherent optical-communication schemes are named by analogy to their electronic counterparts. There are, for example, phase-shift-keying (PSK) homodyne; frequency-shift-keying (FSK) homodyne; amplitude-shift-keying (ASK) homodyne; FSK/PSK/ASK heterodyne; FSK/PSK base-band direct detection; and subcarrier ASK, PSK, and FSK detection systems. Of course, there is the ASK base-band direct detection, which is a fancier name for direct-amplitude modulation (AM) and detection, the scheme that we discussed in the preceding sections. The nomenclature for all these coherent systems might be confusing, but they actually represent very simple things. The FSK, PSK, and ASK all refer to how the source oscillator (laser) is being modulated. For example, in FSK the laser oscillation frequency is being modulated by the signal. In heterodyne detection schemes, the information (usually contained in the frequency/phase of the carrier) is extracted from the incoming optical signal by mixing it with a local laser oscillator, while in homodyne detection the local oscillator is derived from the incoming signal itself. The PSK homodyne detection system is similar to the familiar electronic phase-locked loop. It is also possible to extract the information directly from the incoming signal. In PSK and FSK this is accomplished by using an optical discriminator, which can be an optical balanced bridge (Kaminow, 1964) or by using autocorrelation methods and in ASK by simply using a photodetector.

Numerous experimental and theoretical works have been reported in coherent optical-communication schemes. Yamamoto (1980, 1981) analyzed and compared receiver performance of various digital modulation – demodulation schemes in the $0.5 – 1.0$-μm wavelength region. Experiments on frequency modulation (FM) of a semiconductor laser by current injection were reported by Osterwalder and Rickett (1980). Since the refractive index of the semiconductor material depends on the carrier density, modulation of the carrier density by varying the current flowing into the laser introduces an effective change in the cavity length and produces variations in the oscillation frequency of the laser. The modulation index is typically measured to be $0.1 – 3$ GHz mA^{-1}, depending on the modulation frequency and laser structure. Kishino *et al.* (1982) and Kobayashi *et al.* (1982) have considered theoretically the amount of wavelength variation due to current modulation. Saito *et al.* (1981) reported on an optical FSK heterodyne detection experiment using semiconductor lasers as the transmitter and local oscillator. Figure 47 shows a schematic diagram of the system. In a simulated experiment Kikuchi *et al.* (1981) verified the theoretically predicted bit error rate in a PCM – ASK heterodyne system. Olesen and Jacobsen (1982) calculated the combined AM and FM amplitude and intensity modulation optical spectrum due to dc modulation of a laser. They realized that current modulation of a laser always results in combined intensity and frequency modulation. A deeper relationship between the ratio of the intensity to frequency modulation index was derived by Harder *et al.* (1983). It was shown by a semiclassical analysis of laser dynamics that the ratio of AM to FM index is independent of frequency for a given lasing material and is directly proportional to the α factor, which is defined as the ratio of the real to the imaginary part of the complex dielectric susceptibility of the material. It is also called the linewidth enhancement factor since it also affects the intrinsic linewidth of

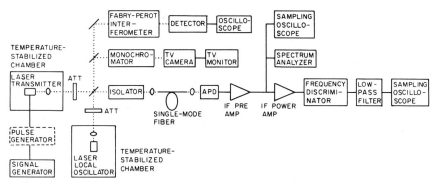

FIG. 47. Schematic diagram of a coherent communication system using an FSK format. [From S. Saito, Y. Yamamoto, and T. Kimura (1981). *IEEE Journal of Quantum Electronics,* **QE-17,** 935. Copyright © 1981 IEEE.]

the lasing mode. This important factor, in fact, has never been measured accurately. By measuring the relative amount of AM and FM under high-frequency current modulation of a semiconductor laser, Harder *et al.* (1983) for the first time produced an accurate determination of α.

In general, coherent detection techniques can potentially provide better performance than a direct-intensity modulated semiconductor laser communication system. The complexity of the coherent schemes (e.g., the stability of the lasing frequency must be very well controlled by enclosing the lasers in a temperature-stabilized chamber and/or by optical feedback techniques) dictates that they be used only in very high-performance systems, in which, for example, a very long repeater spacing must be attained at all cost.

23. Lasing Spectrum under Modulation

The steady-state longitudinal mode spectra of semiconductor lasers have been studied extensively, and major observed features can be understood in terms of modal competition in a common gain reservoir. It was generally agreed that gain saturation in semiconductor lasers is basically homogeneous, and therefore it should oscillate in a single longitudinal mode above the threshold. This has been extensively verified in well-behaved lasers of many different structures. In general, the number of lasing modes increases with increasing spontaneous emission and decreases with increasing optical power. Although most lasers lase in essentially a single longitudinal mode under cw operation, this is not the case during lasing transients. This was predicted by a numerical solution of the multimode rate equations [Eqs. (29)] (Tang *et al.*, 1963). The optical spectrum of a semiconductor laser during excitation transient has been observed by Peterman (1978), Matthews and Steventon (1978), Mengel and Ostoich (1977), Seeway and Goodwin (1976), and others. Further details in transient spectral effects have been elucidated by the numerical calculations of Peterman (1978) and Buus and Danielsen (1977). To explain this phenomenon, it is important to understand the spectral selection process that leads to single-mode oscillation in cw operation. The exact oscillation spectrum of a laser can be calculated from the multimode rate equations [Eqs. (29)]; but a heuristic exposition is given here instead. It must be realized that a laser is essentially a noise-driven oscillator and that no laser really oscillates in one absolute single mode. Mode selection arises from gain clamping and the slight difference in the gain between each mode. In a laser, the optical gain compensates for *most,* but not all, of the losses of the cold cavity so that the Q of each cavity mode is extremely high. All modes in a laser cavity are lossy and would not oscillate if spontaneous emission were not present as a driving force. The amplitude of forced oscillation of each mode is obviously proportional to the individual Q value. Assume that the optical gain g_i of the mode i is given by a Lorentzian distribution as in Eq. (30) and that the loss of each mode is assumed to be identical and equal to $l = g_0 + \delta_1$ (i.e., slightly more than the highest gain of all the modes). Since Q is inversely proportional to the net loss of the mode, it follows that Q_i is proportional to

$$(l - g_i)^{-1} = [\delta + (g_0 bi^2/1 + bi^2)]^{-1}. \tag{62}$$

The factor b is very small ($\sim 2 \times 10^{-4}$) in typical semiconductor lasers, and therefore

$$Q_i \sim (\delta + g_0 bi^2)^{-1}. \tag{63}$$

The oscillation amplitude of mode i is proportional to Q_i. It is obvious from Eq. (63) that the relative strength of the modes depends strongly on the factor δ, which is defined as the excess loss over the gain of mode 0, which is at the center of the Lorentzian spectrum. For example, if $\delta = 10^{-5}g_0$, then the amplitude of the mode 0 is a factor of 20 higher than that of the next higher ones (modes $\pm$ 1), while if δ is $10^{-4}g_0$, the factor falls to 3. In semiconductor lasers, δ decreases monotonically with increasing pump current, and so essentially single-mode operation can be obtained at high bias currents. An analysis of the multimode rate equations in the steady state by Renner and Carroll (1978) gave

$$B = \beta 2\pi/P_0 b, \tag{64}$$

where B is the full width at half-maximum (FWHM) of the spectral envelope, P_0 the normalized photon density, b the spectral discrimination factor as defined in Eqs. (30) and (63).

Understanding that mode selection relies on the very delicate balance of the clamped gain and modal loss, it is not difficult to see that under high-speed modulation, the mode selection process is not operative. It was shown (Thompson, 1982) that during switch-on of lasing emission, the ratio of the power in the ith to that in the jth mode is given by

$$P_i/P_j = \exp(G_i - G_j)t, \tag{65}$$

where

$$G_i = \alpha g_i n(t) \tag{66}$$

is the total mode gain of the ith mode. Equation (65) is derived by assuming that the electron density n remains relatively constant even during lasing transients. This is a good approximation, as is evident from the numerical calculations shown in Section 3. The gain factor g_i varies only slightly from mode to mode (parts in 10^{-4}) and consequently the time it takes the modes to reach their equilibrium power is quite long (on the order of a few nanoseconds), in spite of the fact that the *total* photon density can respond to a drive-current pulse in less than 100 psec. It is also assumed in Eq. (65) that the spontaneous emission factor is zero. As a result, only one mode can oscillate as $t \to \infty$, and Eq. (65) can be applied only in the first moments during switch-on.

In addition to increasing the number of oscillating modes, modulation of an injection laser causes the width of each individual mode to broaden. This is also related to the fluctuation in the electron density during modulation, which causes the refractive index (and therefore the effective length of the laser cavity) to vary in time. A simplified analysis of this effect using rate equations was given by Kishino *et al.* (1982) along with experimental results. An analysis along the same line was used by Henry (1982) to explain linewidth boardening in cw-operated semiconductor lasers. A more rigorous analysis of linewidth broadening is given by Vahala and Yariv (1983), who treated the problem by using the van der Pol model. This analysis yields not only the amount of linewidth broadening, but also the detailed line shape and its dependence on modulation frequency.

Spectral broadening and multimode oscillation in semiconductor lasers under modulation will increase the amount of chromatic dispersion in optical-fiber transmission and therefore significantly reduce the transmission bandwidth in single-mode fiber systems. Means of suppressing multimode oscillation are therefore of particular interest, especially in coherent detection applications discussed in Section 18. This can be achieved by reducing the gain of all other modes except the one closest to the gain spectrum maxima using frequency-selective structures such as a distributed Bragg reflector (Sakakibara *et al.*, 1980; Utaka *et al.*, 1981) or an external cavity controlled laser (Preston *et al.*, 1981). Ebeling *et al.* (1983) accomplished the same effect by using a composite-cavity laser. Lee *et al.* (1982, 1983) and Liu *et al.* (1982) have shown that in short-cavity lasers, in which longitudinal modes are widely separated in frequencies and hence have increased mode discrimination, the time required for the lasing spectrum to settle down to the steady-state, single-mode oscillation was reduced to below 1 nsec during step-current excitation from below the threshold, whereas the corresponding time was ~6 nsec for lasers of conventional lengths. These results can be well explained by the theoretical analysis leading to Eq. (65). On the other hand, Eq. (65) cannot be used in describing the spectral behavior during switch-off or in the case when the laser is biased above the threshold, since spontaneous emission was neglected. A more accurate analysis by Lau *et al.* (1984a) shows that the lasing spectrum of a laser under pseudorandom pulse code modulation will depend on the bit pattern. As mentioned before, the redistribution of the power in the lasing modes during transients is a relatively slow event. Consequently, when modulation current pulses are less than a few nanoseconds apart, the spectral content of the optical pulses is bound to depend on the preceding pulses. Under very high-frequency, continuous-microwave modulation, the spectral content cannot respond to the rapidly changing current, and hence the lasing spectrum will be that of the laser operating continuously at the average optical power. The spectral contents of a BH-on-SI laser under higher-frequency modulation at different frequencies and optical modulation depths are shown in Fig. 48. It can be observed that essentially single-mode oscillation can be maintained at relatively high modulation frequencies (> 1 GHz) at modulation depths up to 70–80%. Further increases in modulation depth drastically alter the lasing spectrum; both the number of oscillating modes and the linewidth of each individual mode increase. In fact, it has recently been observed that single-mode oscillation can be maintained at modulation frequencies up to 4 GHz with optical modulation depths above 90% (Lau *et al.*, 1984c). On the other hand, single-mode lasers are not only unnecessary in multimode fiber systems in which modal instead of chromatic dispersion dominates but are even unsuitable due to excessive noise (modal noise) created by the coherence of the optical signal. In fact, the spectral envelopes of lasers used in some multimode systems were intentionally broadened by superposing a very high-frequency microwave modulation on the signal (Bosch *et al.*, 1980).

24. SELF-PULSATIONS, BISTABILITY, AND INHOMOGENEOUSLY PUMPED INJECTION LASERS

It has been observed that some injection lasers, when driven by a dc current, emit a continuous train of sharp optical pulses at a microwave repetition rate somewhere between 300 MHz and 2–3 GHz. This usually occurs after a laser has been operating for a sufficiently long period and

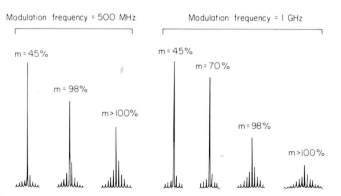

FIG. 48. Measured average spectrum of a BH laser under modulation at various frequencies and modulation depths. [From K. Y. Lau, N. Bar-Chaim, I. Ury, C. Harder, and A. Yariv (1984a). *IEEE Journal of Quantum Electronics* **QE-20,** 71. Copyright © 1984 IEEE.]

marks the beginning of the degradation of the laser. The phenomenon, called self-pulsation, apparently has been observed in all types of III–IV semiconductor lasers; the occurrence is more frequent in some types of lasers than in others. When self-pulsation first develops in a laser, it is in the form of a weak sinusoidal undulation at relatively high frequencies (> 1–2 GHz). However, as the laser continues to age and degrade, the undulation develops into sharp spiking in the output at ever-decreasing frequencies. Some lasers pulsate at a frequency as low as 200 MHz, which severely limits their use as signal transmitters. The self-pulsation frequency increases with increasing bias current. Basov (1968) first attempted an explanation of this peculiar phenomenon by assuming a nonuniformity in the pump current over the length of the laser diode, so that the regions depleted in pumping will form saturable absorption centers. The coexistence of both saturable absorption and saturable gain within a cavity mimics a passively mode-locked laser (Shank and Ippen, 1974). However, the pulsations observed in injection lasers cannot be caused by mode locking, since the pulsation frequency is much lower than the cavity mode separation of ~ 100 GHz. Rather, the pulsation was modeled as a form of undamped relaxation oscillation or repetitive Q-switching, as it is sometimes called. Lee and Roldan (1970) fabricated laser diodes with a purposely built-in nonuniform pump current and produced pulsations as predicted by Basov (1968). However, another theory was advanced by Paoli and Ripper (1969), who speculated that the pulsations, though not produced by the locking of the cavity modes, can be produced instead by the locking of the combination tones (Sargent et al., 1974) of the modes. This is called second-order mode locking. They supported their theory by the experimental observation that when the laser is coupled to a diffraction grating to produce single-mode oscillation, the pulsations are quenched. However, later experiments by Chinone et al. (1978b) and Figueroa et al. (1980) showed that the pulsation can be quenched even with optical feedback from a non-frequency-selective element like a plane mirror. There was no further experimental evidence to support the theory of second-order mode locking. As the quality of injection lasers improved and their commercialization required serious life tests to be undertaken, it became apparent that pulsations were related to the aging and degradation of lasers. The results of Paoli (1977) and Hartman et al. (1979) unmistakably relate self-pulsation to absorbing defects in the active region of the lasers. The exact nature of the defects is still not fully known; some believe them to be deep-level traps (Imai et al., 1978; Copeland, 1979) and microdegradations distributed throughout the active region (Kato, 1977), but there are also suggestions that pulsations are due to mirror facet degradation (Dixon and Joyce, 1979) or proton-induced damage (van der Ziel et al., 1979; Campbell et al., 1980). All of these mechanisms produce saturable absorption inside the laser cavity. The exact cause of self-pulsation in any one particular laser is probably different from others, and there may even be more than one mechanism at work simultaneously.

For self-pulsation to occur, the saturable absorber must possess a certain property, namely, the absorber must saturate more easily than the gain. This condition was spelled out analytically by Dixon and Joyce (1979) and is the same condition as that derived by Haus (1976) for passively mode-locked lasers:

$$|dq/dP| > |dg/dP| \quad \text{at} \quad P = P_s, \tag{67}$$

where q is the total loss introduced by the absorber, g the total gain due to the laser medium, P the optical power inside the cavity, and P_s the optical power at which the total loss of the cavity equals the total gain. The condition prescribed by Eq. (67) is met when the gain, plotted as a function of the optical power, intersects the slope of the loss characteristic such that the slope of the loss characteristic is steeper than the slope of the gain characteristic. This means that at low optical powers, gain exceeds loss, while at high optical powers, loss exceeds gain; the system is unstable and pulsation results.

Since the exact nature of dark-line defects and other microdefects is not fully known, it is difficult to ascertain whether Eq. (67) is satisfied for these absorption centers and hence whether these defects can actually cause self-pulsations in the manner of repetitive Q-switching. One example in which the defect is independently documented and satisfies Eq. (67) is a deep-level trap usually associated with proton-induced damage in GaAs, as proposed by Copeland (1979). This can account for the relatively frequent occurrence of self-pulsation in proton-bombarded stripe lasers even before aging. An example of an absorber that does *not* satisfy Eq. (67) is the intrinsic absorption of GaAs. An inhomogeneously pumped laser diode, in which part of the diode is weakly pumped or not pumped, therefore should not exhibit instability. Instead, analysis (Lasher, 1964) showed that such devices should be stable and exhibit a sizable hysteresis in the light versus current characteristic. However, almost all devices fabricated (Basov, 1968; Lee and Roldan, 1970; Harder *et al.,* 1981; Hanamitsu *et al.,* 1981) that purposely incorporate an inhomogeneous pump *do* exhibit self-pulsations and display no or only a small hysteresis in the light versus current characteristic (more in form of a light jump). To explain these experimental results, several authors suggested the repetitively Q-switching mechanism, which then must assume a sublinear gain dependence on injection carrier concentration. The calculated (Stern, 1973, 1976) and measured (Hakki and Paoli, 1975; Bakker and Acket, 1977; Henry *et al.,* 1980) gain dependences in undoped or lightly doped GaAs, however, do not fulfill this necessary condition.

A more basic understanding of the mechanism governing the behavior of inhomogeneously pumped laser diodes was attempted by Carney and Fonstad (1981, 1983), who fabricated a laser diode consisting of a segmented contact with eight separately pumped sections. They observed some nonlinearities in the light versus current characteristics but observed neither pulsations nor bistability. The range and combination of biasing at the segments were probably not fully explored owing to the complexity of the device. New insights into the statics and dynamics of nonuniformly pumped laser diodes were offered by a two-segment BH laser [whose low threshold makes possible a wide range of biasing conditions and whose stable transverse mode avoids

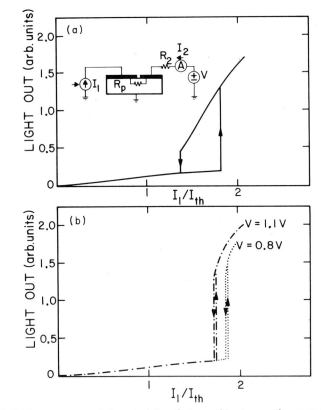

FIG. 49. Light versus current characteristics of a laser with a two-section contact. One of the sections (the absorber section) is reverse biased. The resistivity between the two contacts is ~60 kΩ due to a highly resistive top cladding layer. A large hysteresis results when the absorber section is driven by a high-impedance, pure current source, and a small hysteresis and pulsation result when the absorber section is driven by a low-impedance voltage source, or when a small (1-kΩ) resistor is connected between the two sections. (a) $R_2 = 200$ kΩ, $V = -20$ V; (b) $R_2 = 1$ kΩ. [From Harder *et al.* (1982a).]

effects caused by beam instabilities as described by Lang (1980) and van der Ziel (1981)]. This device, for the first time, produced a sizable hysteresis with *no* pulsations (Harder *et al.,* 1982a,b), as shown in Fig. 49. It was shown that both the size of the hysteresis and the dynamic stability are related to the resistance between the segmented contacts and to the impedance of the current source driving the absorber section. Increasing the value of both of these resistances can result in a nearly ideal situation in which a large hysteresis is manifested with no pulsations. Decreasing these resistances would lead to little or no hysteresis and self-pulsations, *even though the gain dependence on injected carrier density is linear.* It was also shown that a negative resistance reminiscent of a tunnel diode is obtained at the contact over the absorber section, which is optoelectronic in origin. It is this negative resistance interacting with the electrical circuit that leads to instability and pulsations, in much the same way as in a tunnel diode oscillator. A laser diode is after all an optoelectronic device; electrical interactions must be considered for a full understanding of its behavior.

The dynamic switching characteristics of this bistable laser, both by electronic and optical means, were demonstrated subsequently by Lau *et al.* (1982a,b). The switching speed was on the order of a few nanoseconds, which is considerably slower than that attainable by a laser biased continuously above the threshold. Switching a bistable from its off (nonlasing) state to its on (lasing) state is similar to turning on a laser from below threshold, and the basic speed limitations due to the charge storage effect, as discussed in Section 15, apply to a bistable laser as well. The bistable characteristic can prove to be useful in digital optical transmission at moderate speed (a few hundred megahertz), in the same way as an electronic schmidt trigger proves to be indispensable in digital electronics.

VIII. Conclusion and Outlook

Semiconductor lasers are now being used in many optical-fiber field systems, operating at bit rates from tens of megabits per second to around 1 Gbit sec^{-1}. Although higher modulation speeds are now demanded only in specialized systems, they undoubtedly will become more popular once questions regarding high-frequency operation characteristics have been answered and more data have been accumulated. The analysis in this chapter elucidated the basic limits to the high-frequency performance of semiconductor lasers under certain constraints in optical power and current densities. There is still insufficient data to tell where the line should be drawn for these constraints. Specially designed laser structures have stretched the optical constraints to levels previously unattainable, and further improvements can be expected. Even at room temperature, certain lasers fabricated to date already have a direct modulation bandwidth of over 10 GHz.

It is a safe prediction that laser transmitters operating in the 10–20-GHz range will become practically feasible within the next one to two years. This will open up new avenues of application of optical techniques in the fields of microwave transmission technology and very high-speed data transmission.

ACKNOWLEDGMENTS

Much of the work on high-frequency lasers described in this chapter was performed by the authors in collaboration with Dr. Ch. Harder at the California Institute of Technology and Drs. I. Ury and N. Bar-Chaim of Ortel Corporation, under support by the Defense Advanced Research Project Agency, the Naval Research Laboratory, the Army Research Office, and the National Science Foundation under the optical-communication program. Useful comments on the manuscript by Drs. Harder, Ury, Bar-Chaim, and S. Margalit were most appreciated.

REFERENCES

Abbott, S. M., Muska, W. M., Lee, T. P., Dentai, A. G., and Burrus, C. A. (1978). *Electron. Lett.* **14**, 349.

Aiki, K., Nakamura, M., Kuroda, T., Umeda, K., Ito, R., Chinone, N., and Maeda, M. (1978). *IEEE J. Quantum Electron.* **QE-14**, 89.

Amann, M. C., and Boeck, J. (1979a). *Electron. Lett.* **15**, 41.

Amann, M. C., Boeck, J., and Harth, W. (1978). *Int. J. Electron.* **45**, 635.

Amann, M. C., Kuschmider, A., and Boeck, J. (1980). *Electron. Lett.* **16**, 58.

Angerstein, J., and Siemsen, D. (1976). *AEU* **30**, 477.

Auyeung, J. (1981). *Appl. Phys. Lett.* **38**, 308.

Auyeung, J., and Johnston, A. R. (1982). *Proc. Tech. Symp. SPIE* **332**, 47.

Bakker, J., and Acket, G. A. (1977). *IEEE J. Quantum Electron.* **QE-13**, 567.

Bar-Chaim, N., Katz, J., Ury, I., and Yariv, A. (1981). *Electron. Lett.* **17**, 108.

Bar-Chaim, N., Harder, C., Katz, J., Margalit, S., and Yariv, A. (1982). *Appl. Phys. Lett.* **40**, 566.

Bar-Chaim, N., Lau, K. Y., Ury, I., and Yariv, A. (1983). *Appl. Phys. Lett.* **43**, 261.

Basov, N. G. (1968). *IEEE J. Quantum Electron.* **QE-4**, 855.

Blauvett, H., Margalit, S., and Yariv, A. (1982). *Appl. Phys. Lett.* **40**, 1029.

Boeck, J., Amann, M. C., and Stegmuller, B. (1980). *AEU* **34**, 465.

Boers, P. M., and Vlaadingerbroek, M. T. (1975). *Electron. Lett.* **11**, 206.

Bosch, F., Dybwad, G. L., and Swan, C. B. (1980). *CLEOs, San Diego, Calif.* Pap. TuDD7.

Bradley, D. J. (1977). *In* "Ultra Short Light Pulses" (S. L. Shapiro, ed.), Chapter 2, Springer-Verlag Berlin and New York.

Brosson, P., Ruble, W. W., Patel, N. B., and Ripper, J. E. (1981). *IEEE J. Quantum Electron.* **QE-17**, 714.

Buus, J., and Danielsen, M. (1977). *IEEE J. Quantum Electron.* **QE-13**, 669.

Campbell, J. C., Abbott, S. M., and Dentai, A. G. (1980). *J. Appl. Phys.* **51**, 4010.

Carney, J. K., and Fonstad, C. G. (1981). *Appl. Phys. Lett.* **38**, 303.

Carney, J. K., and Fonstad, C. G. (1983). *IEEE J. Quantum Electron.* **QE-19**, 22.

Casey, H. C., Jr., and Panish, M. B. (1978). "Heterostructure Lasers," Part B, Chap. 8. Academic Press, New York.

Casperson, L. W. (1975). *J. Appl. Phys.* **46**, 5194.

Casperson, L. W. (1977). *J. Appl. Phys.* **48**, 258.

Chinone, N., Aiki, K., Nakamura, M., and Ito, R. (1978a). *IEEE J. Quantum Electron.* **QE-14**, 625.

Chinone, N., Aiki, K., and Ito, R. (1978b). *Appl. Phys. Lett.* **33**, 990.

Chinone, N., Saito, K., Ito, R., Aiki, K., and Shize, N. (1979). *Appl. Phys. Lett.* **35**, 513.

Copeland, J. A. (1979). *Electron. Lett.* **14**, 809.

Danielsen, M. (1976). *IEEE J. Quantum Electron.* **QE-12**, 657.

Dixon, R. W., and Joyce, W. B. (1979). *IEEE J. Quantum Electron.* **QE-15**, 470.

Dumant, J. M., Guillausseau, Y., and Monerie, M. (1980). *Opt. Commun.* **33**, 188.

Ebeling, K. J., Coldren, L. A., Miller, B. I., and Rentschler, J. A. (1983). *Appl. Phys. Lett.* **42**, 6.

Fekete, D., Streifer, W., Scrifes, D. R., and Burnham, R. D. (1980). *Appl. Phys. Lett.* **37**, 975.

Figueroa, L., Lau, K. Y., and Yariv, A. (1980). *Appl. Phys. Lett.* **36**, 248.

Figueroa, L., Slayman, C., and Yen, H. W. (1982). *IEEE J. Quantum Electron.* **QE-18**, 1718.

Furuya, K., Suematsu, Y., and Hong, T. (1978). *Appl. Opt.* **17**, 1949.

Gagliardi, R. M., and Karp, S. (1976). "Optical Communication." New York.

Giallorenzi, T. G., Bucaro, J. A., Dandridge, A., Siegel, G. H., Jr., Cole, J. H., Rashleigh, S. C., and Priest, R. G. (1982). *IEEE J. Quantum Electron.* **QE-18**, 626.

Gonda, S. I., and Mukai, S. (1975). *IEEE J. Quantum Electron.* **QE-11**, 545.

Hagimoto, K., Ohta, N., and Nakagawa, K. (1982). *Electron. Lett.* **18**, 796.

Hakki, B. W., and Paoli, T. L. (1975). *J. Appl. Phys.* **46**, 1299.

Hanamitsu, K., Fujiwara, T., and Takusagawa, M. (1981). *Appl. Phys. Lett.* **39**, 14.

Harder, C., Lau, K. Y., and Yariv, A. (1981). *Appl. Phys. Lett.* **39**, 382.

Harder, C., Lau, K. Y., and Yariv, A. (1982a). *Appl. Phys. Lett.* **40**, 124.

Harder, C., Lau, K. Y., and Yariv, A. (1982b). *IEEE J. Quantum Electron.* **QE-18**, 1351.

Harder, C., Katz, J., Margalit, S., Shacham, J., and Yariv, A. (1982c). *IEEE J. Quantum Electron.* **QE-18**, 333.

Harder, C., Vahala, K., and Yariv, A. *Appl. Phys. Lett.* **42**, 328.

Harth, W. (1973). *Electron. Lett.* **9**, 532.

Harth, W. (1975). *AEU* **29**, 149.

Harth, W., and Siemsen, D. (1974). *AEU* **28**, 391.

Hartman, R. L., Logan, R. A., Koszi, L. A., and Tsang, W. T. (1979). *J. Appl. Phys.* **50**, 4616.

Haus, H. A. (1976). *IEEE J. Quantum Electron.* **QE-12**, 169.

Henry, C. H. (1982). *IEEE J. Quantum Electron.* **QE-18**, 259.

Henry, C. H., Logan, R. A., and Merritt, F. R. (1980). *J. Appl. Phys.* **51**, 3042.

Ho, P. T. (1978). *Electron. Lett.* **14**, 725.

Icsevgi, A., and Lamb, W. E. (1969). *Phys. Rev.* **185**, 519.

Ikegami, T., and Suematsu, Y. (1968). *Electron. Commun. Jpn.* **B51**, 51.

Ikegami, T., and Suematsu, Y. (1970). *Electron. Commun. Jpn.* **B53**, 69.

Imai, H., Isozumi, K., and Takusagawa, M. (1978). *Appl. Phys. Lett.* **33**, 330.

Jackson, J. D. (1975). "Classical Electrodynamics." Chapter 8, Wiley, New York.

Kaminow, I. (1964). *Appl. Opt.* **3**, 507.

Kato, D. (1977). *Appl. Phys. Lett.* **32**, 588.

Katz, J., Bar-Chaim, N., Chen, P. C., Margalit, S., Ury, I., Wilt, D. P., Yust, M., and Yariv, A. (1980). *Appl. Phys. Lett.* **37**, 211.

Katz, J., Margalit, S., Harder, C., Wilt, D. P., and Yariv, A. (1981). *IEEE J. Quantum Electron.* **QE-17**, 4.

Kikuchi, K., Okososhi, T., and Kitano, J. (1981). *IEEE J. Quantum Electron.* **QE-17**, 2266.

Kishino, K., Aoki, S., and Suematsu, Y. (1982). *IEEE J. Quantum Electron.* **QE-18**, 343.

Kleinman, D. A. (1964). *Bell Syst. Tech. J.* **43**, 1505.

Kressel, H., and Butler, J. K. (1977). "Semiconductor Lasers and Heterojunction LEDs," Chap. 16. Academic Press, New York.

Kurbatov, L. N., Shakhidzhanov, S. S., Bystrova, L. V., Krapukhin, V. B., and Kolonenkov, S. J. (1971). *Sov. Phys.—Semiconduct. (Engl. Transl.)* **4**, 1739.

Lamb, W. E. (1969). *Phys. Rev.* **8**, 1429.

Lang, R. (1980). *Jpn. J. Appl. Phys. Lett.* **2**, 93.

Lang, R., and Kobayashi, K. (1976). *IEEE J. Quantum Electron.* **QE-12**, 194.

Lasher, G. J. (1964). *Solid-State Electron.* **7**, 707.

Lau, K. Y. (1981). Ph.D. Thesis, Calif. Inst. Technol., Pasadena.

Lau, K. Y., and Yariv, A. (1980). *Opt. Commun.* **35**, 337.

Lau, K. Y., and Yariv, A. (1982a). *Appl. Phys. Lett.* **40**, 452.

Lau, K. Y., and Yariv, A. (1982b). *Appl. Phys. Lett.* **40**, 763.

Lau, K. Y., Figueroa, L., and Yariv, A. (1980). *IEEE J. Quantum Electron.* **QE-16**, 1329.

Lau, K. Y., Harder, C., and Yariv, A. (1981). *Opt. Commun.* **36**, 472.

Lau, K. Y., Harder, C., and Yariv, A. (1982a). *Appl. Phys. Lett.* **40**, 198.

Lau, K. Y., Harder, C., and Yariv, A. (1982b). *Appl. Phys. Lett.* **40**, 369.

Lau, K. Y., Bar-Chaim, N., Ury, I., Harder, C., and Yariv, A. (1983a). *Appl. Phys. Lett.* **43**, 1.

Lau, K. Y., Bar-Chaim, N., Ury, I., Harder, C., and Yariv, A. (1983b). *Appl. Phys. Lett.* **43**, 329.

Lau, K. Y., Bar-Chaim, N., Ury, I., Harder, C., and Yariv, A. (1984a). *IEEE J. Quantum Electron.* **QE-20**, 71.

Lau, K. Y., Bar-Chaim, N., Ury, I., and Yariv, A. (1984b). *Appl. Phys. Lett.* (to be published).

Lee, C. P., Margalit, S., Ury, I., and Yariv, A. (1978). *Appl. Phys. Lett.* **32**, 410.

Lee, T. P., and Derosier, R. M. (1974). *Proc. IEEE* **62**, 1176.

Lee, T. P., and Roldan, R. H. (1970). *IEEE J. Quantum Electron.* **QE-6**, 339.

Lee, T. P., Burrus, C. A., and Miller, B. I. (1973). *IEEE J. Quantum Electron.* **QE-9**, 829.

Lee, T. P., Burrus, C. A., Liu, P. L., and Dentai, A. G. (1982). *Electron. Lett.* **18**, 805.

Lee, T. P., Burrus, C. A., Linke, R. A., and Nelson, R. J. (1983). *Electron. Lett.* **19**, 82.

Lindstrom, C., Scrifes, D. R., and Burnham, R. D. (1983). *Appl. Phys. Lett.* **42**, 28.

Liu, P. L., Daman, T. C., and Elienberger, D. J. (1980). *Top. Meet. Picosecond Phenom., North Falmouth, Mass.*

Liu, P. L., Lee, T. P., Burrus, C. A., Kaminow, I. P., and Ko, J. S. (1982). *Electron. Lett.* **18**, 904.

Lockwood, H. F., Kressel, H., Sommers, H. S., Jr., and Hawrylo, F. Z. (1970). *Appl. Phys. Lett.* **17**, 499.

Lutes, G. (1980). *Jet Propul. Lab., Telecommun. Data Acquis. Prog. Rept.* No. 42, p. 59.

Maeda, M., Nagano, K., Tanaka, M., and Chiba, K. (1978). *IEEE Trans. Commun.* **COM-26**, 1076.

Marshall, P., Scholsser, E., and Wolk, C. (1979). *Electron. Lett.* **15**, 38.

Matthews, M. R., and Steventon, A. G. (1978). *Electron. Lett.* **14**, 649.

Mengel, F., and Ostoich, V. (1977). *IEEE J. Quantum Electron.* **QE-13**, 359.

Moreno, J. B. (1977). *J. Appl. Phys.* **48**, 4152.

Morishita, M., Ohmi, T., and Nishizawa, J. (1979). *Solid-State Electron.* **22**, 951.

Nagano, M., and Kasahara, K. (1977). *IEEE J. Quantum Electron.* **QE-13**, 632.

Nita, S., Namizaki, H., Takamiya, S., and Susaki, W. (1979). *IEEE J. Quantum Electron.* **QE-15**, 1208.

Olesen, H., and Jacobsen, G. (1982). *IEEE J. Quantum Electron.* **QE-18**, 2069.

Olsen, G. H., Hawrylo, F. Z., Channin, D. J., Botez, D., and Ettenberg, M. (1980). *Tech. Dig.—Int. Electron Devices Meet.* Pap. 20.2.

Osterwalder, J. M., and Rickett, B. J. (1980). *IEEE J. Quantum Electron.* **QE-16**, 250.

Ostoich, V., Jeppesen, P., and Slaymaker, N. (1975). *Electron. Lett.* **11**, 515.

Otsuka, K. (1977). *IEEE J. Quantum Electron.* **QE-13**, 520.

Pantell, R. H., and Puthoff, H. E. (1969). "Fundamentals of Quantum Electronics." p. 294, Wiley, New York.

Paoli, T. P. (1977). *IEEE J. Quantum Electron.* **QE-13**, 351.

Paoli, T. P. (1981). *Appl. Phys. Lett.* **39**, 522.

Paoli, T. P., and Ripper, J. E. (1969). *Phys. Rev. Lett.* **22**, 1085.

Paoli, T. P., and Ripper, J. E. (1970). *Proc. IEEE* **58**, 1457.

Peterman, K. (1978). *Opt. Quantum Electron.* **10**, 233.

Peterman, K. (1979). *IEEE J. Quantum Electron.* **QE-15**, 566.

Pierce, J. R., and Posner, E. C. (1980). "Introduction to Communication Science and Systems." Plenum, New York.

Preston, K. R., Wooland, K. C., and Cameron, K. H. (1981). *Electron. Lett.* **17**, 931.

Renner, D., and Carroll, J. E. (1978). *Electron. Lett.* **14**, 781.

Reuven, Y., and Baer, M. (1980). *IEEE J. Quantum Electron.* **QE-16**, 1117.

Russer, P., and Schultz, S. (1973). *AEU* **27**, 193.

Saito, S., Yamamoto, Y., and Kimura, T. (1981). *IEEE J. Quantum Electron.* **QE-17**, 935.

Sakakibara, Y., Furuya, K., Utaka, K., and Suematsu, Y. (1980). *Electron. Lett.* **16**, 456.

Salathe, R., Voumard, C., and Weber, H. (1974). *Opto-electronics (London)* **6**, 451.

Sargent, M., III, Scully, M. O., and Lamb, W. E. (1974). "Laser Physics," Addison-Wesley, Reading, Massachusetts.

Seeway, P. R., and Goodwin, A. R. (1976). *Electron. Lett.* **12**, 25.

Shank, C. V., and Ippen, E. P. (1974). *Appl. Phys. Lett.* **27**, 488.

Statz, H., and deMars, G. (1960). *In* "Quantum Electronics" (C. H. Towns, ed.), p. 235. Columbia Univ. Press, New York.

Stern, F. (1973). *IEEE J. Quantum Electron.* **QE-9**, 290.

Stern, Γ. (1976). *J. Appl. Phys.* **47**, 5382.

Suematsu, Y., and Furuya, K. (1977). *IECE Jpn.,* **E-60**, 467.

Takahashi, S., Kobayashi, T., Saito, H., and Furakawa, Y. (1978). *Jpn. J. Appl. Phys.* **17**, 865.

Takamiya, S., Serwa, Y., Tanaka, T., Sogo, T., Namizaki, H., Susaki, W., and Shirahata, K. (1980). *Int. Semicond. Laser Conf. 7th, London* Pap. 8.

Tang, C. L., Statz, H., and deMars, G. (1963). *J. Appl. Phys.* **34**, 2289.

Tell, R., and Eng, S. T. (1980). *Electron. Lett.* **16**, 497.

Thompson, G. H. B. (1981). *Proc. Inst. Electr. Eng.* **128**, 37.

Thompson, G. H. B. (1982). "Physics of Semiconductor Laser Devices." p. 450 Wiley, New York.

Torphammar, P., Tell, R., Eklund, H., and Johnston, A. (1979). *IEEE J. Quantum Electron.* **QE-1 15**, 1271.

Tsang, D. Z., and Walpole, J. N. (1982). *IEEE Int. Semicond. Conf., 8th, Ottawa* Pap. 26.

Tsang, D. Z., Walpole, J. N., Groves, S. H., Hsieh, J. J., and Donnelly, J. P. (1981). *Appl. Phys. Lett.* **38**, 120.

Tsukada, T. (1974). *J. Appl. Phys.* **45**, 4899.

Ury, I., Lau, K. Y., Bar-Chaim, N., and Yariv, A. (1982). *Appl. Phys. Lett.* **41**, 126.

Utaka, K., Kobayashi, I., and Suematsu, Y. (1981). *IEEE J. Quantum Electron.* **QE-17**, 651.

Vahala, K., and Yariv, A. (1983). *IEEE J. Quantum Electron.* **QE-19**, 1109.

Wakao, K. W., Takagi, N., Shima, K., Hanamitsu, K., Hori, K., and Takusagawa, M. (1982). *Appl. Phys. Lett.* **41**, 1113.

Wilt, D. P., Lau, K. Y., and Yariv, A. (1981). *J. Appl. Phys.* **52**, 4790.

Yamada, J., Machida, S., Mukai, T., Kano, H., and Sugiyama, K. (1979). *Electron. Lett.* **15**, 596.

Yamamoto, Y. (1980). *IEEE J. Quantum Electron.* **QE-16,** 1281.
Yamamoto, Y. (1981). *IEEE J. Quantum Electron.* **QE-17,** 919.
Yen, H. W., and Barnoski, M. (1978). *Appl. Phys. Lett.* **32,** 182.
Yen, H., and Figueroa, L. (1979). Hughes research laboratory internal report.
Yonezu, H. O., Ueno, M., Kamejima, T., and Hayashi, I. (1979). *IEEE J. Quantum Electron.* **QE-15,** 775.
Yu, K. L., Koren, U., Chen, T. R., Chen, P. C., and Yariv, A. (1981). *Electron. Lett.* **17,** 790.

CHAPTER 3

Spectral Properties of Semiconductor Lasers

Charles H. Henry

AT&T BELL LABORATORIES
MURRAY HILL, NEW JERSEY

I. Introduction

Our discussion of the spectral properties of semiconductor lasers is divided into four parts covering the optical properties of the material, mode spectra, and fluctuations. After a brief introduction, we will begin in Part II with a

153

derivation of the Einstein coefficients that relate spontaneous and stimulated emission to optical absorption. These relationships are then used to determine absorption, gain, and refractive index change spectra from spontaneous emission spectra. The measured gain and absorption spectra are compared with the calculated optical properties of a $\mathbf{k} \cdot \mathbf{p}$ band model.

We will discuss mode intensity spectra in Part III, where a large number of phenomena are encountered. Continuously operating index-guided lasers usually have a single dominant mode but become multimode during pulsed operation. In contrast to index-guided lasers, gain-guided lasers are highly multimoded even during continuous operation. Optical nonlinearities enhance the single-mode stability of index-guided lasers and lasing-to-nonlasing ratios as high as 1000 : 1 have been achieved at twice threshold. In many lasers, higher-order transverse modes appear at high power owing to spatial hole burning.

Statistical fluctuations in the intensity and phase of the lasing field are discussed in Part IV. We present a simple model of how phase and intensity noise are caused by the spontaneous emission process. Above threshold, the total light intensity in a laser is stabilized, but, nevertheless, the instantaneous light intensity exhibits a Gaussian distribution that is much greater than that of a fully coherent optical mode with the same power. Mode partition noise consisting of rare but very large fluctuations of individual mode intensities (while the total light intensity remains nearly constant) is an important source of errors in transmission at high bit rates along dispersive fibers. Recent work leading to an understanding of partition noise will be reviewed. Phase noise causes each lasing mode to have a finite linewidth and a nearly Lorentzian shape. The feedback process that stabilizes intensity fluctuations in a laser induces changes in refractive index. These changes greatly enhance the linewidth and produce structure in the tails of the Lorentzian power spectrum. The linewidth requirement for coherent optical communications is discussed.

This chapter is an attempt to review the basic spectral properties of lasers, particularly those that have an impact on optical-communications technology. Our discussions will be restricted to operation near room temperature. The data presented are from measurements of AlGaAs- and quaternary InGaAsP-type lasers, the two lasers that have been studied most widely and are most relevant to optical communications at present. The study of these devices is worldwide and moving at a rapid pace. In many cases, similar studies are carried out by several groups in different laboratories. This chapter is an attempt to give a fairly comprehensive review of spectral phenomena and of the physics necessary for the interpretation of laser spectra, but no attempt is made to review or completely cite the vast literature on these subjects.

II. Absorption, Emission, and Gain

The optical processes in a laser are absorption, spontaneous emission, and stimulated emission. Before the development of a quantum theory of radiation, Einstein (1917) found simple relationships among these processes. The Einstein coefficients, originally derived for atoms, are extendable to the case of a semiconductor (Henry *et al.,* 1980). These relationships allow us to convert spontaneous emission spectra, which are readily measured, into absorption and gain spectra. In this way absorption, spontaneous emission, and gain spectra can be followed from low currents to above threshold.

The gain spectrum of a laser can also be determined from an analysis of the mode spectrum (Hakki and Paoli, 1975). We will only discuss conversion of gain from spontaneous emission spectra here, since this method allows a unified approach to the determination of all optical processes and is not limited in the range of wavelength or current that can be covered.

This part is organized as follows. First, we derive the Einstein coefficients for the semiconductor case. Next, these relationships are applied to the study of absorption and gain spectra. We then compare these spectra with calculations based on a $\mathbf{k} \cdot \mathbf{p}$ band theory model. Finally, we make use of Kramers–Kronig relationships to convert the gain change spectra to refractive index change spectra.

We find that the agreement between theory and experiment is only fair. For this reason, measurement rather than theory is emphasized in this section. The main failing of the theoretical model is that it does not properly account for the band tail.

1. EINSTEIN'S COEFFICIENTS FOR A SEMICONDUCTOR

The photons and electrons of a semiconductor laser are not in thermodynamic equilibrium. Consequently, the light intensity of a laser must be calculated by rate equations rather than by statistical mechanics. However, the rate coefficients that enter into these equations and describe absorption, spontaneous emission, and stimulated emission must be compatible with a thermodynamic equilibrium, which would exist in the absence of optical losses from the laser cavity. This requirement leads to relationships between the rate coefficients that were first derived by Einstein (1917). To derive these relationships, we begin with a discussion of thermodynamic equilibrium.

The electrons in the conduction band and valence band states of a semiconductor are fermions. The average occupation n_k of the kth energy level by fermions at temperature T is given by the Fermi distribution law (Landau and Lifshitz, 1958)

$$n_k = \left[\exp\left(\frac{E_k - \mu}{kT}\right) + 1 \right]^{-1}, \tag{1}$$

where μ is the chemical potential or quasi-Fermi level, defined as the change in free energy F required to add an electron to the conduction band or to the valence band:

$$\mu = \frac{\partial F}{\partial N} = \frac{\partial}{\partial N}(E - TS), \tag{2}$$

where E, S and N are the total values of energy, entropy, and carrier number, respectively. In an excited semiconductor, to a good approximation, the carriers in both the valence band and the conduction band are at the lattice temperature T; the two bands, however, are not in equilibrium with each other, and there is a different chemical potential for each band.

A similar distribution holds for the photons that are bosons. The optical modes are the quantized states of each photon. The average number of photons I_k occupying the kth state with energy $\hbar\omega_k$ is given by the Bose distribution law (Landau and Lifshitz, 1958)

$$I_k = \left[\exp\left(\frac{\hbar\omega_k - \mu}{kT}\right) - 1\right]^{-1}, \tag{3}$$

where μ is the chemical potential of the photons. This quantity is zero for the Planck blackbody distribution but not in general.

The change in sign between Eqs. (1) and (3) for the Fermi and Bose distributions makes an enormous difference. While n_k has a value between 0 and 1, I_k can take on any value. The negative values occurring for $\mu > \hbar\omega_k$ are unphysical. As μ approaches $\hbar\omega_k$, I_k approaches infinity. This is the phenomenon of Bose–Einstein condensation (Landau and Lifshitz, 1958) whereby a single state can acquire a macroscopic occupation. A laser reaching threshold is close to the phenomena of Bose–Einstein condensation, a single mode acquires $\sim 10^5$ photons, but true equilibrium is prevented by cavity losses.

In the absence of optical losses, the photons will come into equilibrium with the electrons. In equilibrium, the net free energy change in emission or absorption of a photon is zero. When an electron drops from the conduction band into the valence band and a photon is created, the loss in free energy by the semiconductor is eV, the separation of the quasi-Fermi levels. In equilibrium, eV must also be the increase in free energy of the photons, that is $\mu = eV$ for photons. Therefore, the equilibrium mode occupancy I_{equil} is given by

$$I_{\text{equil}} = \left[\exp\left(\frac{\hbar\omega - eV}{kT}\right) - 1\right]^{-1}. \tag{4}$$

The rates of absorption, stimulated emission, and spontaneous emission must allow equilibrium to take place in the absence cavity losses. We assume a rate equation of the form

$$\dot{I} = -AI + BI + R - \gamma I, \qquad (5)$$

where AI is the absorption rate given with $A = \alpha v_g$ (where α is the absorption coefficient and v_g the group velocity), BI the rate of simulated emission, R the rate of spontaneous emission, and γI the rate of loss due to scattering, non-band-gap absorption, and facet transmission. In the steady state, with no losses ($\gamma = 0$)

$$I = [(A/R) - (B/R)]^{-1}. \qquad (6)$$

Equating the coefficients in Eqs. (4) and (6) we find

$$B = R = \alpha v_g \exp[(eV - \hbar\omega)/kT] \equiv rv_g. \qquad (7)$$

The net rate of stimulated emission is given by $(A - B)I \equiv GI$, hence

$$G = v_g\alpha \{\exp[(eV - \hbar\omega)/kT] - 1\} \equiv v_g g, \qquad (8)$$

where g is the gain per centimeter.

With these changes Eq. (5) becomes

$$\dot{I} = (G - \gamma)I + R, \qquad (9)$$

where the gain coefficient G and spontaneous emission coefficient R are related to the optical absorption coefficient. Equations (7) and (8) are analogous to the Einstein relationships for an atom with discrete levels. Equation (8) shows that to achieve positive gain, the bias energy eV must be greater than $\hbar\omega$ (Bernard and Duraffourg, 1961).

2. ABSORPTION AND EMISSION AT LOW EXCITATION

We will now show that the measured spontaneous emission spectrum can be converted to an absoprtion spectrum. Spontaneous emission is normally measured in a direction nearly normal to the active layer, where reabsorption effects are small. The density of modes per solid angle is proportional to ω^2 so that using Eq. (7), the spontaneous emission rate S_E is given by

$$S_E \frown \omega^2\alpha \exp(-\hbar\omega/kT). \qquad (10)$$

This relationship can be employed to convert spontaneous emission spectra into absorption spectra. Figures 1 and 2 (Henry et al., 1983a) show an example of this procedure. Figure 1 shows the directly measured absorption spectrum of a 1.3-μm InGaAsP-type quaternary laser active material. The absorption spectrum rapidly increases to 10^4 cm^{-1} and then slowly continues to increase with energy. Figure 2 shows the spontaneous emission spectrum from the same layer under low-level Ar-ion laser excitation. (To

WAVELENGTH (μm)

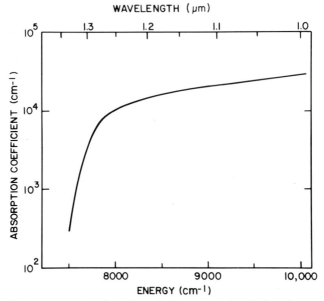

FIG. 1. Directly measured optical absorption spectrum of an InGaAsP quaternary layer grown on InP. $T = 300$ K; $d = 1.3$ μm. [From C. H. Henry, R. A. Logan, H. Temkin, and F. R. Merritt. *IEEE J. Quantum Electron.* **QE-19**, 941 (1983). Copyright © 1983 IEEE.]

facilitate the absorption measurement, the layer was grown 1.3 μm thick, and the spontaneous emission spectrum was corrected for reabsorption of emitted light.) The corrected spontaneous emission spectrum was converted to absorption by using Eq. (10). A temperature of $T = 310$ K gives excellent agreement between the converted absorption spectrum and the directly measured absorption spectrum. The discrepancy between the sample lattice temperature of 300 K and the temperature of 310 K needed to fit the absorption data is probably caused by inaccuracies in the spontaneous emission intensity measurement.

3. SATURATION OF ABSORPTION DURING LASER EXCITATION

Figure 3 illustrates the absorption spectra of the active layer of an AlGaAs-type buried heterostructure laser (Henry *et al.,* 1981a). The laser was fabricated with a window in the p-side contact that allowed the measurement of the spontaneous emission spectrum. The spectra were then converted to absorption spectra, which were made absolute by requiring that they approach the known absorption of GaAs at energies far above the absorption edge.

In the vicinity of the absorption edge, the absorption saturates with increasing excitation, primarily because of the nearly degenerate filling of the

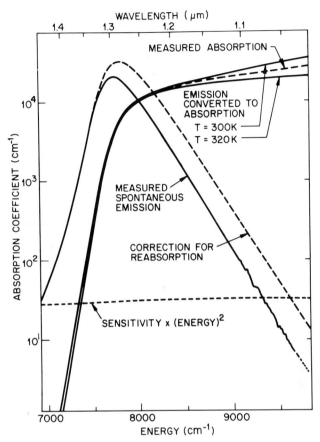

FIG. 2. The spontaneous emission spectrum from the quaternary InGaAs layer of Fig. 1, corrected for reabsorption of emitted light and then converted to an absorption spectrum by using Eq. (10). The converted spectrum at $T = 310$ K and $d = 1.3$ μm is in good agreement with the measured absorption spectrum of Fig. 1. The conversion determines the shape but not the magnitude of the absorption spectrum. [From C. H. Henry, R. A. Logan, H. Temkin, and F. R. Merritt. *IEEE J. Quantum Electron.* **QE-19**, 941 (1983). Copyright © 1983 IEEE.]

conduction band states with electrons. Changes in the band-gap energy and changes in the matrix element due to carrier screening also seem to be observable (Henry *et al.,* 1980). Lasing occurs in the exponential tail. As the laser is brought from low current to threshold, the absorption within GaAs at the laser line drops from about 390 cm^{-1} to 120 cm^{-1}.

4. THE GAIN SPECTRUM

Spontaneous emission spectra can be converted to absorption spectra by using Eq. (10) and then to gain spectra by using Eq. (8). This is illustrated for

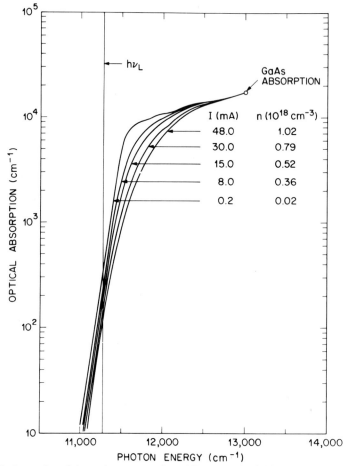

FIG. 3. Saturation of absorption spectra of an AlGaAs-type buried heterostructure laser. The absorption spectra were converted from spontaneous emission spectra by using Eq. (10). [From Henry *et al.* (1981a).]

a 1.3-μm InGaAsP-type quaternary mesa laser (Logan *et al.*, 1982) in Figs. 4 and 5 (Henry *et al.*, 1983a). The low-current and threshold spontaneous emission spectra of Fig. 4 are converted to absorption and gain spectra in Fig. 5.

The energy eV giving the separation of quasi-Fermi levels is determined from the requirement that gain $g(\hbar\omega)$ have a maximum at the laser line $\hbar\omega_L$. The validity of this determination was checked by Henry *et al.* (1981b), who found that eV determined from the spectra were within 8 meV of eV determined by direct measurement of the bias voltage.

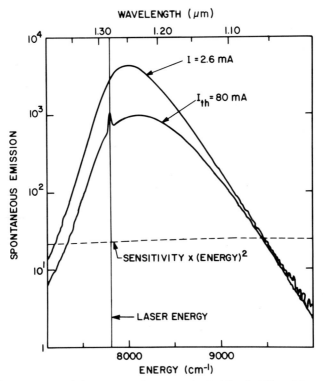

FIG. 4. Spontaneous emission spectra at low current and at threshold for a 1.3-μm InGaAsP-type quaternary mesa laser. $T = 315$ K. [From C. H. Henry, R. A. Logan, H. Temkin, and F. R. Merritt. *IEEE J. Quantum Electron.* **QE-19**, 941 (1983). Copyright © 1983 IEEE.]

The solid curves in Fig. 5 show the spectra at the threshold of the absorption α, the spontaneous emission into a single mode

$$r \equiv \alpha \exp[(eV - \hbar\omega)/kT], \tag{11}$$

and the gain g. Notice that at low energies g and r coincide, but the g abruptly drops to zero at $\hbar\omega = eV$. At higher energies g approaches $-\alpha$.

The ratio of spontaneous emission to gain at the laser line or $r/g = 1.7$ for the 1.3-μm quaternary laser. In GaAs $r/g \approx 2.6$. The smaller value of r/g for the quaternary laser reflects a higher degree of inversion at threshold.

The radiative current at threshold can be found by integrating the spontaneous emission curve (Henry *et al.*, 1983a)

$$J_{rad}/d = (8\pi cen'^2/\phi) \int_0^\infty \alpha \exp[(eV - E)/kT] \, dE, \tag{12}$$

where E is the energy in reciprocal centimeters, n' the refractive index, and ϕ

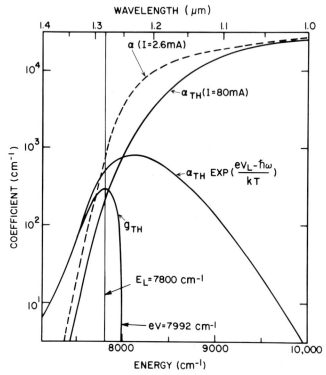

FIG. 5. Conversion of the spontaneous emission spectra of Fig. 4 into absorption and gain spectra. $T = 315$ K. [From C. H. Henry, R. A. Logan, H. Temkin, and F. R. Merritt. *IEEE J. Quantum Electron.* **QE-19,** 941 (1983). Copyright © 1983 IEEE.]

a correction factor equal to the increase in radiative lifetime caused by the reabsorption of emitted radiation (Asbeck, 1977). Henry *et al.* (1983a) estimate $\phi = 1.5$ and find $J_{rad}/d = 4.4$ kA cm^{-2}. This is about 60% of the threshold current density in a good mesa laser. A similar analysis shows that nearly all the threshold current is radiative in low-threshold current density GaAs buried heterostructure lasers (Henry *et al.*, 1981b).

5. CHANGE IN GAIN WITH ENERGY AND CARRIER DENSITY

The shift of the peak gain with excitation is illustrated in Fig. 6 (Kazarinov *et al.*, 1982) and was determined by conversion of spontaneous emission spectra to gain spectra. The point of maximum gain moves to higher energies with increasing excitation because at higher injected carrier densities, levels farther above the band edge become inverted. These levels have greater absorption and gain because of the increase in density of the states with energy.

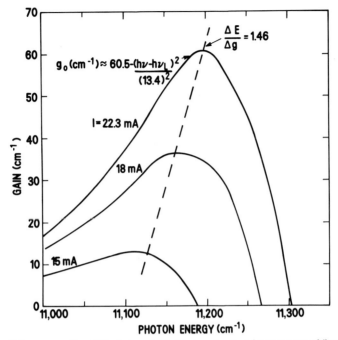

FIG. 6. Gain spectra of an AlGaAs-type buried heterostructure laser converted from spontaneous emission spectra at currents both below and at laser threshold. [From Kazarinov *et al.* (1982).]

A further illustration of this is given in Fig. 7 (Henry *et al.,* 1980), which shows the measured gain versus carrier density at photon energies near the laser line. The gain increases nearly linearly with carrier density, except at low carrier densities. (Many lasers show nearly a linear behavior at all values of carrier densities.) The increase of gain with carrier density becomes progressively steeper at higher energies. The carrier densities in Fig. 7 were determined from the measured values of eV by using a parabolic band model; the absolute carrier density at threshold is 1.1×10^{18} cm^{-3}. This may be low. Recent measurements of carrier density by Manning *et al.* (1983) gave threshold densities of about 3×10^{18} cm^{-3}.

6. COMPARISON WITH $\mathbf{k} \cdot \mathbf{p}$ BAND MODEL

A highly successful and relatively simple model of the energy bands valid near the center of the Brillouin zone is the $\mathbf{k} \cdot \mathbf{p}$ model of Kane (1957). We will now compare absorption, emission, and gain spectra calculated by using this theory with the experimental spectra. The principal parameters needed to describe the energy bands are the energy gap E_g, the spin–orbit splitting Δ,

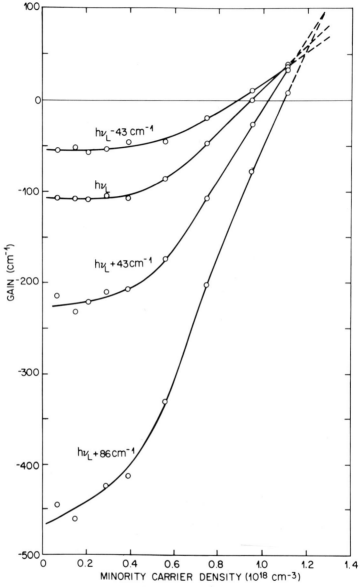

FIG. 7. Gain versus carrier density for an AlGaAs-type buried heterostructure laser with an undoped active layer; Al composition $X_{\text{active}} = 0.08$; differential quantum efficiency $\eta_D = 67\%$. [From Henry et al. (1980).]

and the matrix element of the momentum operator **p** connecting the *s*-like conduction band states with the *p*-like valence band states. These states become mixed by the spin–orbit and **k · p** interactions; the latter interaction arises from computation of the kinetic energy of a Bloch state. These interactions determine the masses and shape of the conduction band, the light hole, and spin–orbit split-off valence bands. The heavy hole band is unaffected, and the mass of this band must be put in as a separate parameter. The calculated bands for 1.3-μm quaternary material are plotted in Fig. 8 (Henry *et al.*, 1983b), which uses the best available band parameters for quaternary material (Pearsall, 1983).

The appeal of the **k · p** model is that the same momentum maxtrix element that determines the band masses also determines the strengths of the optical transitions between states of the same **k** vector. Figure 9 illustrates a calculation of absorption, spontaneous emission, and gain spectra for 1.3-μm quaternary laser material (Henry *et al.*, 1983a). The absorption

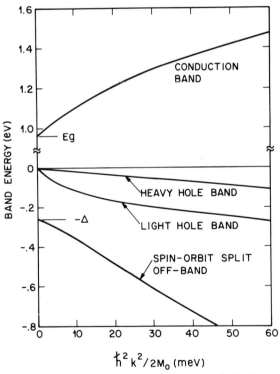

FIG. 8. Energy bands for 1.3-μm InGaAsP-type laser material calculated with a **k · p** band model. $E_g = 0.96$ eV; $\Delta = 0.26$ eV; conductive electron mass $M_e = 0.058M_0$; heavy-hole mass $M_{hh} = 0.55M_0$, where M_0 is the free electron mass. [From C. H. Henry, R. A. Logan, and F. R. Merritt. *IEEE J. Quantum Electron.* **QE-19,** 947 (1983). Copyright © 1983 IEEE.]

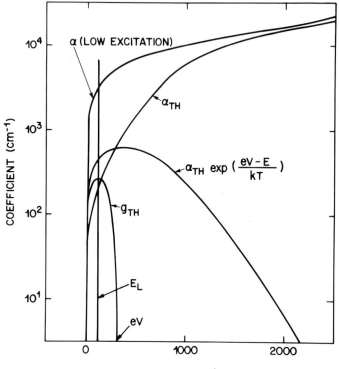

FIG. 9. Absorption, spontaneous emission, and gain calculated by using the **k · p** band model with the carrier density adjusted to give the same peak gain as in Fig. 5. $T = 315$ K. [From C. H. Henry, R. A. Logan, and F. R. Merritt. *IEEE J. Quantum Electron.* **QE-19**, 947 (1983). Copyright © 1983 IEEE.]

spectrum at low excitation can be calculated directly from the formulas of Kane (1957). To calculate absorption under lasing conditions, it is first necessary to compute the quasi-Fermi levels of the conduction band and the valence band at each carrier density and the occupancy factors n_c and n_v of the conduction band and valence band states at each value of k. The absorption coefficient is modified by $n_v(1 - n_c)$, and then the spontaneous emission rate r and the gain g can be calculated by using Eqs. (7) and (8). The carrier density $n = p = 2.15 \times 10^{18}$ cm^{-3} was adjusted to give the same maximum gain as in Fig. 5.

The band calculation accounts well for the magnitude of the absorption far above the band gap and for the shape of the gain and spontaneous emission curves. The main weakness in the theory is that it does not account for the nearly exponential band tail that is found experimentally. In the **k · p** model all spectra begin abruptly at the energy gap. The $\mathbf{k} \cdot \mathbf{p}$ model is a rigorous and accepted model of the energy bands. No comparable model exists for the band tail states. Often Stern's model (Casey and Stern, 1976) for heavily doped semiconductors is used; however, his model is not appropriate to describe normally undoped active layers.

7. SPECTRUM OF REFRACTIVE INDEX CHANGE

The injection of carriers into the active layer of a semiconductor causes a significant decrease in refractive index along with a positive increase in gain. The two changes are related and the spectrum of the change in refractive index can be computed from the spectrum of gain change by making use of the Kramers–Kronig dispersion relations. If the optical field is $E \sim \exp(ikz - \omega t)$, then propagation is described by a complex wave vector $k = k' + ik''$, where $g = -2k''$. The wave vector is associated with a complex refractive index $n = n' + in''$ by means of Maxwell's equations,

$$k = (\omega/c)n = (\omega/c)\epsilon^{1/2}, \tag{13}$$

where $\epsilon = \epsilon' + i\epsilon''$ is the complex dielectric constant. The change in the imaginary part of the refractive index is proportional to the gain change

$$\Delta g = -2(\omega/c)\,\Delta n''. \tag{14}$$

The real and imaginary changes in the dielectric constant are connected by the Kramers–Kronig dispersion relationships. For small changes in refractive index, it can be shown (Henry et al., 1981a) that $\Delta n'$ and $\Delta n''$ are similarly related,

$$\Delta n'(E) = \frac{2}{\pi} P \int_0^\infty \frac{E'\,\Delta n''(E')\,dE'}{E'^2 - E^2}, \tag{15}$$

where P is the principal value of the integral and E is the energy.

Figure 10 (Henry et al., 1981a) shows the spectra of the changes in refractive index occurring when an AlGaAs-type buried heterostructure laser is excited from low current up to threshold. These spectra were calculated from the spectrum of Δg by using Eqs. (14) and (15).

We see that the peak change in $\Delta n''$ occurs on the high energy side of the laser line. This is consistent with Fig. 3. The major changes in gain are decreases in absorption occurring about 400 cm^{-1} above the laser line energy.

An important laser parameter is

$$\alpha \equiv \Delta n'/\Delta n''.$$

It is clear from Fig. 10 that at the laser line $h\nu_L$, α is large compared to unity but depends sensitively on the position of the laser line. This parameter

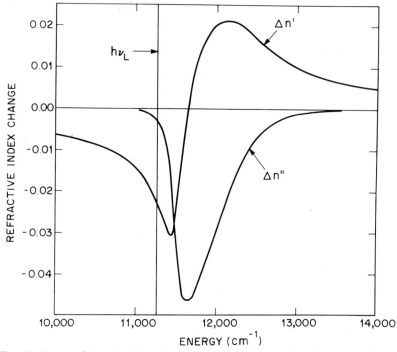

FIG. 10. Spectra of the real and imaginary changes in refractive index of the GaAs active layer of a buried heterostructure laser when it is excited from low currents up to the threshold current. $T = 300$ K. [From Henry *et al.* (1981b).]

determines the spectral width of a single mode (see Section IV). In the theory of gain-guided lasers, α is important in calculations of the laser threshold (Agrawal *et al.*, 1983), kink instabilities (Thompson *et al.*, 1978; Lang, 1979), pulsation instabilities (Lang, 1980; Guo and Wang, 1982) and far field patterns (Asbeck *et al.*, 1979) of gain-guided lasers.

The refractive index is also altered because of the Drude contribution to free carrier absorption. This contribution is about one order of magnitude smaller than the contributions due to changes in the absorption edge, as discussed earlier. A recent and thorough set of measurements of refractive index change at the laser line, as determined from mode shifts, has been made by Manning *et al.* (1983).

III. Mode Intensity Spectra

8. FABRY–PEROT CAVITY MODES

All semiconductor lasers with cleaved faceted mirrors may be regarded as highly resonant Fabry–Perot cavities. Let us begin our consideration of mode intensity spectra with a discussion of Fabry–Perot cavity modes. Consider a cavity of length L along the z axis. A field of the form

$$E \sim \exp[ikz + (g/2)z]$$

will propogate in each direction. A field $E(L)$ at $z = L$, after propagating back and forth in the cavity and making two reflections, must again equal $E(L)$. If the field reflectivity is $R_p^{1/2}$, where R_p is the power reflectivity and α_L is the cavity losses due to scattering, free carrier absorption, etc., then

$$R_p \exp[(2ikL) \exp(g - \alpha_L)L] = 1. \tag{16}$$

The modes satisfying Eq. (16) have

$$k = m\pi/L, \tag{17}$$

where m is an integer and

$$g = \alpha_L + (1/L) \ln(1/R_p) \equiv \alpha_y, \tag{18}$$

where α_y is the average loss per unit length including face losses.

The mode frequencies are given by

$$k = (\omega/c)n', \tag{19}$$

where n' is the effective refractive index of the transverse mode. The mode separations are determined by $\Delta m = 1$ in Eq. (17). Differentiating Eq. (19), we find

$$\Delta E = 1/2n_{\text{eff}}L, \tag{20}$$

where

$$n_{\text{eff}} = n' + \omega \, (dn'/d\omega) = c/v_g \tag{21}$$

and $\Delta E = \Delta\omega/2\pi c$ is the energy separation in reciprocal centimeters. For $L = 250 \, \mu$m and $n_{\text{eff}} = 4.3$, $\Delta E = 4.7 \, \text{cm}^{-1}$.

We are now in a position to relate the number of photons I in a mode to the facet power P_0. Let $u(z)$ be the number of photons per unit length propagating in the z direction. Then

$$u(z) = u(0) \exp[(g - \alpha_L)z]. \tag{22}$$

The total number of photons is then given by

$$I = 2 \int_0^L u(z) \, dz = 2u(0) \, \frac{\exp(g - \alpha_L)L - 1}{g - \alpha_L}.$$

Using Eqs. (18) and $u(0) = R_p u(L)$, we find

$$I = 2u(L)(1 - R_p)L/\ln(1/R_p). \tag{23}$$

The power transmitted through each facet is

$$P_0 = u(L)(1 - R_p)v_g h\nu. \tag{24}$$

Eliminating $u(L)(1 - R_f)$ from Eqs. (23) and (24) we find the relation of P_0 and I to be

$$P_0 = Iv_g h\nu \ln(1/R_p)/2L. \tag{25}$$

Another useful relationship is the formula for the differential quantum efficiency. The net rate of stimulated emission is given by Eq. (8) as $gv_g I$. It is natural to assume that above threshold, all current in excess of the threshold current goes into stimulated emission. (See Section 2 for justification.) If η_i is the fraction of the current passing through the active layer, then

$$gv_g I = \eta_i(C - C_{th}), \tag{26}$$

where C is the current and C_{th} the threshold current. The differential quantum efficiency η is given by $2P_0/h\nu(C - C_{th})$. Using Eq. (25), η becomes

$$\eta = \eta_i \ln(1/R_p)/gL. \tag{27}$$

That is, η is equal to η_i multiplied by the ratio of facet loss to total loss.

9. STEADY-STATE MODE INTENSITIES

In order to discuss mode intensity spectra, we must introduce spontaneous emission into the cavity. This is most easily done by approximating the facet losses as a distributed loss that is uniformly spread across the cavity. Equation (9) is then applicable. The steady-state solution of Eq. (9) is

$$\bar{I} = R/(\gamma - G) = r/(\alpha_\gamma - g), \tag{28}$$

where R is the average spontaneous emission rate, γ the cavity loss rate, and G the gain or net rate of stimulated emission; r, α_γ, and g are the rates per centimeter of these quantities formed by dividing R, γ, and G, respectively, by group velocity v_g.

Equation (9) neglects fluctuations in the spontaneous emission. A more correct derivation of Eq. (28) should take into account correlated fluctuations in the gain and mode intensity $\langle \Delta G \Delta I \rangle$. These fluctuations are the subject of Part IV. Equation (28) is likely to be valid in the description of highly single-moded, index-guided lasers, since the intensity fluctuations of the lasing mode are small and the nonlasing mode intensities are insensitive to gain fluctuations.

The mode intensities given by Eq. (28) are illustrated in Fig. 11 for the case of a 250-μm-long quaternary laser with $P_0 = 2$ mW for the main mode. The mode intensities are completely determined by Eqs. (25) and (28). The curves for $\alpha_\gamma - g$ are also shown. The curvature of $g(E)$ was determined

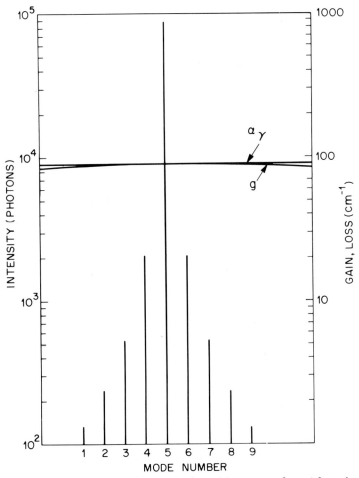

FIG. 11. Calculated mode intensities and gain and loss curves for a 1.3-μm laser with $P_0 = 2$ mW in the lasing mode; $r/g = 1.7$, $L = 250$ μm, $P_{tot} = 2.14$ mW. Note the pinning of g near α_y. The curvature of $g(E)$ is determined from Fig. 5.

from the data in Fig. 5, with the magnitude of the gain reduced by $\Gamma = 0.47$, the mode occupancy factor for a 0.2-μm active layer thickness. Within this reduction,

$$g = g_0 - \left(\frac{E - E_0}{17.4 \text{ cm}^{-1}}\right)^2 = g_0 - \left(\frac{\Delta N}{3.7}\right)^2 \quad \text{cm}^{-1}. \tag{29}$$

For a laser 250 μm long with $\eta = 0.5$ and $\eta_i = 1$, and $R_p = 0.33$, g_0 is determined by Eq. (27) to be 89 cm^{-1}.

The balance between gain and loss is a very delicate one. For the main mode, $\alpha_y - g = 2.0 \times 10^{-3}$ cm^{-1}, and for the most intense nonlasing modes (modes 4 and 6) $\alpha_y - g = 0.074$ cm^{-1}. The gain is a nearly linear function of the carrier density. Spontaneous emission and all other recombination processes except stimulated emission are also functions of carrier density. Therefore, for a laser in steady-state operation, the gain, carrier density, quasi-Fermi level separation, and nonstimulated recombination become pinned when the laser reaches threshold. The prediction of Fig. 11 is that the ratio of the lasing mode to the most intense nonlasing modes will be about 40 at $P_0 = 2$ mW. Mode intensity ratios of this magnitude are commonly observed under dc operation of high-quality, index-guided lasers. The validity of Eq. (27) at the laser threshold has been checked by Kazarinov *et al.* (1982) by constructing the gain curve from the mode intensity spectra and finding it to agree within 10% with the gain curve determined from the analysis of spontaneous emission spectra.

The power in the nonlasing modes is controlled by Eqs. (25) and (28). The denominator in Eq. (28) depends on the mode separation [Eq. (20)]

$$\alpha_y - g = \alpha_y - g_0 + [g''(E)N^2/8(n_{\text{eff}}L)^2]. \tag{30}$$

For powers in the main mode of more than a few tenths of a milliwatt $\alpha_y - g_0$ becomes negligible; consequently, $\alpha_y - g$ is fixed, and the power in the nonlasing modes saturates. From Eqs. (25), (28), and (30) we find that

$$P_{\text{sat}} \sim rL \ln(1/R_p). \tag{31}$$

The saturated power P_{sat} in the nonlasing modes can therefore be reduced by decreasing L and increasing the facet reflectivity. However, as L is reduced, r slowly increases, so that the dependence on L is sublinear. A detailed study of saturated power has been given by Lee *et al.* (1982), who experimentally demonstrated substantial reductions in P_{sat} with both reduced L and high reflecting facets. The predictions of Lee *et al.* are given in Fig. 12.

10. Mode Spectra of Gain-Guided Lasers

Semiconductor stripe geometry lasers of different designs can usually be classified as either index guided or gain guided. The transverse optical mode of index-guided lasers such as the buried heterostructure laser is confined to the active stripe by a positive step in refractive index. No positive refractive index step is provided in gain-guided lasers such as the proton-bombarded or oxide defined stripe geometry lasers. These lasers merely confine the injected current to the active stripe. The stripe of injected carriers produces a complex step in refractive index for which the real part is negative.

The difference between gain-guided and index-guided lasers is illustrated in Fig. 13 for a slab guide model. Modes in both index-guided and gain-

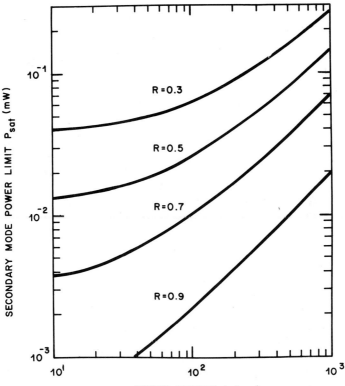

FIG. 12. Saturated power in the nonlasing mode nearest to the laser line for different cavity lengths and reflectivities. [From T. P. Lee, C. A. Burrus, J. A. Copeland, A. G. Dentai, and D. Marcuse. *IEEE J. Quantum, Electron.* **QE-18,** 1101 (1982). Copyright © 1982 IEEE.]

guided lasers can be considered to be plane waves confined to the slab by reflections at the interfaces formed by the refractive index step. The modes are determined by the requirements that a piece of the wave front undergoing two reflections must have the same amplitude and phase change as an unreflected part of the wave front.

In index-guided lasers, the reflection at each interface is total and the mode (aside from evanescent fields) is completely confined to the waveguide. In gain-guided lasers, the reflection at each interface is large (because of the nearly glancing incidence of the rays) but less than unity, and a transmitted wave propagates into the unpumped absorbing medium. This transmission results in the modes of gain-guided lasers having curved wave fronts as illustrated in Fig. 13.

The most striking difference between gain-guided and index-guided lasers is in their longitudinal mode spectra. Figure 14 (Streifer *et al.,* 1982) illus-

(a)

(b)

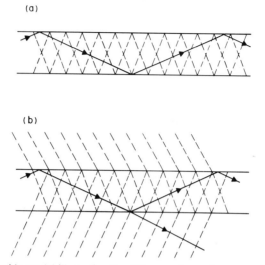

FIG. 13. Slab guide model illustrating the differences in mode propagation for (a) index-guided and (b) gain-guided lasers.

trates the spectrum of a gain-guided laser at different powers. Many modes are present at $P_0 = 4.6$ mW, whereas a well-behaved index-guided laser is highly single moded at half this power.

An explanation for the difference in the spectral behavior of gain-guided and index-guided lasers has been given by Peterman (1979) and Streifer *et al.* (1982). Peterman showed that the rate of spontaneous emission R into each fundamental longitudinal mode of a gain-guided laser is increased compared with index-guided lasers of similar dimensions, and this decreases the longitudinal mode selectivity in accordance with Eq. (28). Peterman made a classical calculation showing that the rate of spontaneous emission is enhanced by

$$K = \frac{(\int_{-\infty}^{\infty} |E(x)|^2 \, dx)^2}{|\int_{-\infty}^{\infty} E(x)^2 dx|^2}, \tag{32}$$

where E is the field of the fundamental mode and x the transverse coordinate. For index-guided lasers, $E(x)$ is real and $K = 1$. In gain-guided lasers, the curved wave fronts produce oscillations in $E(x)$ that enhance K.

Striefer *et al.* (1982) calculated the K factor for 8- and 4-μm-wide stripe guide lasers to be 20 and 30, respectively. They then showed that the spectra of these lasers (Fig.14) could be fitted with these values of K (Fig. 15).

Peterman's calculation has been both attacked (Patzak, 1982) and defended (Yariv and Margolit, 1982; Marcuse, 1982) with regard to its quan-

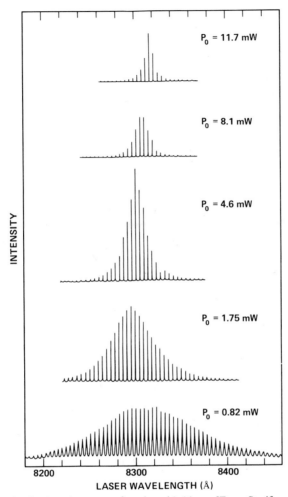

FIG. 14. Longitudinal mode spectra of a gain-guided laser. [From Streifer *et al.* (1982).]

tum mechanical validity. An illuminating discussion of spontaneous emission in loss-guided attenuators and gain-guided amplifiers has been given by Haus and Kawakami (1984). They show that the power coupled from a long amplifier into a single-mode fiber is increased by Peterman's factor. However, as thermal equilibrium is approached in the attenuator, the excess spontaneous emission disappears as required by thermodynamic equilibrium.

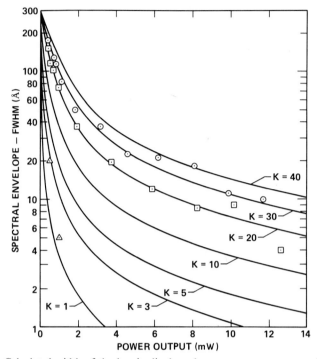

FIG. 15. Calculated width of the longitudinal mode spectrum versus power for different values of K. The spectra of 8- (□) and 4-μm-wide (⊙) stripes are fitted with K = 20 and 30, respectively. The spectrum of CSP (△) laser (thought to be index guided) is fitted with K = 1–3. [From Streifer *et al.* (1982).]

11. TRANSIENT BEHAVIOR OF MODE INTENSITIES

Even index-guide lasers that are highly single-moded under cw operation produce a multiple mode spectrum during the first few nanoseconds of operation when the laser is pulsed from below threshold to above threshold. The delicate balance between gain and loss takes time to develop. During the pulse, the gain g rises until it passes α_y. This causes the mode intensities to increase rapidly and overshoot their steady-state values. The excess light intensity and stimulated emission then drive g below α_y, and a series of damped relaxation oscillations continues until the steady-state intensities are reached. A calculation of the relaxation oscillations of the carrier density and mode intensities by Lee and Marcuse (1983) is shown in Fig. 16. This figure illustrates that suppression of nonlasing modes, which depends on the magnitude of $\alpha_y - g$, is greatly weakened during transient operation. Observations of short-cavity lasers under continuous and short-pulse operation are illustrated in Fig. 17 (Lin and Lee, 1983).

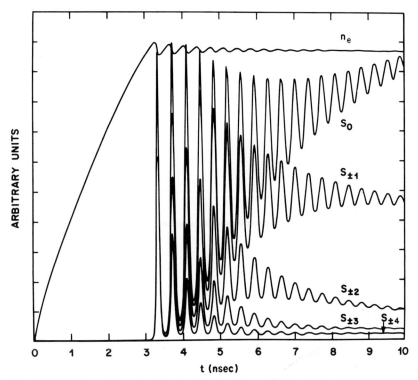

FIG. 16. Mode intensity and carrier density relaxation oscillations following an abrupt current step pulse. Nonlasing modes that are weak during steady-state operation are excited during the relaxation oscillations; $L = 250 \ \mu m$. [From T. P. Lee and D. Marcuse. *IEEE J. Quantum Electron.* **QE-19**, 1397 (1983). Copyright © 1983 IEEE.]

Figure 18 (Lin and Lee, 1983) shows that during pulsed operation the individual mode resonances are greatly broadened. This is due to the change in refractive index with carrier density and the fact that the carrier density is changing during the pulse. The change in refractive index shifts the mode position.

Error-free, high-bit-rate optical communication requires both single-mode operation and rapid pulsing of the laser. Without single-mode operation, errors result from the light of different modes traveling down a dispersive fiber at different velocities that causes pulses of adjacent bits to overlap. Designing lasers for single-mode operation under pulsed operation at high bit rates is an area of active current research (Tsang and Ollson, 1983).

12. LONGITUDINAL MODE SELF-STABILIZATION

Remarkably high ratios of lasing to nonlasing mode intensities have been achieved under cw operation. Figure 19 shows the results of Nakamura *et al.*

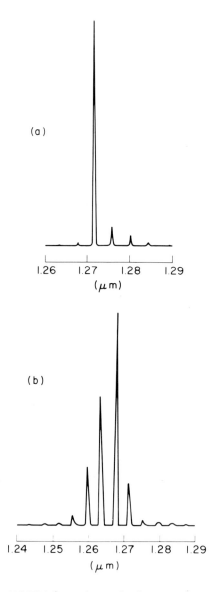

(a)

1.26   1.27   1.28   1.29
              (μm)

(b)

1.24   1.25   1.26   1.27   1.28   1.29
              (μm)

FIG. 17. Longitudinal mode spectra during (a) cw ($c = 1.4c_{th}$) and (b) pulsed (100-psec) operation of a short-cavity laser. [From P. Lin and T. P. Lee. *IEEE J. Lightwave Technol.* **LT-2**, 44 (1984). Copyright © 1984 IEEE.]

(1978a) for a channel substrate planar (CSP) AlGaAs-type laser. At a current of 70 mA above threshold, a ratio of nearly 1000 : 1 was achieved.

The single-mode stability that was observed was greater than expected from the considerations of Section 2. When the temperature is changed, the gain curve (observed by the envelope of nonlasing mode intensities) gradually shifts, but the lasing line holds back and does not switch modes. The

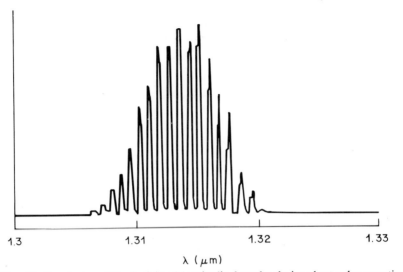

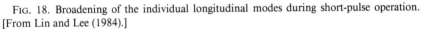

| 1.3 | 1.31 | 1.32 | 1.33 |

λ (μm)

FIG. 18. Broadening of the individual longitudinal modes during short-pulse operation. [From Lin and Lee (1984).]

shifting of the envelope continues until the lasing mode suddenly jumps to a new mode, usually one to four modes away from the previously lasing mode. The holding back causes hysteresis in which the jump occurs at different temperatues depending on whether the temperature is rising or falling. This is illustrated in Fig. 20 (Nakamura *et al.*, 1978a). Intermediate settings with two lasing modes of equal intensity are only observed at low power before mode stabilization sets in. It was expected that spectral hole burning would destroy single-mode stability at high powers, but instead the opposite occurs.

Two explanations have been given for these results. The first, by Yamada and Suematzu (1979) and by Kazarinov *et al.* (1982), invokes stabilization by optical nonlinearities; the second, by Copland (1980), involves saturable absorbing impurities. We will only discuss the theory of Kazarinov *et al.*, which is able to explain the observed structure in the mode spectra that accompanies longitudinal mode stability. The structure is in the form of a dip in the nonlasing mode intensities occurring several modes away from the lasing line shown in Fig. 20. This has been observed both in the spectra of CSP (Nakamura *et al.*, 1978a) and in buried heterostructure (BH) lasers (Kazarinov *et al.*, 1982).

Consider first the optical nonlinearity of spectral hole burning. Electron– hole transitions between discrete levels in the semiconductor have a line width on the order of T_2^{-1} in angular frequency, where T_2 is the time that a polarization associated with the transition takes to decay. The time T_2 is

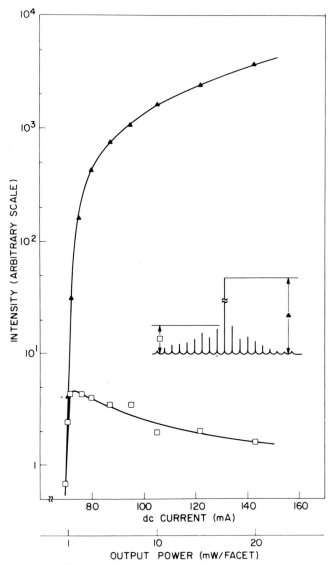

FIG. 19. Intensities of lasing and nonlasing modes as a function of current at 25°C. The total output power from a facet is also shown on the upper abscissa. [From Nakamura *et al.* (1978a).] Here ▲ and □ represent the amplitudes of the lasing and nonlasing modes, respectively, as shown in the inset diagram.

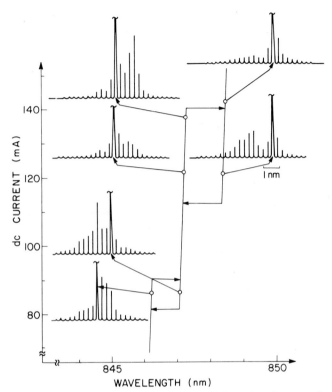

FIG. 20. Hysteresis observed in the lasing wavelength [wavelength versus dc current at constant heat sink temperature (25°C)]. The spectra at various current levels are shown. Note the dips in the nonlasing modes occurring several modes away from the laser line. [From Nakamura *et al.* (1978a).]

essentially an electron scattering time on the order of 10^{-13} sec (Kazarinov *et al.*, 1982; Nishimura, 1974). Consequently, intense stimulated emission at the laser line will depopulate levels contributing to the gain and spontaneous emission that are within T_2^{-1} of the laser line. Spectral hole burning tends to flatten the gain profile and thereby decrease the ratio of lasing to nonlasing mode intensities. This is a small effect, however, since the optical transition linewidth is broad and covers many longitudinal modes due to the smallness of T_2. Spectral hole burning is a nonlinear process in which the gain near the laser line decreases as $|E(\omega_0)|^2$, where $E(\omega_0)$ is the lasing field.

Two other optical nonlinearities that have the same order of magnitude as spectral hole burning arise from the modulation of the inverted population by the beating of lasing and nonlasing mode fields. Consider a lasing mode with angular frequency ω_0 and a nonlasing mode with angular frequency ω_1. Roughly speaking, the rate of stimulated emission of these two modes is $|E(\omega_0) + E(\omega_1)|^2$, which will have a beat term of angular frequency $\Omega = \omega_1 - \omega_0$. Consequently, the inverted population will be modulated at Ω. The modulation is only effective for $\Omega < T_1^{-1}$, where T_1 is the rate at which levels refill after being emptied. Kazarinov et al. (1982) have found $T_1 \approx 10^{-12}$ sec in fitting longitudinal mode spectra of AlGaAs-type buried optical guide lasers.

The modulation of the inverted population at Ω causes a modulation of the dielectric function at this frequency. The lasing field $E(\omega_0)$, acting through the modulated dielectric function, sets up polarizations at $\omega_1 = \omega_0 + \Omega$ and $\omega_2 = \omega_0 - \Omega$. The polarization at ω_1 suppresses the gain of the nonlasing mode and causes single-mode stability. However, the polarization at ω_2 couples the modes that are equally spaced on either side of the laser line and leads to an enhancement of the gain of these modes. The two effects cancel at the laser line, but away from the laser line the net effect is a weak suppression of gain. The effect is greatest several modes from the laser line and this explains the dips in the longitudinal mode spectra in Fig. 20.

The three nonlinear gain changes deduced by Kazarinov et al. (1980) are shown in Fig. 21. They were determined by fitting mode spectra of AlGaAs-type BH lasers. All three functions — spectral hole burning g_1, gain suppression g_2, and mode coupling g_χ — are proportional to the lasing mode inten-

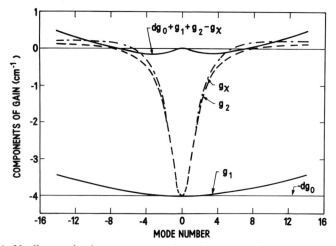

Fig. 21. Nonlinear gain changes g_1, g_2, and g_χ. The change dg_0 is a linear gain change necessary to keep the net change at the laser line equal to zero; $T_1 = 1.0$ psec, $T_2 = 0.1$ psec, $dg_0 = 4.0$ cm^{-1}. The two nonlinear gain changes (g_2 and g_χ) cancel at the laser line. Away from the laser line their net effect is a weak suppression. [From Kazarinov et al. (1982).]

sity. The functions g_2 and g_χ are narrow compared to g_1 because $T_1 \gg T_2$. The changes in the gain profile caused by these nonlinearities is shown in Fig. 22.

Nakamura *et al.* (1978a) found that single-mode stability of CSP lasers was maintained when the laser was modulated at frequencies as high as 2 GHz as long as the modulation dc bias current maintained the laser above threshold throughout the modulation cycle. Such modulation will cause an oscillation of the carrier density and the gain spectrum, but these effects apparently are not great enough to disrupt the single-mode stability in this case.

13. Spectra of Higher-Order Transverse Modes

Near threshold, the higher-order transverse modes of well-designed lasers are suppressed either because the cavity will not support them or because $\alpha_y - g$ for these modes is greater than for the fundamental mode. If higher-order transverse modes exist, a higher carrier density is needed to bring them to threshold than for the fundamental mode.

First-order transverse modes reach threshold through a process of spatial hole burning. At high fundamental mode intensity, the carriers near the center of the active stripe (where stimulated emission is greatest) become

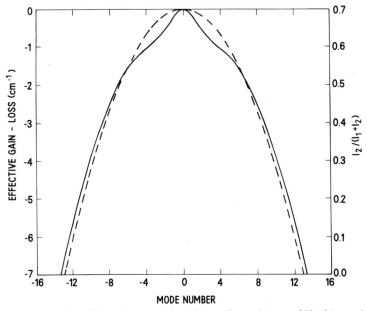

FIG. 22. Modification of the gain curve when the nonlinear changes of Fig. 21 are added to the linear gain curve shown by the dashed line. [From Kazarinov *et al.* (1982).]

depleted. The overall carrier density will increase to maintain $\alpha_\gamma \simeq g$ for the fundamental mode. The net effect is to increase the carrier density near the edges of the stripe, and this increases the gain of the first-order mode. For sufficiently great powers in the fundamental mode, the first-order transverse mode will reach threshold. Spatial hole burning has been directly observed by Nakamura *et al.* (1978b) in CSP lasers.

The laser lines of the first-order modes occur at higher energies than the fundamental mode. This is because a higher carrier density is required for the first-order mode to reach threshold, and the peak of the gain function increases to higher energy with increasing carrier density as shown in Fig. 6. Figure 23 illustrates the onset of the first-order transverse mode in an AlGaAs-type loss-stabilized buried optical guide laser (Henry *et al.*, 1981c). The fundamental and first-order transverse modes are identified by their far-field patterns shown in the insert. The energy of the first-order mode is about 40 cm^{-1} greater than the fundamental mode.

Some simple numerical estimates will show that there will be little correlation between the longitudinal mode positions of fundamental and first-order modes in a narrow-stripe laser. Consider a 1.3-μm BH laser with a stripe width $W = 3$ μm. Assume that the fundamental and first-order modes are strongly confined to the stripe with approximate fields $E_0(x) = \cos(K_0 x)\sin(\beta z)$ and $E_1(x) = \sin(K_1 x)\sin(\beta z)$, where $K_0 \simeq \pi/W$,

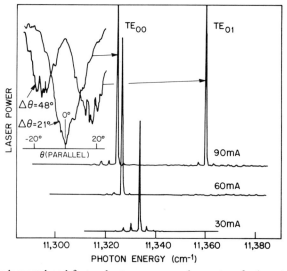

FIG. 23. Fundamental and first-order transverse mode spectra of a loss-stabilized buried optical guide laser. The modes are identified by the far-field patterns shown in the insert. [From [From C. H. Henry, R. A. Logan, and F. R. Merritt. *IEEE J. Quantum Electron.* **QE-17,** 2196 (1981). Copyright © 1981 IEEE.]

$K_1 \simeq 2\pi/W$, and $\beta = N\pi/L$ (where L is the length of the cavity). From the wave equation

$$\left[\frac{\partial^2}{\partial x^2} + \frac{\partial^2}{\partial z^2} + \frac{\omega^2}{c^2} n^2 \right] E(x, z) = 0, \tag{33}$$

we find

$$(M^2/W^2) + (N^2/L^2) = (2En)^2, \tag{34}$$

where $E = \omega/2\pi c$ is the mode energy in reciprocal centimeters, n the refractive index (the mode dependence of n will be neglected), and $M = 1$ and 2 for the fundamental and first-order transverse modes, respectively.

We can estimate N by neglecting $(M/W)^2$ in Eq. (34). From $n = 3.5$, $E = 7700$ cm^{-1} ($\lambda = 1.3$ μm), and $L = 250$ μm, we find

$$N \simeq 2EnL = 1350. \tag{35}$$

For a given value of N, the energy separation of transverse modes is determined by substituting $M = 1$ and $M = 2$ into Eq. (34). For $W = 3$ μm,

$$\Delta E = 3/8En^2W^2 = 44 \quad \text{cm}^{-1}. \tag{36}$$

This is large compared to the longitudinal mode separation determined from Eq. (20) (4.7 cm^{-1}). Hence the fundamental and first-order transverse modes of the same longitudinal mode number are about eight modes apart. At a given energy, the difference in N for the fundamental and first-order transverse modes can be determined from Eq. (34) as

$$\Delta N = 3L^2/2W^2N \simeq 7.7. \tag{37}$$

When ΔN changes by one-half, the energy separation of the fundamental and first-order transverse modes changes by one-half of a mode spacing. This will occur for a 3% variation of either L or W. Such small changes are normally not controlled in laser fabrication. Therefore, the separation of the fundamental and first-order longitudinal modes is not controlled in narrow-stripe lasers.

IV. Spectral Fluctuations

Thus far, our discussion of the spectral properties of lasers has not taken into account noise fluctuations and, consequently, only describe the average behavior of lasers. Amplitude and phase fluctuations of the laser optical field are caused by spontaneous emission and, to a lesser extent, by carrier number fluctuations. Low-frequency total intensity noise is reduced by the lasing process, but intensity fluctuations at high frequencies are substantial. A more serious concern is mode partition noise, in which the individual mode intensities fluctuate in such a way that the total intensity remains constant. This type of noise is not controlled by the lasing process and leads to substantial low-frequency noise fluctuations that can only be suppressed by reducing the intensities of the nonlasing modes.

The phase of the laser field may take on any value and the wandering of the phase is a type of Brownian motion. Phase noise determines the finite width of the laser line. The change in refractive index with carrier density enables intensity fluctuations to induce phase fluctuations that greatly increase the linewidth and produce structure in the line shape (power spectrum) of the lasing mode. The linewidth is a measure of the limited coherence of the semiconductor laser. The avoidance of errors in coherent detection of laser transmission requires highly coherent laser light with a linewidth that is more than 100 times smaller than the bit rate.

14. RATE EQUATIONS AND NOISE SOURCES

In Fig. 24, we give a simple model of how phase and intensity noise are generated by spontaneous emission, which agrees with the results of more formal treatments (Henry, 1982). The laser optical field can be written as the real part of a complex quantity $\beta \exp(-i\omega_0 t)$, where

$$\beta = I^{1/2} \exp(-i\phi). \tag{38}$$

The squared amplitude I is the intensity, and ϕ is the phase of the field. It is

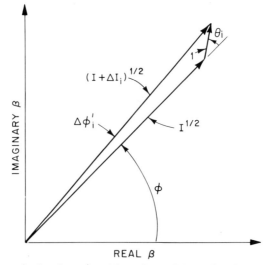

FIG. 24. Changes in the phase ϕ and intensity I of the optical field during spontaneous emission. [From C. H. Henry. *IEEE J. Quantum Electron.* **QE-18**, 259 (1982). Copyright © 1982 IEEE.]

convenient to measure I in units such that the average value of I equals the average number of photons in the mode.

The equation describing the motion of β can be derived by beginning with the time-dependent wave equation (Henry, 1982). It is given by

$$\dot{\beta} = [(G - \gamma)/2] (1 - i\alpha)\beta, \tag{39}$$

where $\alpha = \Delta n'/\Delta n''$, as discussed in Section 7. Equation (39) is more familiar when expressed as the equations for phase and intensity by means of Eq. (38). This yields

$$\dot{\phi} = \tfrac{1}{2} \alpha (G - \gamma) \tag{40}$$

and

$$\dot{I} = (G - \gamma)I. \tag{41}$$

Equation (41) is the same as Eq. (9) in the absence of spontaneous emission. In the derivation of Eq. (39), ω_0 is chosen to be the cavity resonance frequency at a carrier number for which $G = \gamma$. Equation (40) reflects phase changes that result from changes in the cavity resonance frequency when the carrier number changes and $G \neq \gamma$.

To complete the equations for β, I, and ϕ we must add the Langevin noise sources (Lax, 1960) associated with spontaneous emission. Traditionally, this is done by a quantum treatment of both the electromagnetic field and the radiating atoms (Lax, 1966, 1967a,b). A simpler and less rigorous method (but one that agrees completely with the quantized treatment) is given here (Henry, 1982). The effect of spontaneous emission events on the phase and intensity of the field is illustrated in Fig. 24.

The spontaneous emission of an additional photon alters β by $\Delta\beta$ having unit amplitude and random phase θ. Unit amplitude coresponds to $\Delta\beta$ being the field of one additional photon. [Choosing $\Delta\beta$ to have a small random amplitude (not necessarily unity) does not alter our results.] The changes in phase and intensity are

$$\Delta I_i = 1 + 2I^{1/2} \cos(\theta_i), \tag{42}$$

$$\Delta\phi_i = \sin(\theta_i)/I^{1/2}. \tag{43}$$

During a short time t, there are an average of Rt spontaneous emission events, where R is the average rate of spontaneous emission ($R \approx 10^{12}$ sec^{-1}).

The appropriate changes in intensity and phase spontaneous emission events will be correctly predicted by the equations of motion by adding an average spontaneous emission rate R and random Langevin forces $F_I(t)$ and $F_\phi(t)$ to Eqs. (40) and (41), as

$$\dot{\phi} = \tfrac{1}{2} \alpha (G - \gamma) + F_\phi(t), \tag{44}$$

$$\dot{I} = (G - \gamma)I + R + F_I(t), \tag{45}$$

where

$$F_\phi(t) = \sum_i \frac{\sin(\theta_i)}{I^{1/2}} \delta(t - t_i), \tag{46}$$

$$F_I(t) = \sum_i 2I^{1/2} \cos(\theta_i)\, \delta(t - t_i). \tag{47}$$

In general, the average values Langevin forces are zero:

$$\langle F_\phi(t) \rangle = \langle F_I(t) \rangle = 0. \tag{48}$$

The random angles θ_i ensure that this is the case in Eqs. (46) and (47).

Most calculations require only a knowledge of the average of two Langevin forces (Lax, 1960), which takes the form

$$\langle F_i(t)F_j(u) \rangle = 2D_{ij}\, \delta(t - u). \tag{49}$$

The delta function in Eq. (49) arises because of the assumption that the Langevin forces are Markovian or uncorrelated in time (Lax, 1960). From Eqs. (46) and (47) it is easily shown (Henry, 1982) that

$$D_{\phi\phi} = R/4I, \tag{50}$$

$$D_{II} = RI, \tag{51}$$

$$D_{I\phi} = 0. \tag{52}$$

To describe a laser with several longitudinal modes, a separate intensity equation (45) is needed for each mode. The Langevin forces for different modes are uncorrelated, $D_{I_i I_j} = 0$ for $i \neq j$, where i and j label different modes.

The description of the laser can be completed by a Langevin rate equation for the carrier number N, determined by the current C, spontaneous emission rate S, and the net stimulated emission into each mode $G_i I_i$ as

$$\dot{N} = C - S - \sum_i G_i I_i + F_N(t). \tag{53}$$

Carrier noise arises from shot noise resulting because electron levels have occupations of either 1 or 0. Lax (1960) has shown that twice the diffusion coefficient for shot noise is given by the sum of in and out rate processes

$$D_{NN} = C + S + \sum_i (R_i + A_i)I_i. \tag{54}$$

In computing D_{NN} we use the fact that the net stimulated emission $G_i I$ is the difference between stimulated emission $R_i I$ and absorption $A_i I$ (see Subsection II.1).

Processes that increase the numbers of photons in a mode also decrease the number of carriers by the same amount. This leads to a correlation of $F_N(t)$ and $F_{I_i}(t)$ and

$$D_{NI_i} = -RI_i = -D_{I_iI_i}. \tag{55}$$

In summary, Equations (44), (45), and (53) and non-zero diffusion coefficients given by Eqs. (50), (51), (54), and (55) completely describe the fluctuation properties of lasers.

15. FREQUENCY SPECTRUM OF INTENSITY FLUCTUATIONS

The frequency spectrum of intensity noise can be measured by detecting the laser light with a wide-bandwidth photodiode, amplifier, and spectrum analyzer. The frequency spectrum of a single mode can be displayed by selecting the light from a single mode with a grating monochromator. The quantity usually measured is the relative intensity noise $\Delta I(\omega)^2/I^2$ where $\Delta I(\omega)$ is the Fourier component of the intensity fluctuations.

A detailed study of the frequency spectra of intensity noise of AlGaAs-type semiconductor lasers has been made by Jäckel (1980) (see also Guekos and Jäckel, 1977). Fig. 25 shows the frequency spectrum of a CSP laser in which the average intensity of the strongest nonlasing modes is less than 5% of the intensity of the dominant mode. The spectrum of the total intensity noise displays a strong resonance at the relaxation oscillation frequency and little low-frequency noise.

In contrast to the total light intensity, the frequency spectrum of the dominant mode contains a great amount of low-frequency noise. This low-frequency noise is mode partition noise in which the nonlasing and lasing modes fluctuate in such a way as to leave the total intensity constant.

Jäckel (1980) compared his experimental results with the theory of McCumber (1966) and found that this theory can account for the shapes of the frequency spectra of both the total intensity and individual modes. McCumber's theory solves Eqs. (45) (one for each mode considered) and (53) in the quasilinear approximation, in which small oscillations of mode intensity and carrier number are assumed, and nonlinear terms involving $\Delta I_i \Delta N$ and the dependence of the Langevin forces on mode intensity are neglected. These approximations are not negligible, and, consequently, McCumber's theory underestimates the mode partition noise. To obtain agreement with his experimental curves, Jäckel had to assume average intensities of nonlasing modes that were considerably greater than he had measured. A qualitative explanation of the spectra in Fig. 25 will be given in Section 18.

16. INSTANTANEOUS DISTRIBUTION OF INTENSITY NOISE

A second way to characterize intensity noise is by measurement of the distribution of instantaneous light intensity. The instantaneous distribu-

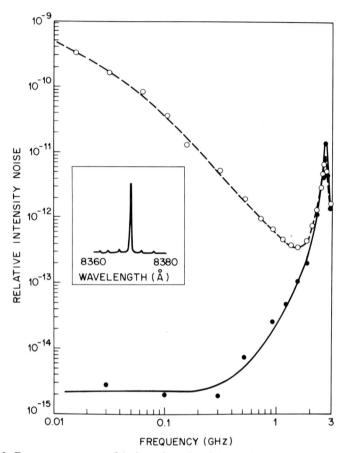

FIG. 25. Frequency spectrum of the intensity noise of the dominant mode (dashed line) and the total intensity (solid line) of a CSP laser; $T = 20°C$; $J_L = 1.2\,J_{th}$. The average mode intensities are shown in the insert. [From Jäckel (1980).]

tions of total light intensity and of individual modes have been measured by Liu *et al.* (1983). This was accomplished by detecting laser light with a fast detector, repetitively sampling the output of the detector for a short time, and displaying the distribution of signal heights. Figure 26 shows the distributions of the total intensity of a 1.3-μm quaternary laser at powers of 1.2 and 3 mW. The distributions are nearly Gaussian with some asymmetry favoring fluctuations at high intensity. The noise distributions are more than an order of magnitude wider than expected for a fully coherent lasing mode that obeys a Poisson distribution (Glauber, 1963).

The distributions of instantaneous total intensity noise involve mainly high-frequency fluctuations, and, consequently, the distribution rapidly narrows with increasing sampling time as well as with increasing power.

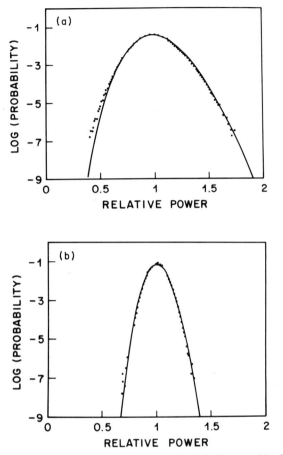

FIG. 26. Distribution of the total light intensity of an InGaAsP laser at (a) 1.2 and (b) 3 mW. Time resolution = 120 psec. [From P. Liu, L. E. Fencil, J.-S. Ko, I. P. Kominow, T. P. Lee, and C. A. Burrus. *IEEE J. Quantum Electron.* **QE-19,** 1348 (1984). Copyright © 1984 IEEE.]

Consequently, this source of noise is expected to be a serious source of errors in transmision only at rates of several gigabits per second or higher.

Figure 27 shows the data of Liu *et al.* (1984) that contrasts total intensity and partition noise fluctuations. The data are taken on a transverse junction AlGaAs laser having a large ratio of lasing to nonlasing average intensities. The distribution of total intensity is Gaussian. The distribution is narrower than in Fig. 26 because of the larger sampling time (500 psec).

While total light intensity distributions are Gaussian, the distributions of the individual modes are nearly exponential. This is illustrated in Fig. 27b showing the distribution of the lasing mode and that of a nonlasing mode that on the average has 1.5% the intensity of the lasing mode. The distribu-

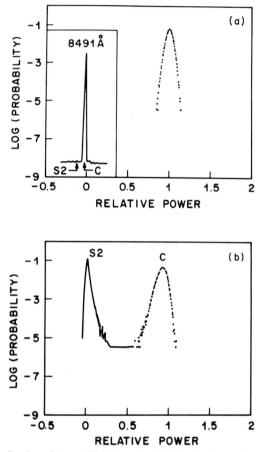

FIG. 27. (a) Distribution of the total light intensity. (b) Distributions of the dominant mode C and a nonlasing mode S_2 having an average intensity of 1.5% of C. The relatively narrow distribution in (a) is due to the use of 500-psec time resolution. The inset in (a) shows the mode spectrum. Measurements were made on an AlGaAs-type transverse-junction laser. [From P. Liu, L. E. Fencil, J.-S. Ko, I. P. Kaminow, T. P. Lee, and C. A. Burrus. *IEEE J. Quantum Electron.* **QE-19,** 1348 (1984). Copyright © 1984 IEEE.]

tion of the nonlasing mode appears to be nearly exponential. The low-intensity side of the lasing mode distribution also appears to be exponential. Evidently, increases in the nonlasing mode intensities are compensated for by drops in the lasing mode intensity, thereby causing the exponential tails to be absent in the distribution of total intensity.

Linke *et al.* (1983) observed large but rare drops in lasing mode intensity and found that they are correlated with equally large increases in nonlasing mode intensities. Furthermore, Linke *et al.* showed that the rate of drops in lasing mode intensity decreases exponentially with $\langle I_0 \rangle / \langle I_1 \rangle$. They found that the "dropouts" last 1–8 nsec and that a ratio $\langle I_0 \rangle / \langle I_1 \rangle \approx 50$ is required to reduce drops in lasing mode intensity of more than 50% to below to one per second. If the drops primarily last for about 1 nsec, this rate would then correspond to an error rate of about 10^{-9} in a system operating at 1 Gbit sec^{-1}.

17. A SIMPLE MODEL OF MODE-PARTITION NOISE

In this section, we give an analysis of mode partition noise that explains in simple terms all of the observed features of mode partition noise discussed in Sections 15 and 16 (Henry *et al.*, 1984). These features are the shapes of the frequency spectra, the time scale of a few nanoseconds for partition noise events, the exponential intensity distribution of nonlasing modes, the exponential dependence of the dropout rate of the lasing mode on $\langle I_0 \rangle / \langle I_1 \rangle$, and the fact that $\langle I_0 \rangle / \langle I_1 \rangle \approx 50$ is necessary to reduce the probability of a 50% dropout to about 10^{-9}.

Our model rests on two features satisfied by index-guided lasers operating at several milliwatts power and having a large value of $\langle I_0 \rangle / \langle I_1 \rangle$ (e.g. on the order of 50 : 1). First, the power in the nonlasing modes is saturated, and therefore these modes are insensitive to carrier density fluctuations. Second, the relaxation oscillation resonance is several gigahertz, and, consequently, the lasing mode intensity responds in subnanosecond times to fluctuations in carrier density.

In principle, the instantaneous distribution of any set of variables a_i obeying the Langevin rate equations

$$\dot{a}_i = A_i(a_i) + F_i(t), \tag{56}$$

$$\langle F_i(t) \rangle = 0, \tag{57}$$

$$\langle F_i(t)F_j(u) \rangle = 2D_{ij}\,\delta(t - u), \tag{58}$$

with random Gaussian noise sources $F_i(t)$ can be determined by solving the Fokker–Planck equation

$$\frac{\partial P}{\partial t} = -\sum_i \frac{\partial}{\partial a_i}(A_i P) + \sum_{ij} \frac{\partial^2}{\partial a_i \partial a_j}(D_{ij}P) \tag{59}$$

for the probability distribution $P(a_1, a_2, \ldots, t)$ (Lax, 1960). In practice, this equation is only solvable when there are one or two variables or in simple situations when all variables are quasilinear or when detailed balance is obeyed (Lax, 1960). A multimode laser does not satisfy these restrictions, and the simplest multimode laser has three variables N, I_0, and I_1.

The Fokker–Planck equation can be solved for the intensity distribution $P(I_1)$ (a single nonlasing mode) and gives an exponential intensity distribution. Since this mode is not sensitive to the small fluctuations in the carrier number N occurring in a laser above threshold, we will neglect these fluctuations. With this assumption, the Langevin rate equation for I_1 is not coupled to the equations for other mode intensities, and the Fokker–Planck equation for $P(I_1)$ contains only the single variable I_1. In the steady state, with $\partial P/\partial t = 0$, this equation is

$$-\frac{\partial}{\partial I_1} \{[(G - \gamma)I_1 + R]P(I_1)\} + \frac{\partial^2}{\partial I_1^2} [RI_1 P(I_1)] = 0, \tag{60}$$

where G is the average value of $G(N)$. The solution of Eq. (60) is

$$P(I_1) = \langle I_1 \rangle^{-1} \exp(-I_1/\langle I_1 \rangle), \tag{61}$$

where

$$\langle I_1 \rangle \equiv R/(\gamma - G). \tag{62}$$

Thus, the nonlasing mode has an exponential distribution.

Most noise fluctuations are Gaussian. Lax (1960) shows that when a system of variables obeys linear rate equations and the Gaussian Langevin forces are independent of the variables, $P(a_1, a_2, \ldots)$ is a Gaussian function of the system variables. This is not the case for light intensity since according to Eqs. (47) and (51) $F_I(t)$ is proportional to $I^{1/2}$ and D_{II} is proportional to I. The random force $F_\beta(t)$ is independent of β, and for a nonlasing mode, β can be shown to have a Gaussian distribution. Equation (61) can also be derived by beginning with a Gaussian distribution for β and then transforming to $I_1 = |\beta|^2$ (Henry et al., 1984).

As a consequence of the exponential distribution, there is a nonnegligible probability that the nonlasing mode intensities can equal a significant fraction of the average lasing mode intensity I_0. For example, consider two nonlasing modes I_1 and I_2 on either side of the laser line with equal average intensities $\langle I_1 \rangle$. Their combined probability is

$$P(I_1, I_2) = \langle I_1 \rangle^{-2} \exp[-(I_1 + I_2)/\langle I_1 \rangle]. \tag{63}$$

The probability that the sum $I_1 + I_2$ is more than a fraction f of the lasing mode intensity is

$$P_f = \int \int P(I_1, I_2)\, dI_1\, dI_2, \tag{64}$$

where integration is over the region for which I_1 and I_2 are positive and $I_1 + I_2 > f\langle I_0 \rangle$. Integration gives

$$P_f = [1 + (f\langle I_0 \rangle/\langle I_1 \rangle)) \exp[-(f\langle I_0 \rangle/\langle I_1 \rangle)]. \tag{65}$$

For $P_f < 10^{-9}$,

$$\langle I_0 \rangle/\langle I_1 \rangle > 24/f. \tag{66}$$

If increases in nonlasing mode fluctuations induce equal and opposite changes in the lasing intensity, then Eq. (65) is also the probability for the lasing mode to drop below $f \langle I_0 \rangle$. The probability decreases exponentially with $\langle I_0 \rangle / \langle I_1 \rangle$, in agreement with the observations of Linke et al. (1983). Errors known as dropouts occur when the laser intensity drops below $\frac{1}{2} I_0$. According to Eq. (66) with $f = \frac{1}{2}$, to keep the probability of dropouts to less than $\frac{1}{2} \langle I_0 \rangle$ at a value less than 10^{-9},

$$\langle I_0 \rangle / \langle I_1 \rangle > 48, \tag{67}$$

which is in close agreement with the value of approximately 50 as determined by Linke et al. (1983).

Increases on the order of $\frac{1}{2} I_0$ in the nonlasing mode intensity will increase the rate of stimulated emission by 50%. This will decrease the carrier number and abruptly alter the light in the lasing mode but will have little effect on the nonlasing mode. If the lasing mode can adiabatically follow the change in the nonlasing mode, it will decrease in intensity by 50% in order to keep the total light intensity constant.

The response time of the modes can be inferred from their eigenfrequencies for small oscillations. At a facet power of 2 mW, the relaxation oscillation frequency that arises from the coupling of I_0 and N is about 3 GHz. The response time of the lasing mode is about one-half of a relaxation oscillation period (~ 170 psec).

In order to understand the response time of the nonlasing mode, we return to Eq. (39) for the field β. We will neglect the $i\alpha$ term, which can be removed by changing the angular frequency ω_0 of the mode by $\frac{1}{2}(G - \gamma)\alpha$. The Langevin force equation is

$$\dot{\beta} = -(\beta/T) + F_\beta(t), \tag{68}$$

where

$$T = 2/(\gamma - G) = 2 \langle I_1 \rangle / R. \tag{69}$$

We can evaluate T with the aid of Fig. 11, calculated for a 250-μm-long quaternary laser with $P_0 = 2$ mW. We see that $\langle I_1 \rangle = 2000$, $g = 90$ cm^{-1}, $R = 1.7 r v_g = 1.0 \times 10^{12}$ sec^{-1}; hence $T = 3.7$ nsec.

When β is driven by a sinusoidal noise at Ω, $\beta \sim [(1/T) + i\Omega]^{-1}$ and $I_1 = |\beta|^2 \sim [(1/T^2) + \Omega^2]^{-1}$. The response will drop to one-half the maximum at a frequency of $(2\pi T)^{-1} = 43$ MHz. This explains the low-frequency cutoff in the partition noise spectrum in Fig. 25.

Consider the response of the nonlasing mode to a large step in $F_\beta(t) = A$ beginning at $t = 0$. Solving Eq. (68) gives

$$\beta = AT(1 - e^{-t/T}), \tag{70}$$

$$I_1 = |\beta|^2 = A^2 T^2 (1 - e^{-t/T})^2. \tag{71}$$

The slow increase of I_1 with time (which begins quadratically in t) reaches $0.4A^2T^2$ at $t = T$ and $0.75A^2T^2$ at $t = 2T$. This explains the long time values (1 – 8 nsec) for partition noise fluctuations observed by Linke et al. (1983). In view of the fact that the response time of the lasing mode is over an order of magnitude faster than the time of partition noise fluctuations, it is reasonable that the lasing mode is able to follow adiabatically the nonlasing mode fluctuations and maintain a nearly constant rate of total stimulated emission.

18. LINEWIDTH OF A SINGLE MODE

The laser power spectrum $G(\omega)$ is the line shape measured when the energy of a single-mode laser is passed through a dispersive instrument such as a scanning Fabry–Perot interferometer. We will first relate $G(\omega)$ to phase noise and then use a model of phase noise to calculate the mode linewidth. If $E(\omega)$ is the Fourier transform of the optical field $E(t)$, then the power spectrum $G(\omega)$ is proportional to $|E(\omega)|^2$. It is readily shown (Lax, 1960) that

$$G(\omega) = \int dt \, \langle E(t)E(0) \rangle \, e^{i\omega t}. \tag{72}$$

That is, $G(\omega)$ is the Fourier transform of the correlation of the optical field at times 0 and t.

To evaluate Eq. (72), we use the expression for the optical field

$$E(t) = I(t)^{1/2} \exp[-i(\omega_0 t + \phi)] + \text{c.c.}. \tag{73}$$

Above threshold, the intensity of a highly single-moded laser becomes stabilized, and therefore, for simplicity, we will neglect the small intensity fluctuations in computing $\langle E(t)E(0) \rangle$. This is not entirely correct. Vahala et al. (1983) have shown that the correlation of intensity and phase fluctuations causes a slight asymmetry in the power spectrum structure. Substituting Eq. (73) into Eq. (72) and neglecting high-frequency terms with $(\omega_0 + \omega)$ yield

$$G(\omega) = \int dt \, e^{i(\omega - \omega_0)t} \langle e^{-i\Delta\phi(t)} \rangle, \tag{74}$$

where

$$\Delta\phi(t) \equiv \phi(t) - \phi(0). \tag{75}$$

Equation (74) can be simplified because $\Delta\phi$ is a Gaussian variable. The phase of the optical field of a laser can take on any value. We saw (Fig. 24) that $\phi(t)$ is continually undergoing a random walk because of spontaneous emission events. It is not surprising, therefore, that the probability distribution $P(\Delta\phi)$ is a Gaussian. This was proved by Lax (1967c) and established experimentally for semiconductor lasers by Diano *et al.* (1983). For a Gaussian distribution it is simple to show (Lax, 1967c) that

$$\langle \exp(-i\Delta\phi) \rangle = \exp(-\tfrac{1}{2}\langle \Delta\phi^2 \rangle). \tag{76}$$

Therefore

$$G(\omega) = \int dt \; e^{i(\omega - \omega_0)t} \; e^{-1/2 \langle \Delta\phi^2(t) \rangle} \tag{77}$$

We will now derive an expression for $\langle \Delta\phi^2(t) \rangle$. The phase of the optical field changes abruptly during each spontaneous emission event according to Eq. (44). The intensity change during each spontaneous emission event [Eq. (45)] results in an even larger *delayed* phase change. This change is brought about by a gain change that follows the spontaneous emission event and restores the laser intensity back to the steady state. Equations (40) and (41) show that this gain change is accompanied by a phase change due to the shift of the cavity resonance when G departs from the steady-state value. From these equations and Eq. (42) we find the delayed phase change to be

$$\Delta\phi = (\alpha/I^{1/2}) \cos(\theta_i). \tag{78}$$

Therefore, the complete phase change due to a number of emission events is

$$\Delta\phi = \sum_{i=1}^{n} [\sin(\theta_i) + \alpha \cos(\theta_i)] I^{-1/2}. \tag{79}$$

Squaring and averaging $\Delta\phi$ and using $n = Rt$, where R is the average spontaneous emission rate, we find (Henry, 1982)

$$\langle \Delta\phi^2(t) \rangle = (R/2I)(1 + \alpha^2)t. \tag{80}$$

It can be shown that $\Delta\phi^2(t)$ is even in t (Lax, 1960). Substitution of Eq. (80) into Eq. (77) gives a Lorentzian power spectrum with a full width at half-maximum (FWHM) Δf of

$$\Delta f = R(1 + \alpha^2)/4\pi I. \tag{81}$$

This equation accounts for the linewidth data of Welford and Mooradian (1982) that exhibit a nearly I^{-1} dependence and a linewidth at 300 K of about 80 MHz at 1 mW. Their data are shown in Fig. 28. The small residual linewidth found by extrapolating to $I = \infty$ is not yet accounted for. The magnitude of the linewidth was explained by Henry (1982) as due to the $1 + \alpha^2$ correction with $\alpha \approx 5$. A value of $\alpha \approx 6$ was measured for BH lasers (see Fig. 10). The variation of α with energy shown in Fig. 10 makes it understandable that this parameter will be slightly different in different lasers.

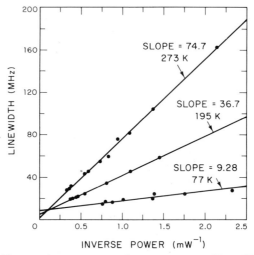

FIG. 28. Linewidth versus inverse power at three temperatures. [From Welford and Mooradian (1982).]

19. RELAXATION OSCILLATION REPLICAS IN THE POWER SPECTRUM

The simple theory of the preceding section, which predicts a Lorentzian line shape and quantitatively accounts for the observed linewidth, is based on an approximation. Removing this approximation does not alter the theory of the laser linewidth but gives rise to additional structure in the power spectrum in the form of weak replicas of the laser line evenly spaced by separations equal to the relaxation oscillation frequency. This structure has been observed by Osterwalder and Rickett (1979), Diano *et al.* (1983), and Vahala *et al.* (1983). The data of Vahala *et al.* are shown in Fig. 29. They explain the slight asymmetry in the replica intensities as due to the correlation of phase and intensity noise. This correlation is neglected in our discussion.

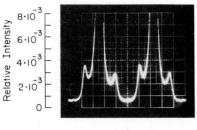

FIG. 29. Fabry–Perot scan of a laser line showing slightly asymmetric side peaks separated from the main peak by the relaxation oscillation frequency. [From Vahala *et al.* (1983).]

The power spectrum is determined by the Fourier transform of $\exp(-\frac{1}{2} \langle \Delta\phi^2(t) \rangle)$. To determine the structure in the mode power spectrum occurring several gigahertz away from the line center ω_0, we must accurately know the time dependence of $\langle \Delta\phi^2(t) \rangle$ at short times. However, the calculation leading to Eq. (80) is only valid for times long compared to the relaxation oscillation damping time since the deviation of Eq. (80) assumes that the relaxation oscillations following spontaneous emission are able to die out in the times of interest.

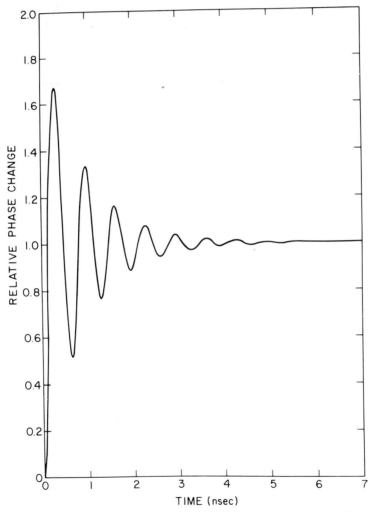

FIG. 30. Oscillating delayed phase change following an abrupt intensity change; $P_0 = 1$ mW, $\Gamma = 1.1 \times 10^9$ sec^{-1}, $\Omega/2\pi = 1.52$ GHz. [From C. H. Henry. *IEEE J. Quantum Electron.* **QE-19**, 1391 (1983). Copyright © 1983 IEEE.]

Figure 30 (Henry, 1983c) shows the time dependence of the delayed phase change following a spontaneous emission event. It is oscillatory and approaches the value given by Eq. (79) after the oscillations die away. Figure 31 shows a calculation of $\langle \Delta\phi^2(t) \rangle$ taking into account this oscillatory behavior (Henry, 1983c). The mean-square phase change increases initially as t^2 and then oscillates and finally approaches the linear time dependence with a slope given by Eq. (80) after a relaxation oscillation decay time. These features have been directly measured by Diano et al. (1983).

The power spectra in Fig. 32 are calculated from the Fourier transform of $\exp(-\tfrac{1}{2}\langle \Delta\phi^2(t) \rangle)$ for the three values of $\exp(\langle \Delta\phi^2(t) \rangle)$ of Fig. 31. The main peak and linewidth are unchanged from the Lorentzian prediction. The oscillatory behavior gives rise to replicas of the main peak that are spaced by the relaxation oscillation frequency. The strength of these peaks diminishes with increased damping of the relaxation oscillations.

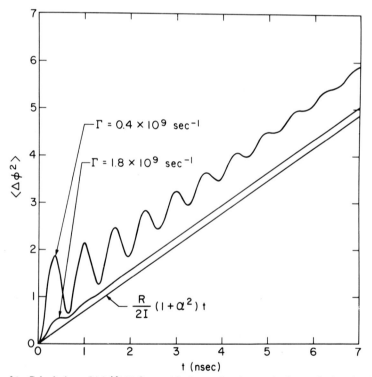

FIG. 31. Calculation of $\langle \Delta\phi^2(t) \rangle$ for weakly damped and strongly damped relaxation oscillations; $P_0 = 1$ mW, $I = 3.1 \times 10^4$, $\Omega/2\pi = 1.52$ GHz, $\alpha = 5.3$, $R = 1.51 \times 10^{12}$ sec^{-1}. [From C. H. Henry. *IEEE J. Quantum Electron.* **QE-19**, 1391 (1983). Copyright © 1983 IEEE.]

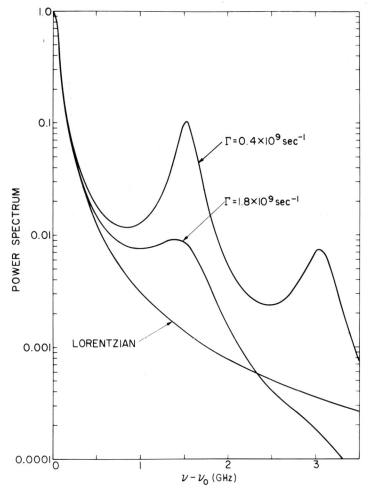

FIG. 32. Structure in the power spectrum of a single mode. The three power spectra were calculated from the three curves for $\langle \Delta\phi^2(t) \rangle$ of Fig. 31. [From C. H. Henry. *IEEE J. Quantum Electron.* **QE-19,** 1391 (1983). Copyright © 1983 IEEE.]

20. LINEWIDTH REQUIREMENTS FOR COHERENT OPTICAL COMMUNICATIONS

Just as in radio and microwave transmission, increased sensitivity and a high degree of wavelength selection are obtainable by coherent detection of the received radiation. In superheterodyne reception, the incoming optical field is mixed coherently with the field of a local oscillator. These improvements are at the expense of a much more difficult technology. Both the light source and the local oscillator must be highly coherent. The laser linewidth is a measure of optical coherence. We will make a simple estimate here of the linewidth required to avoid errors in coherent optical communications.

Various schemes for digital transmission with coherent detection have been discussed (Yamamoto and Kimura, 1981) including optical amplitude shift keying, frequency shift keying, and phase shift keying. We will only calculate the linewidth requirements for phase shift keying, which is conceptually simple and is considered to be the most sensitive scheme.

In phase shift keying, a bit of information is transmitted by shifting the phase of the optical signal by π. The method of detection must be insensitive to slow drifts in phase and only sense the phase shift during the bit time T. Errors will be encountered if random phase shifts as large as $\pm\pi/2$ occur during T in either the source or the local oscillator. The probability of the phase shifting randomly by $\Delta\phi$ during time T is a Gaussian

$$P(\Delta\phi) = \frac{\exp[-\Delta\phi^2/2\langle\Delta\phi^2(T)\rangle]}{[2\pi\,\Delta\phi^2(T)]^{1/2}}. \tag{82}$$

The probability of error is found by integrating $P(\Delta\phi)$ from $+\pi/2$ to ∞ and from $-\pi/2$ to $-\infty$. The approximate value of this integral can be obtained by substituting $\Delta\phi^2 = [(\pi/2) + x]^2 \approx (\pi^2/4) + \pi x$ and integrating x from 0 to ∞. The result is

$$P_e \approx \left(\frac{8\langle\Delta\phi^2(T)\rangle}{\pi^3}\right)^{1/2} \exp\left(\frac{-\pi^2}{8\langle\Delta\phi^2(T)\rangle}\right). \tag{83}$$

For rapidly damped relaxation oscillations,

$$\langle\Delta\phi^2(T)\rangle \approx 2\pi\,\Delta f\,T, \tag{84}$$

and

$$P_e \approx (4/\pi)(\Delta f\,T)^{1/2}\exp(-\pi/16\,\Delta f\,T). \tag{85}$$

Equation (85) shows that the error probability rapidly decreases with decreasing linewidth and increasing bit rate. Evaluation of Eq. (85) shows that to achieve a error rate of less than 10^{-9}, a laser linewidth of $\Delta f\,T \approx 0.01$ is required. That is Δf (megahertz) equals the bit rate (megabits) divided by 100; i.e., at 100 Mbits, $\Delta f = 1$ MHz and at 1 Gbit, $\Delta f = 10$ MHz. Therefore coherent detection will become more feasible at high bit rates, where the linewidth requirements are less stringent. It should be remembered that these are minimum requirements that become more stringent if the relaxation oscillations are not rapidly damped (see Fig. 31).

References

Agrawal, G. P., Joyce, W. B., Dixon, R. W., and Lax, M. (1983). *Appl. Phys. Lett.* **43**, 11.
Asbeck, P. M. (1977). *J. Appl. Phys.* **48**, 820.

Asbeck, P. M., Commack, D. A., Daniele, J. J., and Klabanoff, V. (1979). *IEEE J. Quantum Electron* **QE-15**, 727.

Bernard, M. G., and Duraffourg, G. (1961). *Phys. Status Solidi* **1**, 699.

Casey, H. C., and Stern, F. (1976). *J. Appl. Phys.* **47**, 631.

Copland, J. A. (1980). *IEEE J. Quantum Electron* **QE-16**, 721.

Diano, B., Spano, P., Tambarini, M., and Piazolla, S. (1983). *IEEE J. Quantum Electron* **QE-19**, 226.

Einstein, A. (1917). *Phys. Z.* **18**, 121.

Glauber, R. J. (1963). *Phys. Rev.* **131**, 2766.

Guekos, G., and Jäckel, H. (1977). *Opt. Quantum Electron* **9**, 233.

Guo, C.-z., and Wang, K.-g. (1982). *IEEE J. Quantum Electron.* **QE-18**, 1728

Hakki, B. W., and Paoli, T. L. (1975). *J. Appl. Phys.* **46**, 1299.

Haus, H. A., and Kawakami, S. (1984). *IEEE J. Quantum Electron.* To be published.

Henry, C. H. (1982). *IEEE J. Quantum Electron.* **QE-18**, 259.

Henry, C. H. (1983). *IEEE J. Quantum Electron.* **QE-19**, 1391.

Henry, C. H., Logan, R. A., and Merritt, F. R. (1980). *J. Appl. Phys.* **51**, 3042.

Henry, C. H., Logan, R. A., and Bertness, K. A. (1981a). *J. Appl. Phys.* **52**, 4457.

Henry, C. H., Logan, R. A., and Bertness, K. A. (1981b). *J. Appl. Phys.* **52**, 4453.

Henry, C. H., Logan, R. A., and Merritt, F. R. (1981c). *IEEE J. Quantum Electron.* **QE-17**, 2196.

Henry, C. H., Logan, R. A., Temkin, H., and Merritt, F. R. (1983a). *IEEE J. Quantum Electron.* **QE-19**, 941.

Henry, C. H., Logan, R. A., Merritt, F. R., and Luongo, J. P. (1983b). *IEEE J. Quantum Electron.* **QE-19**, 947.

Henry, C. H., Henry, P. S., and Lax. M. (1983). *IEEE J. Lightwave Technol.* **LT2**, 209.

Jäckel, H. (1980). Ph.D. Thesis, BTH, Zurick, Switzerland.

Kane, E. O. (1957). *J. Phys. Chem. Solids* **1**, 249.

Kazarinov, R. F., Henry, C. H., and Logan, R. A. (1982). *J. Appl. Phys.* **53**, 4631.

Landau, L. D., and Lifshitz, E. M. (1958). *in* "Statical Physics," Chapter V, Addison-Wesley, Reading, Massachusetts.

Lang, R. (1979). *IEEE J. Quantum Electron* **QE-15**, 718.

Lang, R. (1980). *Jpn. J. Appl. Phys.* **19**, L93.

Lax, M. (1960). *Rev. Mod. Phys.* **32**, 25.

Lax, M. (1966). *Phys. Rev.* **145**, 110.

Lax, M. (1967a). *Phys. Rev.* **157**, 213.

Lax, M. (1967b). *IEEE J. Quantum Electron.* **QE-3**, 37.

Lax, M. (1967c). *Phys. Rev.* **160**, 290.

Lee, T. P., and Marcuse, D. (1983). *IEEE J. Quantum Electron.* **QE-19**, 1397.

Lee, T. P., Burrus, C. A., Copland, J. A., Dentai, A. G., and Marcuse, D. (1982). *IEEE J. Quantum Electron.* **QE-18**, 1101.

Linke, R. A., Burrus, C. A., Kominow, S. P., Ko, J. S., and Lee, T. P. (1983). *Dig. Tech. Pap. Top. Meet. Opt. Fiber Commun., 6th,* pap. PD-4.

Lin, P., and Lee, T. P. (1984). *J. Lightwave Tech.* **LT-2**, 44.

Liu, P., Fencil, L. E., Ko, J.-S., Kominow, I. P., Lee, T. P., and Burrus, C. A. (1983). *IEEE J. Quantum Electron.* **QE-19**, 1348.

Logan, R. A., Herry, C. H., van der Ziel, J. P., and Temkin, H. (1982). *Electron Lett.* **18**, 782.

McCummber, D. E. (1966). *Phys. Rev.* **141**, 306.

Manning, J., Olshansky, R., and Su, C. B. (1983). *IEEE J. Quantum Electron.* **QE-19**, 1525.

Marcuse, D. (1982). *Electron Lett.* **18**, 920.

Nakamura, M., Aiki, K., Chinone, N., Ito, R., and Umeda, J. (1978a). *J. Appl. Phys.* **49**, 4644.

Nakamura, M., Aiki, K., and Umeda, J. (1978b). *Appl. Phys. Lett.* **32**, 322.

Nishimura, Y. (1974). *Jpn. J. Appl. Phys.* **13**, 109.

Osterwalder, J. M., and Rickett, B. J. (1979). *Proc. IEEE* **67,** 1671.

Patzak, E. (1982). *Electron. Lett.* **18,** 278.

Pearsall, T. P. (1983). "GaInGaP Alloy Semiconductor." Wiley, New York.

Peterman, K. (1979). *IEEE J. Quantum Electron.* **QE-15,** 566.

Streifer, W., Scifres, D. R., and Burnham, R. D. (1981). *Electron. Lett.* **17,** 934.

Streifer, W., Scifres, D. R., and Burnham, R. D. (1982). *Appl. Phys. Lett.* **40,** 305.

Thompson, G. H. B., Lovelace, D. F., and Turley, S. E. H. (1978). *IEE J. Solid-State Electron. Devices* **2,** 12.

Tsang, W. T., and Olsson, N. A. (1983). *Appl. Phys. Lett.* **42,** 650.

Vahala, K., Harder, C., and Yariv, A. (1983). *Appl. Phys. Lett.* **42,** 211.

Welford, D., and Mooradian, A. (1982). *Appl. Phys. Lett.* **40,** 865.

Yamada, M., and Suematzu, Y. (1979). *IEEE J. Quantum Electron.* **QE-15,** 743.

Yamamoto, Y., and Kimura, T. (1981). *IEEE J. Quantum Electron.* **QE-17,** 919.

Yariv, A., and Margolit, S. (1982). *IEEE J. Quantum Electron.* **QE-18,** 1831.

SEMICONDUCTORS AND SEMIMETALS, VOL. 22, PART B

CHAPTER 4

Dynamic Single-Mode Semiconductor Lasers with a Distributed Reflector

Yasuharu Suematsu, Katsumi Kishino, Shigehisa Arai, and Fumio Koyama

DEPARTMENT OF PHYSICAL ELECTRONICS
TOKYO INSTITUTE OF TECHNOLOGY
TOKYO, JAPAN

I. Introduction

Single-wavelength operation or single-longitudinal-mode operation of a rapidly modulated injection laser in the wavelength range from 1.5 to 1.6 μm is very attractive as a light source, especially for single-mode fiber communication systems in the minimum-loss wavelength region (Miya *et al.,* 1979; Utaka *et al.,* 1980a, 1981a,b,c; Kobayashi *et al.,* 1981; Mikami, 1981; Koyama *et al.,* 1981a; Tanbun-ek *et al.,* 1981; Abe *et al.,* 1981, 1982a; Matsuoka *et al.,* 1982; Akiba *et al.,* 1982; Yamamoto *et al.,* 1982), since the

205

chromatic dispersion (Payne and Gambling, 1975) caused by dynamic spectral broadening of the light source (Ikegami, 1975) severely limits the transmission bandwidth. Single-longitudinal-mode operation in a rapidly modulated 1.5-μm GaInAsP-type laser with a distributed Bragg reflector (DBR) structure was first reported by Utaku et al. (1981a). These lasers are called dynamic single-mode (DSM) lasers (Suematsu et al., 1983). Single-mode operation under rapid direct modulation has been reported as shown in Table I. Single-mode operation was observed in 0.85-μm GaAlAs-type distributed-feedback (DFB) lasers at a pulse width of 5 nsec and in 1.3-μm GaInAsP/InP-type distributed-Bragg-reflector (DBR) lasers at a pulse width of 1.5 nsec (Sakakibara et al., 1980). Single-longitudinal-mode dc operation is frequently observed in narrow-striped conventional lasers (Namizaki, 1975; Aiki et al., 1977), but spectral broadening occurs when they are modulated rapidly, and the dynamic spectral width increases up to about 10 nm in the 1.6-μm GaInAsP-type laser (Kishino et al., 1982). This is why the 1.5–1.6-μm wavelength region has not yet been used for fiber communications, even though the transmission loss is minimal in this wavelength region. However, 1.3-μm wavelength fiber-communication systems are now in use even though the loss is larger than in the 1.5–1.6 μm region, because the chromatic dispersion is small at this wavelength (Payne and Gambling, 1975) and thus the effect of dynamic spectral broadening can be minimized. Investigations aimed at shifting the wavelength of zero chromatic dispersion of single-mode fiber into the 1.5–1.6-μm wavelength region have been made (see, e.g., White and Nelson, 1979; Cohen et al., 1979; Miya et al., 1981), but this work is still under development. Thus, a DSM laser is required in order to establish high-capacity and low-loss optical-fiber communications systems. Another advantage of the DSM laser is that the jump in lasing mode

TABLE I

SINGLE-MODE OPERATION OF 1.55-μM DYNAMIC SINGLE-MODE
SEMICONDUCTOR LASERS UNDER RAPID DIRECT MODULATION

Wavelength	Pulse width	Rate	Reference
1.55 μm	1 nsec	—	Utaka et al. (1981b)
1.55 μm	—	280 Mbit/sec	Yamamoto et al. (1982)
1.55 μm	—	400 Mbit/sec	Ikegami et al. (1982)
1.55 μm	450 psec	—	Ebeling et al. (1983)
1.55 μm	—	1.6 Gbit/sec	Nakagawa et al. (1983)
1.55 μm	—	1.0 Gbit/sec	Tsang et al. (1983)
1.55 μm	—	2.0 Gbit/sec	Lin et al. (1983)

that is normally observed in the conventional semiconductor laser is suppressed over wider temperature and current ranges of operation. The excellent low-noise property thus obtained is expected to be advantageous in optoelectronic applications of DSM lasers.

The DSM laser is a tightly mode-controlled laser in which axial, and transverse, and TE or TM modes are maintained as fixed modes during very rapid direct modulation. In principle, axial-mode control can be obtained by the use of a wavelength-selective resonator, such as a multicavity structure (Suematsu *et al.,* 1975a; Kishino *et al.,* 1978; Coldren *et al.,* 1981; Choi and Wang, 1982; Tsang *et al.,* 1983), external mirror (Chinone *et al.,* 1978; Preston *et al.,* 1981; Arnold *et al.,* 1978; Fleming and Mooradiani, 1981), short cavity (Matsumoto and Kumabe, 1979; Burrus *et al.,* 1981; Soda *et al.,* 1979), distributed feedback (DFB) and distributed Bragg reflector (DBR) (Kogelnik and Shank, 1971; Kaminow and Weber, 1971; Nakamura *et al.,* 1973a,b, 1974, 1975a,b; Yen *et al.,* 1973; Shank *et al.,* 1974; Scifres *et al.,* 1974, 1975, 1976; Stoll and Seib, 1974; Anderson *et al.,* 1974; Alferov *et al.,* 1975; Zory and Comerford, 1975; Reinhart *et al.,* 1975; Casey *et al.,* 1975; Tsang and Wang, 1976; Tsang *et al.,* 1979a,b,c; Burnham *et al.,* 1976; Aiki *et al.,* 1976a,b; Kawanishi *et al.,* 1977, 1980; Yamanishi *et al.,* 1977; Namizaki *et al.,* 1977; Shams *et al.,* 1978a,b; Shams and Wang, 1978; Kawanishi and Suematsu, 1978; Umeda *et al.,* 1977), or a light injection technique (Kobayashi *et al.,* 1980; Yamada *et al.,* 1981; Hodgkinson *et al.,* 1982). On the other hand, single transverse-mode operation can be obtained by a narrow stripe configuration, such as a buried heterostructure (BH). Polarization control can be obtained by the use of differences in reflectivity or loss for the TE and TM modes (Utaka *et al.,* 1980a, 1981a,b; Kobayashi *et al.,* 1981; Streifer *et al.,* 1975b, 1976; Suematsu *et al.,* 1972; Takano and Hamasaki, 1982). External modulation techniques (Kubota *et al.,* 1980) are, in principle, also applicable for the elimination of spectral broadening. Developments in monolithic integration of optical devices have also contributed to the formation of DSM lasers. Such a laser is sometimes called an integrated laser (Suematsu *et al.,* 1975a,b; Kishino *et al.,* 1978; Coldren *et al.,* 1981; Aiki *et al.,* 1976b; Hurwitz *et al.,* 1975; Campbell and Bellavance, 1977; Merz and Logan, 1977). Due to larger mode stability and simpler fabrication, GaInAsP/InP-type DBR integrated lasers (Kawanishi *et al.,* 1978, 1979; Utaka *et al.,* 1980b) and DFB lasers (Doi *et al.,* 1979) in both 1.3-μm and 1.5–1.6-μm wavelengths have been developed. There has been considerable work done on the theory of DFB and DBR lasers (Kogelnik and Shank, 1972; Wang, 1973, 1974, 1974–1975, 1977; DeWames and Hall, 1973; Yariv, 1973; Yariv and Nakamura, 1977; Subert, 1974; Chinn and Kelly, 1974; Nakamura and Yariv, 1974; Iga and Kawabata, 1975; Streifer *et al.*, 1975a,b, 1976, 1977; Elachi and Evans, 1975; Okuda and Kubo, 1975; Haus and Shank, 1976; Stoll, 1979; Suzuki and Tada, 1980).

It was found that the lasing wavelength of a DSM laser shifts during one period of direct modulation (Koyama *et al.*, 1981a). This wavelength shift is called dynamic wavelength shift and it gives dynamic spectral broadening to the DSM laser (Kishino *et al.*, 1982). The maximum value of the dynamic wavelength shift was observed to be about 0.27 nm at the resonance-like modulation frequency (1.8 GHz) (Koyama *et al.*, 1981a). This observation makes it possible to estimate the transmission bandwidth of a conventional, single-mode fiber driven by a DSM laser at a wavelength of 1.55 μm to be 185 GHz km, which is about 37 times larger than that driven by a conventional semiconductor laser (Koyama *et al.*, 1981a). Recently, cw operations of DBR integrated lasers (Kobayashi *et al.*, 1981; Tanbun-ek *et al.*, 1981; Abe *et al.*, 1982a) and DFB lasers (Utaka *et al.*, 1981c; Matsuoka *et al.*, 1982; Akiba *et al.*, 1982), in the wavelength region of 1.5 to 1.6 μm have been reported with an output power of more than several milliwatts. The reliability tests performed very recently estimate the operation life to be comparable to that of conventional lasers (Ikegami, 1983; Tanbun-ek *et al.*, 1984).

In this chapter, the progress of DSM lasers is reviewed and the basic principles and conditions of DSM operation are given. The structure and basis of lasing properties are detailed in Section II, in which the coupling problems between the active and passive regions, the coupling parameters, the reflectivity of the distributed reflector, the lasing threshold condition, and the theoretical static- and dynamic-lasing properties are discussed. The fabrication and experimental lasing properties of the DSM laser are detailed in Section III. Possible further improvements, future applications, and a discussion of monolothic integration of optical devices are presented in Section IV.

II. Structure and Theoretical Considerations of DSM Lasers

1. STRUCTURE OF DBR AND DFB LASERS

In conventional Fabry–Perot lasers, a cleaved, faceted mirror is used in order to obtain the feedback for laser operation. The feedback can be also obtained by a periodic, equivalent, refractive index variation of the corrugated optical waveguide. Such a periodic structure is utilized in DBR and DFB lasers. Figure 1 shows the basic structures of a DBR laser in which the corrugation is fabricated separate from the active region on the surface of the low-loss, external waveguide and a DFB laser, in which the corrugation is fabricated on the active waveguide. In a DBR laser, a low-loss waveguide for the DBR region is necessary in order to obtain the high reflectivity and

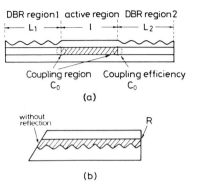

FIG. 1. Schematic diagram of (a) a DBR integrated laser and (b) a DFB integrated laser. Parameters C_0 and C_0' denote the coupling efficiency between the active and passive regions within a certain coupling scheme, and R is the power reflectivity of the faceted mirror.

fine-mode selectivity. The power coupling efficiency between the active and passive DBR regions is an essential factor for high-performance operation, as will be discussed in next section. Since such a coupling to the external region is not essential in a DFB laser, the fabrication process becomes relatively simple. The integration of the external waveguide, however, enables us to prepare an integrated laser for monolithic integration of an optical circuit. The external waveguides provide a TE-mode filter that eliminates the TM mode for a DBR laser completely. They also provide for a possible fine-wavelength control mechanism as discussed in Section III.

The structure difference between DBR and DFB lasers gives some variations in the lasing properties. In a DFB laser without the faceted mirror, the resonant-mode characteristics are symmetric with respect to the Bragg wavelength so that longitudinal modes with the same minimum threshold gain exist in pairs on both sides of the Bragg wavelength. This theoretically allows two-mode operation (Kogelnik and Shank, 1972). In this symmetric case, the lasing oscillation at the Bragg wavelength is essentially forbidden. In order to obtain the stable single-mode operation, some means of making the resonant-mode characteristics asymmetric have been studied (Haus and Shank, 1976; Suzuki and Tada, 1980). One of the simplest solutions, using the effect of the one-sided faceted mirror (Streifer *et al.*, 1975a), is being used successfully in these lasers.

In DBR lasers, the resonant-mode characteristics determined by the phase shifts of the active and passive DBR regions are usually asymmetric. The deviation of the lasing wavelength from the Bragg wavelength is changed by adjusting the length of active region. Therefore, if the phase shift of the active region is finely controlled, the lasing wavelength can be matched to the Bragg wavelength. However, this may be difficult in the actual fabrication process. A DBR laser with a wavelength control mechanism, in which the external phase-controlled waveguide was integrated, has been developed for high-performance operation (Tohmori *et al.*, 1983).

The lasing modes of DBR and DFB lasers also exhibit different temperature dependencies. The DBR lasing mode jumps to another DBR mode as
the temperature rises. In contrast, the DFB lasing mode remains constant
over a large temperature region. This will be discussed further in Section 7.

2. Coupling Structure

For DBR lasers, the low-loss DBR waveguide is required to get a sharp
wavelength selectivity of reflectivity (Utaka *et al.,* 1981b; Wang, 1974).
Simultaneously, a highly efficient coupling between the active and external
waveguides is required in order to obtain a higher differential quantum
efficiency (Utaka *et al.,* 1981b).

Five kinds of coupling schemes that are used for integrated lasers are
shown in Fig. 2. In the phase coupling scheme, both layers of the active and
output waveguides are coupled using a directional coupler. It is also required
that the phase velocities of both waveguides be nearly equal (Suematsu *et al.,*
1975a; Kishino *et al.,* 1978). With taper coupling, the optical absorption at
the taper portion is produced (Reinhart *et al.,* 1975). The direct coupling
scheme is shown in Fig. 2c (Hurwitz *et al.,* 1975). The large-optical-cavity
(LOC) laser uses the evanescent coupling scheme in which the coupling

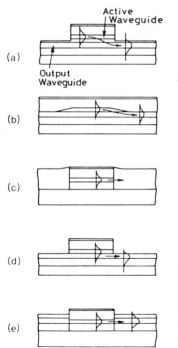

(a)

(b)

(c)

(d)

(e)

Fig. 2. Various coupling schemes used for integrated
lasers. (a) Phase coupling (Suematsu *et al.,* 1975a, (b)
taper coupling (Reinhart *et al.,* 1975), (c) direct coupling
(Hurwitz *et al.,* 1975), (d) evanescent coupling (Campbell and Bellavance, 1977; Merz and Logan, 1977), and
(e) direct coupling with optimized coupling (Abe *et al.,*
1981). [From Suematsu *et al.* (1983). Copyright © 1983
IEEE.]

efficiency is rather low. However, it is expected theoretically that if a finely adjusted external waveguide is used, the coupling efficiency increases. (Chang and Garmire, 1980).

Monolithic integration of six GaAlAs/GaAs-type DFB lasers was demonstrated (Aiki *et al.*, 1976b) using a directly coupled external waveguide (Hurwitz *et al.*, 1975), as shown in Fig. 2(c). The coupling efficiency, however, was relatively low. In the novel integrated laser with a butt-jointed, built-in (BJB) external waveguide, a high coupling efficiency between the two butt-jointed waveguides (namely, the active and external guides) is realized by matching both the equivalent refractive indexes and the field profiles of two waveguides (Abe *et al.*, 1981; Kishino *et al.*, 1982). In this subsection, we will examine the dependence of the coupling efficiency of the BJB structure on the displacement and width difference of the two waveguides.

A schematic diagram of the joint part of this laser is shown in Fig. 3. The field distribution in the active and external waveguides can be expressed as

$$E(x, y) = \overline{E}_{ix}(x)\overline{E}_{iy}(y)e^{-j\beta_i z}, \tag{1}$$

where i stands for a or e (a: active waveguide, e: external waveguide), and β_i is the propagation constant of region i.

Then the coupling efficiency from the active waveguide to the external waveguide C_0 is expressed as

$$C_0 = \left(\frac{2\beta_e}{\beta_a + \beta_e}\right)^2 \left| \frac{\int \overline{E}_{ax}(x)\overline{E}_{ex}(x)\, dx \int \overline{E}_{ay}(y)\overline{E}_{ey}(y)\, dy}{\int \overline{E}_{ex}(x)^2\, dx \int \overline{E}_{ey}(y)^2\, dy} \right|^2. \tag{2}$$

Figure 4 shows the coupling efficiency C_0 as a function of the displacement

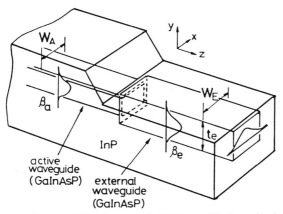

FIG. 3. Schematic diagram of the joint part of a BJB laser. W_i, β_i, and t_i denote the width, propagation constant, and thickness, respectively, of the waveguide in the region i.

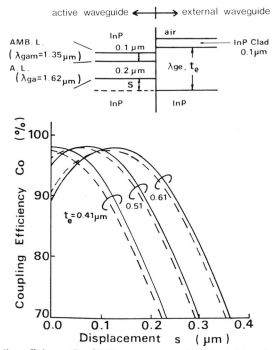

FIG. 4. Coupling efficiency C_0 of BJB structure as a function of the displacement s. The parameter t_e is the thickness of the external waveguide and λ_{ge} the corresponding bandgap wavelength. [After Abe *et al.*, (1981).] ——, $\lambda_{ge} = 1.40$ μm; – – –, $\lambda_{ge} = 1.30$ μm. The refractive indexes of the quaternary and InP cladding layers at a wavelength of 1.62 μm were estimated from the SEO method (Utaka *et al.*, 1981b). [From Utaka *et al.* (1980b).]

s of the active and external waveguides for a 1.6-μm GaInAsP/InP-type BJB laser with the assumption that the widths of both the active and external waveguides are the same. A maximum coupling efficiency of 98% is calculated at $s = 0.08$ μm.

Figure 5 also shows the coupling efficiency C_0 as a function of the width of the external waveguide W_E. For this calculation the active region was assumed to be 4 μm wide and the lasing wavelength was 1.62 μm with TE$_{00}$-mode oscillation.

The coupling efficiency remains greater than 95% for an external waveguide width of 4 ∓ 1 μm ($\mp 2.5\%$ deviation) or for the thickness of 0.51 ± 0.1 μm as shown in Figs. 4 and 5. This tolerance is considered to be achievable with a good reproducibility in the fabrication processes.

3. PARAMETERS OF A DISTRIBUTED REFLECTOR

In this subsection, the fundamental properties of a distributed Bragg reflector are analyzed by the use of the coupled-wave equations (Kogelnik and

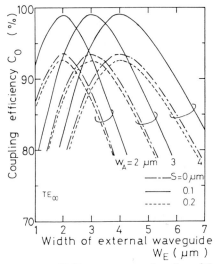

FIG. 5. Coupling efficiency C_0 of BJB structure as a function of the width of the external waveguide W_E. The parameter W_A is the width of the active waveguide where $t = 0.5$ μm, $b = 0.3$ μm, $\lambda = 1.62$ μm, $\lambda_g = 1.37$ μm. ---, $s = 0$ μm; ——, $s = 0.1$ μm; - - -, $s = 0.2$ μm. [From Tabun-ek et al. (1984). Copyright © 1984 IEEE.]

Shank, 1972), in preparation for the analysis of the DBR and DFB laser properties. The coupled-wave analysis of a plane-wave propagation in a periodic structure was given by Kogelnik and Shank (1972). Since the coupled-wave theory has been treated in many books, only a short description is given here. The based-wave equation is given as

$$\nabla^2 E + k^2 E = 0, \qquad k^2 = k_0^2 n^2 - j k_0 n \alpha_g, \qquad (3)$$

where $k_0 = 2\pi/\lambda$ is the propagation constant, α_g the absorption coefficient in the waveguide, and n the refractive index.

A corrugated periodic structure, shown schematically in Fig. 6, is generally described by the periodic refractive index distribution along the dielectric waveguide. Thus, the refractive index function can be expressed by the sum of the refractive index of the waveguide without corrugations and the purturbation produced by corrugations as

$$n^2(x, z) = n_0^2 + \Delta n^2(x, z). \qquad (4)$$

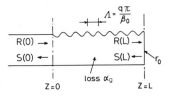

FIG. 6. Schematic of a corrugated optical waveguide.

In such a periodic structure with period Λ, the forward and backward propagating waves in the z direction are coupled to each other. Therefore, the total electric field is represented as the sum of two propagating guided waves with the amplitude of the function with respect to z and is given as

$$E(x, z) = [R(z)e^{-j\beta_0 z} + S(z)e^{j\beta_0 z}]E_x(x), \tag{5}$$

where $\beta_0 = q\pi/\Lambda$ for the qth order of Bragg reflection, $R(z)$ and $S(z)$ the amplitudes of the forward and backward propagating waves, respectively, and $E_x(x)$ the electric field distribution in the x direction of the guided wave of the waveguide without corrugations. We now substitute Eqs. (4)–(5) into Eq. (3) and neglect the second derivatives $\partial^2 R/\partial z^2$ and $\partial^2 S/\partial z^2$ because the power transfer between the counterrunning waves occurs slowly. As a result, the coupled-wave equations between the forward and backward waves $R(z)$ and $S(z)$ can be obtained as (Kogelnik and Shank, 1972)

$$
\begin{aligned}
-(dR/dz) - [(\alpha_g/2) + j\,\Delta\beta]R &= j\kappa e^{-j\psi}S \\
(dS/dz) - [(\alpha_g/2) + j\,\Delta\beta]S &= j\kappa e^{j\psi}S,
\end{aligned}
\tag{6}
$$

where $\Delta\beta = \beta - \beta_0$ shows the deviation of the propagation constant β from the Bragg wavelength, ψ shows the phase of corrugation at $z = 0$, and κ is the coupling coefficient between the counterrunning waves (Streifer et al., 1975b), and is given by

$$\kappa = \frac{k_0^2}{2\beta N^2} \int_{\text{corrugation}} \Delta n^2(x, z)\, E_x(x)^2 \, dx \tag{7}$$

for the TE mode, where $\Delta n^2(x, z)$ is the pertubation in refractive index produced by corrugations and N a normalized constant given by

$$N^2 = \int_{-\infty}^{\infty} E_x(x)^2 \, dx.$$

If the refractive index perturbation term $\Delta n(z)$ is given as

$$\Delta n \cos[2\pi z/\Lambda] + \psi]. \tag{8}$$

the coupling coefficient κ is easily obtained from Eq. (7) and is expressed by

$$\kappa = (k_0 \,\Delta n)/2. \tag{9}$$

The electric field distribution, $S(z)$ and $R(z)$ in the distributed Bragg reflector (of length L), is obtained by solving the coupled wave equation at the base of the boundary condition. The boundary condition of the DBR waveguide shown in Fig. 6 is

$$S(L)/R(L) = r_0,$$

where r_0 is the amplitude reflectivity at the end of the Bragg reflector. $S(z)$ and $R(z)$ are obtained by

$$S(z) = R(0) \frac{\gamma \cosh[\gamma(z - L)] - \{[(\alpha_g/2) + j\,\Delta\beta] + j\kappa r_0 e^{-j(\psi + 2\beta_0 L)}\} \sinh[\gamma(z - L)]}{\gamma \cosh(\gamma L) + \{[(\alpha_g/2) + j\,\Delta\beta] + j\kappa r_0 e^{-j(\psi + 2\beta_0 L)}\}}, \tag{10}$$

$$R(z) = R(0) \frac{\begin{array}{c}\gamma r_0 e^{-j2\beta_0 L} \cosh(\gamma(z - L)) \\ + \{[(\alpha_g/2) + j\,\Delta\beta]r_0 e^{-j2\beta_0 L} + j\kappa e^{j\psi}\} \sinh[\gamma(z - L)]\end{array}}{\begin{array}{c}\gamma \cosh(\gamma L) \\ + \{[(\alpha_g/2) + j\,\Delta\beta] + j\kappa r_0 e^{-j(\psi + 2\beta_0 L)}\} \sinh(\gamma L)\end{array}} \tag{11}$$

where the dispersion relation is given by

$$\gamma^2 = (\alpha_g/2 + j\,\Delta\beta)^2 + \kappa^2. \tag{12}$$

Therefore the amplitude reflectivity r of the distributed Bragg reflector is expressed as

$$r = S(0)/R(0) = \frac{\gamma r_0 e^{-j(2\beta_0 L + \psi)} \cosh(\gamma L) - \{[(\alpha_g/2 + j\,\Delta\beta]r_0 e^{-j(2\beta_0 L + \psi)} + j\kappa\} \sinh(\gamma L)}{\gamma \cosh(\gamma L) + \{[(\alpha_g/2 + j\,\Delta\beta] + j\kappa r_0 e^{-j(2\beta_0 L + \psi)}\} \sinh(\gamma L)}$$

$$= |r|\, e^{-j\phi}. \tag{13}$$

The amplitude reflectivity r and transmittivity t of the Bragg reflector without the end mirror (that is, $r_0 = 0$) is easily introduced from Eq. (13),

$$r = \frac{-j\kappa \tanh(\gamma L)}{\gamma + [(\alpha_g/2) + j\,\Delta\beta] \tanh(\gamma L)} \tag{14}$$

$$t = \frac{\gamma \exp(-j\beta_0 L)}{\gamma \cosh(\gamma L) + [(\alpha_g/2) + j\,\Delta\beta] \sinh(\gamma L)} \tag{15}$$

Figure 7 shows an example of the coupling coefficient κ for the TE mode calculated from Eq. (7) for three kinds of typical grooves (Streifer et al., 1975b), as a function of the corrugation depth g in which first-order diffraction is assumed. The expression of the coupling coefficient for the TE mode is somewhat different from that for the TM mode (Streifer et al., 1976). However, the value of coupling coefficient for the two modes is almost same, so that polarization-control mechanism such as TE-mode filters in the resonator are necessary for DBR or DFB lasers.

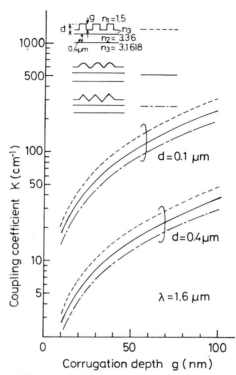

FIG. 7. Coupling coefficient k of three kinds (rectangular, sinusoidal, and triangular) of gratings as a function of the corrugation depth g for different thicknesses d of the InP cladding layer (Streifer *et al.*, 1975b).[From Suematsu *et al.* (1983). Copyright © 1983 IEEE.]

Figure 8 shows the power reflectivity $R = |r|^2$ and the phase shift ϕ as a function of the normalized wavelength deviation $\Delta\beta L$, when we assume that the reflectivity of the end of DBR is zero. At $\Delta\beta = 0$, i.e., the Bragg wavelength, the reflectivity is a maximum and is given by

$$R = \tanh^2(\kappa L), \qquad (16)$$

where $\alpha_g = 0$.

Figure 9 shows the numerated power reflectivity R and power transmittivity $T = |t|^2$ at the Bragg wavelength as a function of κL (Utaka, 1981). The phase shift ϕ of the DBR is approximately expressed as a linearly varying function of $\Delta\beta$ around the Bragg wavelength (see Fig. 8). The proportionality

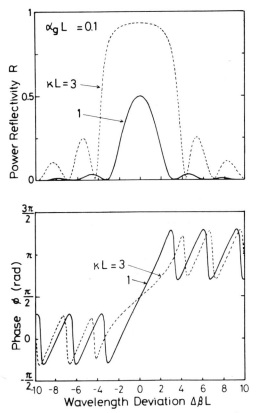

FIG. 8. Power reflectivity R and phase shift ϕ of a distributed Bragg reflector as a function of a wavelength deviation $\Delta\beta L$ from the Bragg wavelength.

constant can be defined as

$$\left.\frac{\partial\phi}{\partial(\Delta\beta)}\right|_{\Delta\beta=0} \equiv 2L_{\text{eff}}, \qquad (17)$$

where L_{eff} is the effective length of the DBR region. Just as the phase shift is obtained by Eq. (13), L_{eff} is given by

$$L_{\text{eff}} = \frac{1}{2}\frac{(\alpha_g/2)L\{[\tanh(\gamma_0 L)]/(\gamma_0 L) - 1/\cosh^2(\gamma_0 L)\} + \tanh^2(\gamma_0 L)}{(\alpha_g/2)\tanh^2(\gamma_0 L) + \gamma_0\tanh(\gamma_0 L)}, \qquad (18)$$

where $\gamma_0 = \sqrt{\kappa^2 + (\alpha_g/2)^2}$. The relationship between the effective length L_{eff} of the DBR region as given by Eq. (18) and the length L of the DBR region is shown in Fig. 10. The effective length is used for the approximate solution of the lasing condition and dynamic properties as described later.

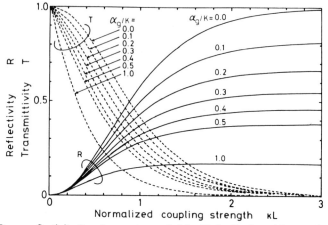

FIG. 9. Power reflectivity R and power transmittivity T of a distributed Bragg reflector at the Bragg wavelength as a function of a normalized coupling strength κL (Utaka, 1981d). [From Suematsu *et al.* (1983). Copyright © 1983 IEEE.]

4. LASING THRESHOLD CONDITIONS

a. DBR Laser

The threshold conditions for the threshold gain g_{th} and lasing wavelength of a DBR laser, shown schematically in Fig. 1, are given by (Utaka, 1981)

$$(\sqrt{C_0})^4 |r_1||r_2| \exp\{\tfrac{1}{2} (\xi_t g_{th} - \alpha)2l\} \exp\{j(-\phi_1 - \phi_2 - 2\bar{\beta}l)\} = 1, \quad (19)$$

$$\alpha = \xi_t \alpha_{ac} + (1 - \xi_t)\alpha_{ex}, \quad (20)$$

where r_i $(i = 1, 2)$ is the reflectivity r of DBR at region i given by Eq. (13), α the energy absorption loss, ξ_t the energy confinement factor of the active waveguide of length l ($C_0 = C_0'$ is assumed), α_{ac} the absorption loss of the active layer, and α_{ex} the absorption loss of the cladding layers (Asada *et al.*, 1981). From the imaginary part of Eq. (19), the phase condition of the laser oscillation is given by

$$\bar{\beta}l + (\phi_1 + \phi_2)/2 = \bar{\beta}l + (L_{eff,1} + L_{eff,2}) \Delta\beta + \pi = m\pi, \quad (21)$$

where m is an integer and $\bar{\beta} = 2\pi\bar{n}_{eq}/\lambda$ is the propagation constant in the active region. By using Eq. (21), the mode spacing $\Delta\lambda_m$ between the neighboring modes is obtained approximately as

$$\Delta\lambda_m = \lambda^2/2[\bar{n}_{eff}l + n_{eff}(L_{eff,1} + L_{eff,2})], \quad (22)$$

where

$$\bar{n}_{eff} = \bar{n}_{eq} - \lambda (\partial\bar{n}_{eq}/\partial\lambda), \qquad n_{eff} = n_{eq} - \lambda (\partial n_{eq}/\partial\lambda),$$

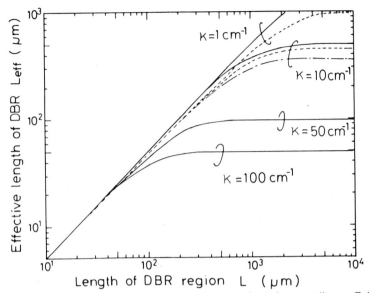

FIG. 10. Calculated effective length L_{eff} of the DBR region for various coupling coefficients κ and loss α_g as a function of the real length of the DBR region ——, $\alpha_g = 1$ cm^{-1}; – – –, $\alpha_g = 5$ cm^{-1}; – · –, $\alpha_g = 10$ cm^{-1} (Koyama *et al.*, 1981b, 1983). [From Suematsu *et al.* (1983). Copyright © 1983 IEEE.]

and $\bar{n}_{eq}$ and $\bar{n}_{eq}$ are the equivalent refractive indexes of the active and DBR regions, respectively. From the real part of Eq. (19), the threshold gain g_{th} is given by

$$g_{th} = (\alpha_{in} + \alpha_{ac} + \frac{1 - \xi_t}{\xi_t} \alpha_{ex}) + \frac{1}{\xi_t l} \ln \frac{1}{|r_1||r_2| C_0^2}. \tag{23}$$

The second term of the right-hand side of Eq. (23) corresponds to the mirror loss per unit length. Figure 11 shows the mirror loss as a function of the normalized wavelength deviation $\Delta\beta L$ calculated on the assumption of zero reflectivity at the end of the DBR. Owing to such wavelength-dependent mirror loss, the dynamic single-mode operation of a DBR laser can be obtained as detailed in Subsection 5. The mirror loss for a DBR laser with a faceted mirror at the end is shown in Fig. 12, where $\theta = 2\beta_0 L + \psi$ denotes the position of the faceted mirror relative to the corrugations. Note that the wavelength characteristics of the mirror loss around the Bragg wavelength are deformed by the faceted mirror and show a drastic change with the relative position of the faceted mirror. The reflectivity of the external faceted mirror should be reduced for stable, single-mode operation.

Assuming that the total carrier lifetime τ_t is inversely proportional to the

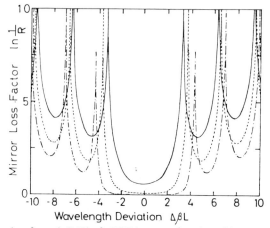

FIG. 11. Mirror loss factor $\ln(1/R)$ of a DBR laser as a function of the normalized wavelength deviation $\Delta\beta L$. $\alpha_g L = 0$. κL is the normalized coupling strength. ——, $\kappa L = 1.0$; – – –, $\kappa L = 2.0$; – · –, $\kappa L = 3.0$ (Utaka, 1981d).

carrier density (Stubkjaer *et al.*, 1981; Asada *et al.*, 1981), the threshold current density J_{th} is written as (Utaka, 1981d)

$$J_{th} = \frac{ed}{A_0 \tau_t} \left(\alpha_{in} + \alpha_{ac} + \frac{1 - \xi_t}{\xi_t} \alpha_{ex} + \frac{1}{\xi_t l} \ln \frac{1}{|r_1||r_2| C_0^2} \right)^2, \qquad (24)$$

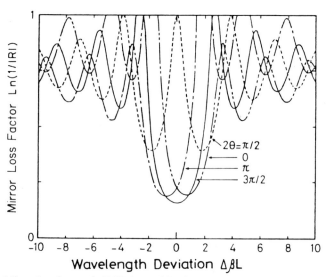

FIG. 12. Mirror loss factor of a DBR laser with a facet mirror at the end as a function of the wavelength deviation. $\kappa L = 1$; $\alpha L = 0.1$; $1r_0 1^2 = 0.32$. [From Kawanishi *et al.* (1979). Copyright © 1983 IEEE.]

where e, d, and τ_t are the electronic charge, the active layer thickness, and the total carrier lifetime, respectively, and A_0 and α_{in} are the proportionality constants of gain and residual internal absorption loss. The one-sided differential quantum efficiency of the DBR laser is expressed approximately as (Utaka *et al.*, 1981b)

$$\eta_d = \frac{1 - R_2}{(1 + \sqrt{R_2/R_1})(1 - C_0^2 \sqrt{R_1 R_2})}$$
$$\times \frac{(1/l) \ln(1/\sqrt{R_1 R_2})}{\xi_t \alpha_{ac} + (1 - \xi_t) \alpha_{ex} + (1/l) \ln(1/C_0^2 \sqrt{R_1 R_2})}. \qquad (25)$$

Figure 13 shows the calculated threshold current density J_{th} and the differential quantum efficiency η_d as a function of the normalized coupling strength $\kappa_2 L$ of the left hand side DBR, in which Bragg wavelength oscillation is assumed. The coupling strength of the right-hand side DBR is set to be two, which means $R_1 = 0.9$ for $\alpha_g = 0$, as shown in Fig. 9. We notice here that a quantum efficiency of greater than 50% can be achieved in such an asymmetric DBR laser.

b. DFB Laser

The threshold conditions of a DFB laser can be obtained by using the coupled-wave equation [Eq. (6)]. Since the corrugated waveguide has a positive gain $\xi_t g$, Eqs. (10), (11), and (13) can be applied to a DFB laser with one reflector at the end by exchanging the loss coefficient α_g for the net gain term $(\alpha - \xi_t g)$. The reflection power $S(0)$ from the corrugated waveguide be-

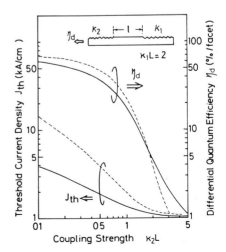

FIG. 13. Calculated threshold current density J_{th} and differential quantum efficiency η_d of a DBR laser as a function of the normalized coupling strength $\kappa_2 L$ of the left-hand-side DBR. ——, $l = 300\ \mu m$; – – –, $l = 100\ \mu m$.

comes infinite, as understood from Eq. (10), when the condition of

$$\{[(\xi_t g_{th} - \alpha)/2] - j[\Delta\beta + \kappa r_0 e^{-j(\phi+2\beta_0 L)}]\} \sinh(\gamma L) = \gamma \cosh(\gamma L) \quad (26)$$

is satisfied, where

$$\gamma = \{[(\xi_t g_{th} - \alpha)/2] - j\,\Delta\beta\}^2 + \kappa^2. \quad (27)$$

In this condition we can get a finite light output $S(0)$ from the corrugated waveguide without regard to the amount of the incident light field $R(0)$, and this defines the lasing oscillation. The corrugated waveguide, therefore, becomes an asymmetric DFB laser with a faceted reflector at one end, and the threshold condition is given by Eq. (26). By solving Eq. (26), deviations of both the threshold gain and the wavelength of the axial modes around the Bragg wavelength caused by the position of the faceted mirror on one period of the corrugation can be obtained (Streifer $et\ al.$, 1975a). Figure 14 shows examples of the relationship between the threshold gain and the wavelength of axial modes for two phase positions of the faceted mirror, $\pi/2$ and $3\pi/2$. At the special phase condition of $3\pi/2$ there are two axial modes with the same threshold gain that cause two-mode operation of the DFB laser with the faceted reflector. We must note here that the difference in threshold gains among the neighboring resonant modes is very sensitive to changes in the position of the faceted mirror. It may introduce deterioration of the yield of DFB lasers. Such an effect of the faceted reflector on axial-mode selectivity will be discussed in the next subsection.

The threshold gain of a DFB laser without the faceted reflector is obtained from Eq. (26) assuming relatively small gain $[(\xi_t g_{th} - \alpha)/\xi t \ll 1]$

$$g_{th} = [\pi^2/(\kappa^2 L^3) + \alpha]/\xi_t. \quad (28)$$

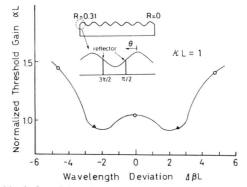

FIG. 14. Threshold gain for axial modes around the Bragg wavelength in an asymmetric DFB laser. θ is the position of the facet mirror relative to the corrugations. $\bigcirc, \theta = \frac{\pi}{2}$; $\blacktriangle, \theta = \frac{3\pi}{2}$. $\kappa L = 1$ (Streifer $et\ al.$, 1975a).

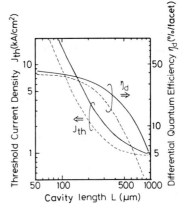

FIG. 15. Calculated threshold current density J_{th} and differential quantum efficiency η_d of a symmetric DFB laser as a function of cavity length L. ——, $\kappa = 50$ cm^{-1}; – – –, $\kappa = 100$ cm^{-1}.

The threshold current density is calculated in the same manner as Eq. (24)

$$J_{th} = \frac{ed}{A_0 \tau_t} \left(\alpha_{in} + \alpha_{ac} + \frac{1 - \xi_t}{\xi_t} \alpha_{ex} + \frac{\pi^2}{\xi_t \kappa^2 L^3} \right)^2. \qquad (29)$$

The mirror loss of a DFB laser with first-order grating can be defined as

$$\alpha_m = \xi_t g_{th} - \alpha. \qquad (30)$$

By using Eq. (30), the differential quantum efficiency of the DFB laser as well as a conventional Fabry–Perot laser is expressed as

$$\eta_d = \frac{1}{2} \frac{\alpha_m}{\alpha + \alpha_m}. \qquad (31)$$

Figure 15 shows the calculated threshold current density and the differential quantum efficiency of a DFB laser as a function of cavity length. The differential quantum efficiency can be increased by decreasing the cavity length just as in a conventional laser.

5. DYNAMIC SINGLE-MODE OPERATING CONDITIONS

In order to clarify the conditions of single-mode operation under rapid direct modulation, a simplified analysis evaluating the intensity of the sub-mode is carried out. The multimode rate equations are used for this purpose. These equations include the effect of lateral carrier diffusion (Furuya *et al.*, 1978) and are written as follows

$$\frac{dS_i}{dt} = \int F_i^2 g(N - N_g) S_i \, dv - \frac{S_i}{\tau_{pi}} + \frac{C_i}{\tau_s} \int F_i^2 N \, dv, \qquad (32)$$

$$\frac{\partial N}{\partial t} = \frac{I}{eV} - \sum_i V F_i^2 g(N - N_g) S_i - \frac{N}{\tau_s} + D \frac{\partial^2 N}{\partial x^2}, \qquad (33)$$

where S_i is the averaged photon density of mode i, N the injected carrier density, τ_{pi} the photon life time of mode i, τ_s the carrier lifetime, e the electron charge, V the volume of the active layer, C_i the spontaneous emission factor of mode i, $D = L_D^2/\tau_s$ the diffusion constant with the diffusion length L_D, and F_i is the field distribution of the light intensity of mode i.

A laser cavity consisting of the rectangular waveguide structure as shown in Fig. 16 is assumed for the theoretical analysis. The field distribution $F(x, y, z)$ is extended from the active DBR region into the passive region, as shown in Fig. 16. Therefore, the optical confinement factor that is a part of light energy confined in the active region is different from that in the conventional semiconductor laser. The optical confinement factor ξ of a DBR laser is defined as

$$\xi \equiv \int_{-\overline{W}/2}^{\overline{W}/2} \int_{-d/2}^{d/2} \int_{-l/2}^{l/2} F_i(x, y, z)^2 \, dx \, dy \, dz, \tag{34}$$

where $F_i(x, y, z) = F_t(x, y) \cdot F_1(z)$, $F_t(x, y)$ is the transverse field distribution, and $F_1(z)$ the longitudinal field distribution. The transverse field distribution $F_t(x, y)$ is determined by the passive structure of the waveguide. Assuming that the reflectivity of DBR is nearly equal to unity, $F_1(z)$ is given by

$$F_1(z) = \begin{cases} 2F_{10} \cos[\bar{\beta}(z - z_0)] & |z| < l/2, \\ F_{10}\{R(z) \exp[-j\beta_0(z - z_0)] + S(z) \exp[j\beta_0(z - z_0)]\} \\ & (l/2) \leq |z| < L_i + (l/2), \end{cases} \tag{35}$$

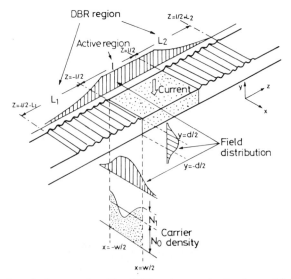

FIG. 16. Rectangular laser cavity with a DBR region, carrier density, and field distribution.

where F_{10} is the normalized constant, z_0 indicates the axial boundary between the active region and the passive DBR region, and $R(z)$ and $S(z)$ are the field amplitudes derived by the coupled-mode equation (Kogelnik and Shank, 1972). The optical confinement factor ξ is expressed as the product of transverse confinement factor ξ_t (which is equal to the conventional confinement factor) and the longitudinal confinement factor ξ_1, i.e., $\xi = \xi_t \xi_1$, where ξ_t and ξ_1 are given by

$$\xi_t = \int_{-\overline{W}/2}^{\overline{W}/2} \int_{-d/2}^{d/2} F_t(x, y)^2 \, dx \, dy, \tag{36}$$

$$\xi_1 = \int_{-l/2}^{l/2} F_1(z)^2 \, dz = l/(l + L_{\text{eff},1} + L_{\text{eff},2}). \tag{37}$$

The integral in Eq. (32) covers the entire volume of the active layer. In the present analysis, the gain of each mode at the active layer is assumed to be the same and is expressed as $g(N - N_g)$ (Asada et al., 1981), where g is the gain coefficient and N_g the carrier density above which the positive gain appears. This assumption is reasonable for a laser with a mirror whose loss difference for each mode is much larger than the gain difference, such as a DBR laser. Equations (32)–(33) are complicated, and we simplify them by employing a reasonable assumption that the intensities of the submodes are fairly small compared to the main mode so that the rate equation of main mode is not influenced by these submodes. Under such a condition, Eq. (33) is approximated by

$$\frac{\partial N}{\partial t} = \frac{I}{eV} - VF_0^2 g(N - N_g) S_0 - \frac{N}{\tau_s} + D \frac{\partial^2 N}{\partial x^2}, \tag{38}$$

where S_0 is the photon density of the main mode ($i = 0$). Here, we consider the laser diode to be modulated by a sinusoidal current superimposed on the dc bias current as follows

$$I = I_b + I_m \sin(\omega t), \tag{39}$$

where I_m is the modulation current amplitude and I_b the bias current. Following Kishino et al. (1982) and Furuya et al. (1978), we assume that only the fundamental transverse mode exists and the carrier is completely confined within the rectangular waveguide. We also assume that the carrier distributes in the x direction as shown in Fig. 16. The space-averaged carrier density N_{eff} is easily obtained from Eqs. (32) and (38) under the assumption of small signal as follows (Kishino et al., 1982)

$$N_{\text{eff}} = \int F_0^2 N \, dv = \xi N_{\text{tho}} + \Delta N_{\text{eff}}, \tag{40}$$

where

$$N_{tho} = 1/(\xi g \tau_{p0}) + N_g$$

$$\Delta N_{eff} = \frac{\xi \sqrt{1 + (h/\omega\tau_s)^2}\ m(I_b - I_{th})}{e\omega_{r0}V}$$

$$\times \left([(\omega_{r0}/\omega)^2 - 1]^2 (h/\omega_{r0}\tau_s)^2 \right.$$

$$+ \left\{ (\omega_{r0}/\omega) \left[1 + \frac{\eta}{\xi} + h\frac{(\omega_{r0}^2\tau_s\tau_p + 1)}{(\omega_{r0}\tau_s)^2} \right] \right.$$

$$\left. \left. - (\omega/\omega_{r0}) \right\}^2 \right)^{-1/2} \sin(\omega t + \theta), \tag{41}$$

where ω_{r0} is the resonancelike angular frequency without taking into account the effect of lateral carrier distribution and is expressed as (Kishino *et al.*, 1982)

$$\omega_{r0} = (I_b/I_{th} - 1)/[\tau_s\tau_p(1 - N_g/N_{tho})], \tag{42}$$

where I_{th} is the threshold current and $N_g/N_{tho} = u$ is usually about 0.6 (Asada *et al.*, 1981). The modulation depth m is defined as $m = I_m/(I_b - I_{th})$. The parameter is η, and $h = (2\pi L_D/W)^2$ is the square of diffusion length L_D relative to stripe width W. Since the intensity of the nonlasing submodes ($i = 0$) is maximum at the resonancelike frequency, we wish to discuss the magnitude of the submode intensity at the resonancelike frequency. We assume that the stripe width W is equal to the carrier diffusion length L_D which gives an optimum width at which the resonancelike peak of modulation is minimized (Furuya *et al.*, 1978). The carrier density variation at the resonancelike frequency $\Delta N_{eff,max}$ is a maximum and is given from Eq. (41) by

$$\Delta N_{eff,max} = \frac{mB\omega_{r0}}{2g} \sin(\omega_{r0} + \theta)$$

$$= \frac{\xi\sqrt{1 - u}\sqrt{\tau_s\tau_p}mBI_{th}}{2eV}\sqrt{(I_b/I_{th} - 1)}\sin(\omega_{r0} + \theta), \tag{43}$$

where $B \simeq 8$ and is a constant dependent on the stripe width. By substituting Eqs. (40) and (43) into Eq. (32), the rate equation of the photon density for submode S_1 is

$$\frac{dS_1}{dt} = \left[\frac{mB\omega_{r0}}{2}\sin(\omega_{r0}t + \theta) - \Delta\left(\frac{1}{\tau_p}\right) \right] S_1 + \frac{C_1\xi}{\tau_p}N_{tho}, \tag{44}$$

where $\Delta(1/\tau_p) = (1/\tau_{p1}) - (1/\tau_{p0})$ is the mirror loss difference between the submode and the main lasing mode. From Eq. (44), the ratio of the time-averaged photon density of the nonlasing submode to that of the main lasing mode in a stationary condition is obtained as

$$\frac{\overline{S}_1}{\overline{S}_0} = \frac{\omega_{r0} C_1 \xi}{2\pi \tau_{p0}(I_b/I_{th} - 1)}$$

$$\times \int_{t_1}^{t_1 + 2\pi/\omega_{r0}} \int_0^t \exp\left\{\frac{mB}{2}[\cos(\omega_{r0}u + \theta)\right.$$

$$\left. - \cos(\omega_{r0}t + \theta)] - \Delta\left(\frac{1}{\tau_p}\right)(t - u)\right\} du\, dt, \qquad (45)$$

where $t_1 \gg 2\pi/\omega_{r0}$. Figure 17 shows the relative intensity $\overline{S}_1/\overline{S}_0$ of the submode to the main mode under direct modulation at the resonancelike frequency as a function of mirror loss difference $\Delta\alpha_m = \Delta(1/\tau_p)/v$ between the submode and the main mode. A large signal analysis of the carrier density N at $m = 100\%$ shows that the carrier density analyzed from the small signal analysis deviates from the result of the large signal analysis by 30% at the resonancelike frequency. Therefore, the small signal analysis given here is a good measure even for $m = 100\%$.

We can now define the DSM condition that the time-averaged intensity of the nearest neigboring submode $\overline{S}_1$ be q times less than the intensity of the

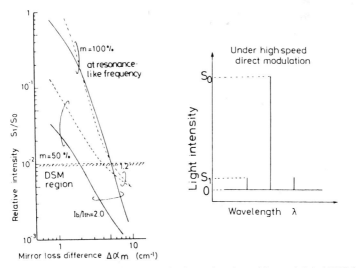

FIG. 17. Normalized submode photon relative intensity of a rapidly modulated DBR laser as a function of mirror loss difference at the resonancelike frequency (Koyama *et al.*, 1981b, 1983), where the spontaneous emission factor $C_1 \xi$ is 5×10^{-5}.

main mode $\bar{S}_0$, i.e., $\bar{S}_1/\bar{S}_0 \le q$. We could define q as, say, 1%. A mirror-loss difference of several reciprocal centimeters is required to maintain the relative intensity of submode to main mode at less than 1%, as shown in Fig. 17.

6. AXIAL-MODE SELECTIVITY

To maintain the DSM operation, the mirror loss of the neighboring submode should be larger than main mode by a certain amount as shown earlier. In this section, the mirror-loss difference between axial modes for both DBR and DFB lasers is discussed.

The mirror-loss difference $\Delta\alpha_m$ between the main mode and submode per unit length in a DBR laser with wavelengths λ_0 (main mode) and λ_1 (submode) is given by

$$\Delta\alpha_m = 1/(l + L_{eff,1} + L_{eff,2}) \ln[R(\lambda_0)/R(\lambda_1)], \qquad (46)$$

where $R(\lambda)$ is the power reflectivity at the wavelength λ. Figure 18 shows the mirror-loss difference between the main mode and the neighboring submode for a symmetric DBR laser in the axial direction, assuming that the main mode coincides with the Bragg wavelength. The mirror-loss difference depends on the coupling coefficient κ and the active region length l. The

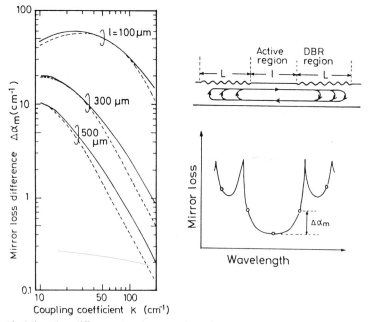

FIG. 18. Mirror loss difference $\Delta\alpha_m$ between the main mode and the neighboring submode in a symmetric DBR laser as a function of coupling coefficient κ. ——, $\kappa l = 1$; – – –, $\kappa l = 2$.

value of κ is limited in order to obtain a greater amount of mirror-loss difference for the DSM operation. It is effective for DSM operation to reduce the active region length. The wavelength deviation of the main mode from the Bragg wavelength reduces the mirror-loss difference and causes the unstable operation of the DBR laser. In order to solve this problem, a DBR laser was fabricated with the tuning mechanism that adjusts the ratio of the lasing wavelength to the Bragg wavelength. This will be discussed further in Section III.

In a DFB laser without a faceted reflector, there are two axial modes with the same threshold gain that are equidistant from the Bragg wavelength (Kogelnik and Shank, 1972). This indicates two-mode operation and is considered to be undesirable for DSM operation. On the other hand, the addition of a faceted mirror to one side gives an asymmetric spectrum with respect to the Bragg wavelength and results in a single-mode lasing condition (Streifer *et al.*, 1975a). However, the axial-mode selectivity depends on the position of the faceted mirror relative to the corrugations.

Figure 19 shows the mirror-loss difference between the main mode and the

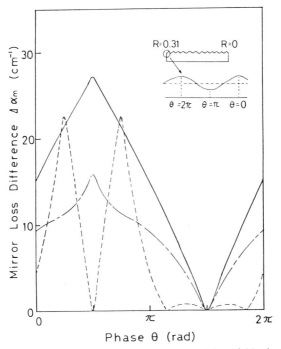

FIG. 19. Mirror loss difference between the main mode and the neighboring submode in an axially asymmetric DFB laser as a function of the phase ϕ. $L = 300 \ \mu m$. ——, $\kappa L = 0.5$; – · –, $\kappa L = 1.0$; – – –, $\kappa L = 2.0$ (Streifer *et al.*, 1975a).

neighboring submode of such an axially asymmetric DFB laser as a fuction of the phase θ, which denotes the position of the faceted mirror relative to the corrugations. As can be seen, the axial-mode selectivity is sensitive to the position of the mirror, and in a certain case ($\theta = 3\pi/2$), there are two axial modes with the same threshold gain. This is also understood from Fig. 14.

7. TEMPERATURE DEPENDENCE OF LASING WAVELENGTH OF DBR LASERS

In conventional, single-mode lasers such as BH lasers, the lasing mode jumps from one axial mode to another by increasing the temperature only a few degrees, since the temperature coefficient of the gain peak wavelength is much larger than that of the lasing wavelength. In contrast, DBR and DFB lasers operate without mode jumps over much larger temperature ranges.

The temperature coefficient of the lasing wavelength of a DBR laser can be obtained by differentiation of Eq. (21) (Arai and Suematsu, 1981),

$$\frac{d\lambda}{dT} = \lambda \frac{l\left[(\partial\bar{n}_{eq0}/\partial T) + \xi_t\,(\partial\bar{n}/\partial N)\,(dN_{th}/dT)\right] + (L_{eff,1} + L_{eff,2})\,(\partial n_{eq}/\partial T)}{l\bar{n}_{eff} + (L_{eff,1} + L_{eff,2})n_{eff}}, \tag{47}$$

where n_{eq0} is the equivalent refractive index of the active region without carrier injection, $\bar{n}$ the refractive index of the active layer, and N_{th} the threshold carrier density. On the other hand, the temperature coefficient of the Bragg wavelength is given by

$$\frac{d\lambda_B}{dT} = \frac{\lambda_B}{n_{eff}}\frac{\partial n_{eq}}{\partial T}. \tag{48}$$

From the experimental results of the temperature dependence of the lasing wavelength of DBR lasers, $\partial n_{eq}/\partial T$ was estimated to be approximately 3×10^{-4} deg^{-1}, and $\partial\bar{n}/\partial N$ was estimated to be $-7-12 \times 10^{-21}$ cm^3 from the measurement of a 1.6-μm wavelength BH laser (Kishino et al., 1982; Stubkjaer et al., 1981). The threshold carrier density variation due to temperature variation is estimated to be approximately 3×10^{16} cm^{-3} deg^{-1}, taking account of the temperature dependence of the threshold current density and the carrier lifetime (Stubkjaer et al., 1981; Asada et al., 1981). Thus, the effect of the injection carrier on the temperature coefficient of the equivalent refractive index amounts to a few tenths of $\partial n_{eq0}/\partial T$ whereas the equivalent refractive index is less than 1%.

When dN_{th}/dT is positive, $d\lambda/dT$ is always smaller than $d\lambda_B/dT$ and the lasing mode moves to the shorter-wavelength side in relation to the Bragg wavelength with increasing temperature. The DBR lasing mode then jumps to the next DBR mode when the wavelength shifts to the shorter wavelength side of λ_B by one half of the mode spacing $\Delta\lambda_m/2$. A schematic diagram of the

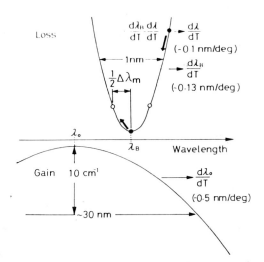

FIG. 20. Schematic diagrams of the mirror loss factor of a DBR, a gain profile, and movements of the lasing mode, Bragg wavelength λ_B and peak gain wavelength λ_0. [From Suematsu *et al.* (1983). Copyright © 1983 IEEE.]

movements of the lasing mode, Bragg wavelength, and peak gain wavelength is shown in Fig. 20.

By using Eqs. (22), (24), (27), (47), and (48), the temperature dependences of the lasing wavelength and the threshold current density of a DBR laser were calculated. The calculated temperature range T_s necessary to maintain a fixed single mode at ambient temperature is shown in Fig. 21 (Arai and Suematsu, 1981). The value of T_s increases when both the optical confinement factor ξ of the active layer and active region length l are reduced, and when the normalized coupling strength L of DBR region is increased. If the device structure is designed properly, a value of T_s of approximately 100 degrees may be realized in DBR lasers. It is, however, understood that the temperature range for fixed, single-mode operation will be larger in DFB lasers, though the flexibility of the device design is larger in DBR lasers.

8. DYNAMIC WAVELENGTH SHIFT

The lasing spectrum of a conventional Fabry–Perot, single-mode, semiconductor laser is broadened under the rapid direct modulation (Ikegami, 1975). The lasing mode of these lasers sometimes hops from one mode to another, even if it operates in single mode. Thus, mode partition noise, which limits the transmission bit rate, arises in fiber transmission (Nakagawa *et al.*, 1979). Dynamic spectral broadening as wide as 10 nm was observed in a conventional 1.5–1.6-μm wavelength GaInAsP/InP-type laser (Kishino *et al.*, 1982). Whereas dynamic single-mode operation is attained

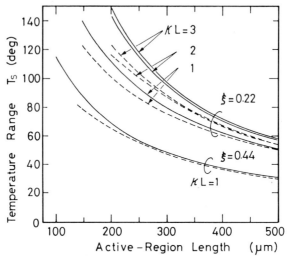

FIG. 21. Calculated temperature range T_s for fixed, single-mode operation of a symmetrical DBR laser as a function of the active-region length. The initial condition of $\lambda = \lambda_B = \lambda_0 = 1.58\ \mu m$ at $T = 60°C$ was used. ——, $\kappa = 100\ cm^{-1}$; – – –, $\kappa = 50\ cm^{-1}$ (Arai and Suematsu, 1981). [From Suematsu et al. (1983). Copyright © 1983 IEEE.]

in DSM lasers, the lasing wavelength shifts periodically within a period of modulation owing to variations of the refractive index that is originated by oscillation of the carrier density. This phenomenon is known as the dynamic wavelength shift (Koyama et al., 1981a; Kishino et al., 1982). Therefore, the dynamic spectral broadening of a DSM laser is equivalent with the dynamic wavelength shift (Koyama et al., 1981a).

The dynamic wavelength shift $\Delta\lambda(t)$ caused by the refractive index change $\Delta N(x, t)$ is given as (Koyama and Suematsu, 1983)

$$\Delta\lambda(t) = \xi_l\,\xi_t\,(\lambda/n_e)\,(\partial\bar{n}/\partial N)\,\Delta N(t),$$
$$n_e = \xi_l\,\bar{n}_{eff} + (1 - \xi_l)\,n_{eff},$$

(49)

where $\Delta N(t)$ is the carrier-density variation (Kishino et al., 1982) and $\partial\bar{n}/\partial N$ shows the coefficient of carrier-induced refractive index change. The carrier-density variation becomes maximum when the modulation frequency is at the resonancelike frequency (Kishino et al., 1982). From Eqs. (43) and (49), the maximum dynamic wavelength shift appearing at the resonancelike frequency is (Koyama and Suematsu, 1984)

$$\Delta\lambda_{s,max} = (\lambda B/n_e\,\sqrt{geV})\,(\partial\bar{n}/\partial N)\,m\,\sqrt{\xi_t\xi_l}\,(I_b - I_{th}),$$

(50)

where the stripe width of the active region is assumed to be equal to the carrier diffusion length (in that case $B = 8$), g is the gain coefficient, e the

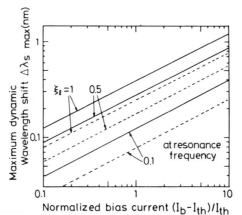

FIG. 22. Maximum dynamic wavelength shift of a DBR laser as a function of the normalized bias current where ξ_l is the longitudinal optical confinement factor and ξ_t the transverse confinement factor. Theoretical parameters are $g = 1.2 \times 10^{-6}$ cm^3/sec and $I_{th}/V = 10$kA/cm^2/μm. Modulation depth = 100%. ——, $\xi_t = 0.5$; – – –, $\xi_t = 0.2$ (Koyama *et al.*, 1981b, 1983). [From Suematsu *et al.* (1983). Copyright © 1983 IEEE.]

electronic charge, V the volume of the active medium, I_b the bias current, I_{th} the threshold current, and m the modulation depth. Figure 22 shows the maximum dynamic wavelength shift of DBR lasers as a function of the normalized bias current $(I_b - I_{th})/I_{th}$ and the longitudinal light confinement factor ξ, where $\partial \bar{n}/\partial N$ is assumed to be -7×10^{21} cm^3, and the transverse confinement factor ξ_t is assumed to be 0.5 and 0.2, as indicated in the figure by solid and dashed lines, respectively (Koyama and Suematsu, 1984).

III. Fabrication and Lasing Properties of DSM Lasers

9. FABRICATION

Figure 23 shows typical schematic structures of reported GaInAsP/InP-type BH–DBR and BH–DFB lasers; a phase coupling and improved direct coupling methods were used for BH–DBR–ITG (Kobayashi *et al.*, 1981; Utaka *et al.*, 1981a; Koyama *et al.*, 1981a; Tanbun-ek *et al.*, 1981), and BH–DBR–BJB (Abe *et al.*, 1981, 1982a), structures. Buried heterostructures were used to reduce the threshold current as well as to maintain transverse single-mode operation. The first-order gratings are formed on a thin InP layer just as on a low-loss output waveguide of GaInAsP-type crystal with a corresponding bandgap wavelength of 1.4 μm. To activate the TM mode filters, Cr and Au were evaporated on a thin SiO$_2$ film in the DBR region. Figure 23c shows a BH–DFB laser where one facet was cleaved and the one on the opposite side was sawed obliquely in order to suppress the

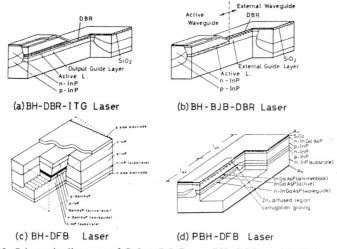

(a) BH-DBR-ITG Laser

(b) BH-BJB-DBR Laser

(c) BH-DFB Laser

(d) PBH-DFB Laser

FIG. 23. Schematic diagrams of GaInAsP/InP-type BH–DBR, and DFB lasers with the emission wavelength of 1.5–1.6 μm. (a) BH–ITG–DBR laser (Kobayashi *et al.*, 1981), (b) BH–BJB–DBR laser (Abe *et al.*, 1982a), (c) BH–DFB laser (Matsuoka *et al.*, 1982), and (d) PBH–DFB laser (Utaka *et al.*, 1981c). [From Suematsu *et al.* (1983). Copyright © 1983 IEEE.]

Fabry–Perot mode oscillation (Matsuoka *et al.*, 1982). Figure 23d shows a planer BH–DFB laser where the Fabry–Perot mode was suppressed by the sufficiently long unpumped region (Utaka *et al.*, 1981c; Akiba *et al.*, 1982). The pitch of the grooves of the first-order grating of 1.5–1.6-μm wavelength GaInAsP/InP-type lasers is 240–250 nm.

From Fig. 7, we can estimate the coupling coefficient κ of the first-order grating in relation to the groove depth. It is assumed that the refractive indexes of the DBR waveguide layers are 3.36 for the core and 3.1618 for the cladding, and the core layer is 0.4 μm thick. Then a grating with a depth of 50 nm is formed on the waveguide, the coupling coefficient of the sinusoidal grating is estimated to be 100 cm^{-1}.

Fine gratings were fabricated by employing holographic lithography and a chemical etching technique. First, a sample wafer was coated with AZ-1350 photoresist that had been diluted 2:1 (v/v) with AZ-thinner. The coating was applied at a spin velocity of 8000 rpm for 15 sec. This resulted in a coating thickness of approximately 100 nm. The wafer was then prebaked at 90°C for 15 min and the holographic exposure using the 325-nm line of a He–Cd laser with 7-mW output power was carried out. The exposure time was typically 5 sec for an exposure area of 1 × 2 cm. After developing it in AZ-developing for 5 sec, it was postbaked at 100°C for 5 min and then the grating was transcribed on the wafer by KKI etchant (HCl:CH$_3$COOH:H$_2$O$_2$ = 1:1:1) at 30°C for 3–5 sec (Iga *et al.*, 1979).

This etchant is suitable for both InP- and GaInAsP-type crystals. Uniform and bridgeless gratings for both crystals were obtained. An example of the cross-sectional SEM photograph of the grating transcribed on InP is shown in Fig. 24. As can be seen, smooth and sinusoidal gratings were obtained with a pitch and depth of 246.5 and 80 nm, respectively. Other etching techniques such as a dry etching and an ion etching will be preferable for rectangular gratings (Hu and Howard, 1980; He *et al.*, 1981). The Bragg wavelength is determined not only by the pitch of the grooves but also by the equivalence refractive index as determined by the material composition and geometric structure of waveguide. The coupling coefficient of the DBR shown in Fig. 24 is calculated to be about 80 cm^{-1} from Fig. 7. The effective length L_{eff} of the DBR from Fig. 10 is about 70 μm, assuming a loss coefficient $\alpha_g = 10$ cm^{-1}. The loss coefficient of the passive region of a BJB laser wihout the grooves of a DBR was measured by observing the threshold current with different positions of the cleaved facet along the passive region. From such measurements, the loss of the passive region was estimated to be less than

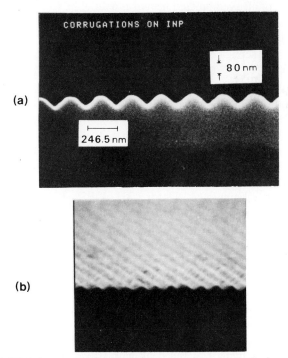

FIG. 24. (a) SEM photograph of the first-order corrugation transcribed on an InP substrate (b) with a pitch and depth of 246.5 and 80 nm, respectively. [From Tanbun-ek *et al.* (1984). Copyright © 1984 IEEE.]

10 cm^{-1} (Abe *et al.*, 1982b). The reflectivity of such a DBR is estimated from Fig. 9 to be 0.8.

10. STATIC LASING PROPERTIES

A low threshold current of less than 100 mA was obtained for the BH–DBR–ITG, BH–DBR–BJB, and BH–DFB lasers (Tanbun-ek *et al.*, 1981; Matsuoka *et al.*, 1982; Akiba *et al.*, 1982), emitting in the 1.5–1.6-μm region. Figure 25 shows the typical current and light output characteristics and lasing spectra at various injection current levels for BH–DBR–BJB (Tanbun-ek *et al.*, 1984), PBH–DFB (Akiba *et al.*, 1982), BH–DFB (Matsuoka *et al.*, 1982), and DC–PBH–DFB lasers (Kobayashi, 1982; Kitamura, 1983). A threshold current of 37 mA and a differential quantum efficiency of 16.3% were obtained at a temperature of 248 K for a BH–DBR–ITG laser with a stripe width of 3 μm and an active laser thickness of 0.2 μm (Tanbun-ek *et al.*, 1981). The reflectivity of the DBR was estimated using Eq. (25) to be 0.15 with $\alpha = 38$ cm^{-1} (Asada *et al.*, 1981), $l = 350$ μm, $R_2 = 0.42$, and $\eta_d = 0.163$. A larger effective reflectivity of the DBR compared with that of cleaved faceted mirror is available. In a DC–PBH–DFB laser at a wavelength of 1.3 μm, an output power of more than 40 mW was obtained as shown in Fig. 25d (Kobayashi *et al.*, 1982; Kitamura *et al.*, 1983). As can be seen, completely stable single-mode operation with an injection level up to 1.7 times the threshold current and with an output power level up to 3 mW was observed. The saturationlike characteristics as seen in I–L curves in Fig. 25a–b at high injection current levels are considered to be caused by a junction temperature rise and current leakage through the blocking layer of the BH structure, since the lasing-mode jump was not observed at any injection current level but rather a slight shift of lasing wavelength toward longer side. This wavelength shift will be relaxed by reducing the threshold current and the thermal resistance. An external differential quantum efficiency of 14% for a BH–DBR–BJB laser was obtained at the temperature of 20°C. It will be increased significantly by device optimization such as the improvement of coupling efficiency C_0 and the reduction of leakage current. A low threshold current of $I_{th} = 50$ mA was reported for a DFB laser at room temperature cw (Matsuoka *et al.*, 1982; Akiba *et al.*, 1982). The life test has been carried out for a few thousand hours without any degradation (Matsuoka *et al.*, 1982; Akiba *et al.*, 1982).

Another intrinsic characteristic in these DSM lasers is the small temperature coefficient of the wavelength shift. In conventional Fabry–Perot-type lasers, the temperature coefficient of the lasing wavelength is dominated by the temperature coefficient of the peak gain wavelength $d\lambda_0/dT$, which is measured to be approximately 0.5 nm/deg for 1.5–1.6-μm wavelength GaInAsP/InP-type double heterostructure lasers (Li *et al.*, 1980). However,

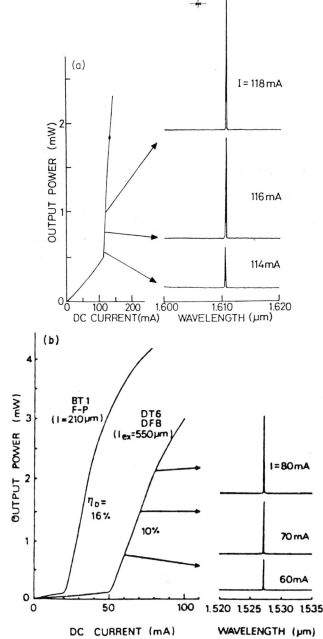

FIG. 25. Injection-current and light-output characeristics of various lasers. (a) BH–DBR–
BJB laser; mount B–74; CW at 18°C, I_{th} = 110 mA.[From Tanbun-ek *et al.* (1984). Copyright
© 1984 IEEE.] (b) PBH–DFB laser; CW at 22°C and I_{th} = 50 mA.[From Akiba *et al.* (1982).
Copyright © 1982 IEEE.] (c) BH–DFB laser (after Matsuoka *et al.,* 1982), and (d) DC–PBH–
DFB laser; η_d = 26% (after Kobayashi, 1982; Kitamura *et al.,* 1983). (*Continued*)

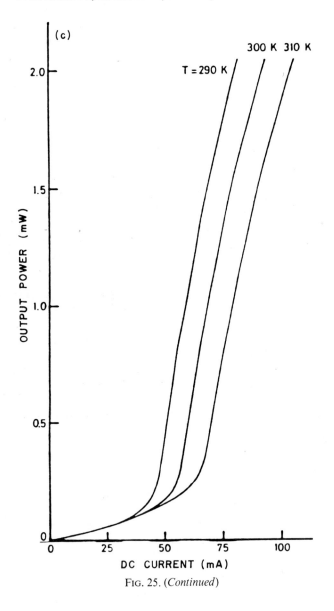

FIG. 25. (*Continued*)

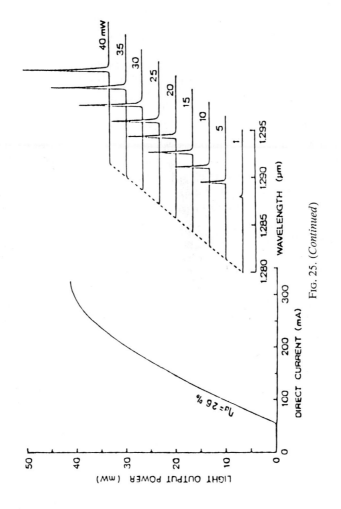

Fig. 25. (*Continued*)

in DBR and DFB lasers, it depends mainly on the temperature dependence of the reactive index and is approximately 0.1 nm/deg. Figure 26a shows the temperature dependence of lasing wavelength for typical BH – DBR – BJB lasers. The temperature range for a fixed-mode oscillation of a BH – DBR – BJB laser is approximately 50 – 65 deg (Abe *et al.,* 1982a). A large wavelength change at T ≃ 240 K, as seen in Fig. 26a (Tanbun-ek *et al.,* 1984), was caused by a transverse mode change, since the TE-mode oscillation was confirmed in this temperature range. As for the DFB laser, the temperature range was measured to be more than 100 deg, as shown in Fig. 26b (Sakai *et al.,* 1982). The temperature dependence of the threshold current of a DFB laser is different from that of a Fabry – Perot laser as shown in Fig. 26c (Akiba *et al.,* 1982).

An example of a spontaneous emission spectrum of a BH – BJB – DBR laser is shown in Fig. 27 (Tanbun-ek *et al.,* 1984). In contrast to that of the conventional laser, the one round-trip gain of the neighboring modes is very small compared to the main mode with its large emission intensity. On the other hand, the spontaneous emission spectrum with a stop band, in which no resonance is allowed, was observed in a DFB laser as shown in Fig. 28 (Itaya *et al.,* 1984).

A schematic diagram of a 1.5-μm GaInAsP/InP-type BH – BJB – DBR laser with a monolithically integrated wavelength-tuning region is shown in Fig. 29 (Tohmori *et al.,* 1983). In order to tune the lasing wavelength to the Bragg wavelength, the optical path length between the distributed reflector and the cleaved mirror is controlled by changing the refractive index of the tuning region. The refractive index change in the tuning region is attained by use of the plasma effect of the injected carrier.

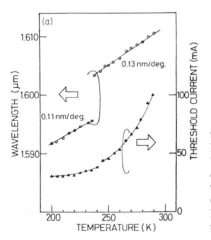

FIG. 26. Temperature dependence on (a) the threshold current and lasing wavelength of a BH – DBR – BJB laser; *CW* operation; mount B – 74; $l = 300$ μm; $L = 500$ μm [From Tanbun-ek *et al.* (1984). Copyright © 1984 IEEE.], (b) the wavelength [From Sakai *et al.* (1982). Copyright © 1982 IEEE.], and (c) the threshold current of a PBH – DFB laser in comparison that of a conventional BH laser [From Akiba *et al.* (1982).]

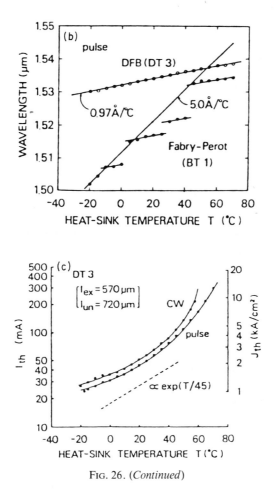

FIG. 26. (*Continued*)

Figure 30 shows the measured wavelength shift as a function of the injected tuning current (Tohmori *et al.*, 1983). The lasing wavelength was shifted by 4 Å with a tuning current of 4.1 mA. Such techniques of fine-wavelength tuning would be applied to improve the temperature behavior of a DSM laser, to increase the degree of the wavelength division multiplexing (WDM) transmission, and to obtain precise optical measurement.

11. DYNAMIC LASING PROPERTIES

The typical spectral properties of a conventional single-mode laser are shown in Fig. 31. Single-mode operation is observed in under dc operating conditions, but mode hops from one mode to another are observed with a

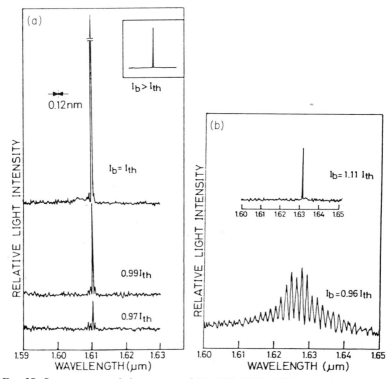

FIG. 27. Spontaneous emission spectra of (a) a BH–DBR–BJB laser below the threshold current (mount B–74; CW operation; I_{th} = 106 mA) and (b) a conventional BH laser fabricated from the same wafer. [After Tanbun-ek *et al.* (1984). Copyright © 1984 IEEE.]

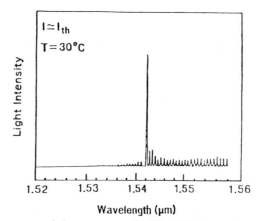

FIG. 28. Spontaneous emission spectrum of a BH–DFB laser with stop band below the threshold current. $I \simeq I_{th}$; $T = 30°C$. [From Itaya *et al.* (1984). Copyright © 1984 IEEE.]

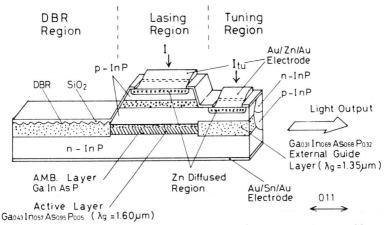

FIG. 29. Schematic diagram of a GaInAsP/InP-type BH–BJB–DBR integrated laser with wavelength tuning mechanism due to carrier injection [From Tohmori *et al.* (1983).]

variation of the temperature or bias current. The lasing spectrum tends to be multimode when the laser is operated under rapid, direct modulation.

The modulation sensitivity of a BH–DBR–ITG laser was measured and was almost linear without any appreciable resonancelike peaks, as shown in Fig. 32 (Koyama and Suematsu, 1983), and the 6-dB cutoff modulation

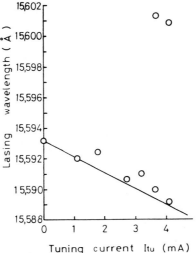

FIG. 30. Lasing wavelength versus tuning current. O, measured values; ——, theoretical values. $I = 1.1\, I_{th}$; $T = 294$ K. [From Tohmori *et al.* (1983).]

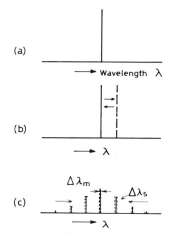

FIG. 31. Schematic diagrams of the lasing spectrum of a conventional Fabry–Perot-type laser (a) under dc operation, (b) with the variation of temperature or bias current, and (c) under a rapid, direct modulation. In case (c), the dynamic spectral broadening $\Delta\lambda_{br}$ and dynamic wavelength shift $\Delta\lambda_s$ are observed. [From Suematsu *et al.* (1983). Copyright © 1983 IEEE.]

frequency was approximately 2 GHz for $I_b/I_{th} = 1.1$ and 2.5 GHz for $I_b/I_{th} = 1.2$.

The spectra of a GaInAsP/InP-type BH–DBR–ITG laser with an emitting wavelength of 1.58 μm under direct rapid modulation were observed by use of a monochromator in which the modulation depth was 100% and the bias current level I_b was 1.2 times the threshold current. The modulation frequency was increased from 0.5 GHz to 3.05 GHz. The time-averaged lasing spectrum of a DSM laser under rapid direct modulation is compared with that of a conventional laser, as shown in Fig. 33 (Koyama and Suematsu, 1983). A stable, dynamic, single-mode operation was observed over all of the modulation frequencies as shown in Fig. 34 (Koyama *et al.*, 1981a). Magnified spectra are shown in the upper half of the figure. Note that the line

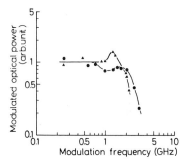

FIG. 32. Modulation sensitivity of a BH–DBR–ITG laser as a function of the modulation frequency. CW at 258 K; $m = 50\%$. ▲, $I/I_{th} = 1.1$; ●, $I/I_{th} = 1.2$. [From Suematsu *et al.* (1983). Copyright © 1983 IEEE.]

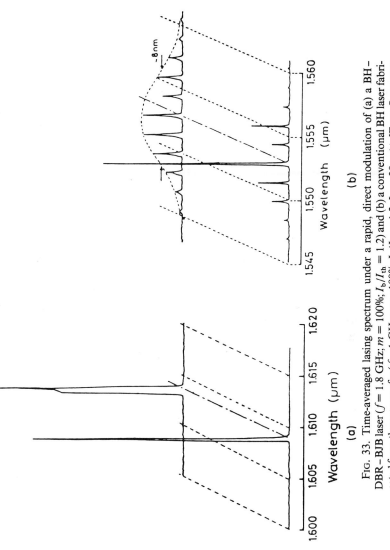

Fig. 33. Time-averaged lasing spectrum under a rapid, direct modulation of (a) a BH–DBR–BJB laser ($f = 1.8$ GHz; $m = 100\%$; $I_b/I_{th} = 1.2$) and (b) a conventional BH laser fabricated from the same wafer ($f = 1$ GHz; $m = 100\%$; $I_b/I_{th} = 1.2$; $I_{th} = 30$ mA). [From Suematsu et al. (1983). Copyright © 1983 IEEE.]

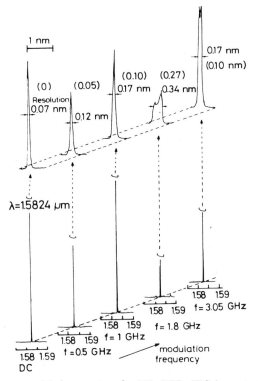

FIG. 34. Time-averaged lasing spectra of a BH–DBR–ITG laser at various modulation frequencies. Magnified spectra are shown in the upper side, where the dynamic wavelength shift (nm) is indicated in the parentheses. CW at 253 K; $I_{th} = 43$ mA; $I/I_{th} = 1.2$; $m = 100\%$. [From Koyama et al. (1981a).]

width of the lasing mode was broadened due to the dynamic wavelength shift, and it became maximum at the resonancelike frequency of 1.8 GHz. The dynamic wavelength shift of the BH–DBR–ITG laser increased appreciably when the modulation depth m increased over 50% at a frequency of 1.8 GHz (Koyama et al., 1981b, 1983). The width at half maximum intensity of the spectrum without modulation was observed here to be about 0.07 nm, which was the resolution of monochromator. Subtracting this spectral resolution from the line width under direct modulation, the full width of the dynamic wavelength shift $\Delta\lambda_{s,max}$ was 0.27 nm at a resonancelike frequency of 1.8 GHz with $m = 100\%$ and was about $\frac{1}{37}$ of the dynamic spectral width of conventional lasers emitting at same wavelength (Koyama et al., 1981a).

The frequency dependence of the dynamic wavelength shift was numerically calculated by using the rate equations (Koyama et al. 1983), as shown in

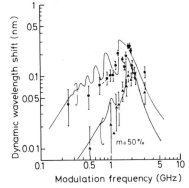

Fig. 35. Dynamic wavelength shift of a BH–DBR–ITG laser as a function of the modulation frequency. CW at 253 K; $I/I_{th} = 1.2$. ▲, $m = 50\%$; ●, $m = 100\%$; ——, theoretical values. [From Suematsu *et al.* (1983). Copyright © 1983 IEEE.]

Fig. 35. As can be seen, the experiment is in agreement with theory. A maximum dynamic wavelength shift $\Delta\lambda_{s,max}$ of 0.35 nm was observed for 1.6-μm wavelength GaInAsP/InP-type BH–DBR–BJB lasers (Koyama *et al.*, 1983). In a similar way, dynamic single-mode operation was observed in GaInAsP/InP-type DFB lasers under direct modulation of 0.4–1.6 Gbit sec^{-1} (Utaka *et al.*, 1981c; Matsuoka *et al.*, 1982; Nakagawa *et al.*, 1983).

IV. Discussion

12. LASING PROPERTIES

The details of DSM lasing properties have been investigated for a few years, and only a few fundamental properties have been measured. Therefore, interesting properties at the extreme or optimized conditions are not yet known. However, very satisfactory results such as very narrow spectral width as well as the significant reliability of DSM lasers were observed. Output powers of up to 40 mW from a DFB laser were measured as shown in Fig. 25d. The output power could be significantly increased because the mirror surface is large and it is located at the passive region in DBR lasers. The threshold current could be reduced further by increasing the coupling coefficient or the reflectivity of the distributed reflector as given by Eq. (14) and simultaneously by reducing the length of active region of the DFB or DBR lasers. A reliability test has been undertaken, and there is no indication that the reliability of DSM lasers is inferior to that of a conventional laser (Matsuoka *et al.*, 1982; Ikegami, 1982; Tanbun-ek *et al.*, 1984). As the problem of the mirror facets for the case of a conventional laser is eliminated for DSM lasers, the operational lifetime could be significantly increased by a reduction in the threshold current density.

The modulation sensitivity is essentially same as that of conventional lasers; therefore, the resonancelike peak appearing at the sinusoidal modulation or the relaxation oscillation appearing at the pulse modulation can be minimized by narrowing the stripe width down to the carrier diffusion length of a few micrometers (Furuya *et al.,* 1978). Since it is possible to increase the mirror reflectivity of DSM lasers, the problem of reflection noise can also be reduced further (Hirota *et al.,* 1981).

13. APPLICATION TO SINGLE-MODE FIBER COMMUNICATION

A very typical future application is to wideband optical-fiber transmission in the minimum loss wavelength region of 1.5–1.6 μm. The transmission bandwidth of single-mode (SM) fibers is limited by the chromatic dispersion τ_c (Payne and Gambling, 1975). The transmission bandwidth B of a conventional single-mode fiber of the length L with the chromatic dispersion τ_c per unit length and unit light-source spectral width is (Payne and Gambling, 1975)

$$B = 1/(\tau_c \, \Delta\lambda_s L), \qquad (51)$$

where $\Delta\lambda_s$ is the spectral width of the light source and is assumed to be larger than that of the corresponding modulation frequency. When the spectral width of the light source is smaller than that of the corresponding modulation bandwidth $\Delta\lambda_{mod}$, the transmission bandwidth B is (Furuya *et al.,* 1979)

$$B = 1/(\tau_c \, \Delta\lambda_{mod} L) = [c/(2\lambda^2 \tau_c L)]^{1/2}. \qquad (52)$$

This equation gives the theoretical limit of the transmission bandwidth of a single-mode fiber. The spectral width corresponding to $B = 1.6$ GHz is 0.025 nm at the a wavelength of 1.55 μm. The narrowest dynamic spectral width observed in a DSM laser is 0.27–0.35 nm at the resonance-like modulation frequency (Koyama *et al.,* 1981a, 1983), which is about ten times larger than the corresponding modulation spectral width at the modulation frequency of 1.6 GHz.

The bandwidth-distance product of a single-mode fiber transmission system using a conventional laser at a wavelength of 1.55 μm is calculated from Eq. (51) to be BL = 6.25 GHz km assuming a chromatic dispersion τ_c of 20 psec/nm km and a dynamic spectral width of 10 nm. However, the bandwidth is calculated to be more than 50 GHz km at a wavelength of 1.3 $\pm$ 20 nm, which is about 10 times larger than that at 1.55 μm (Li, 1980). This is due to small material dispersion of silica at 1.3 μm and is a reason why existing long-wavelength fiber communications adopt the 1.3-μm wavelength even though the fiber loss is larger than that at 1.55 μm. However, the use of DSM lasers with the maximum dynamic spectral width of 0.27 nm gives a bandwidth of 185 GHz km for conventional single-mode silica fibers. Figure 36 shows the transmission bit-error rate and the eye diagrams for various signals transmitted through a single-mode fiber with a 1.55-μm DFB laser light source (Yamamoto *et al.,* 1982; Nakagawa *et al.,* 1983; Ichihashi *et al.,* 1983). The effects of the spectral broadening and mode hopping were not observed. Several transmission experiments using a DSM laser were reported in the 1.5-μm wavelength region (Ikegami *et al.,* 1982; Nakagawa *et al.,* 1983; Linke *et al.,* 1983; Ichihashi *et al.,* 1983; Kasper *et al.,* 1983). Figure 37 shows the transmission bandwidth B of a conventional single-mode fiber 100 km long as a function of the light source wavelength. The solid lines were drawn from Eqs. (51) and (52). The dotted line gives the fiber transmission loss.

We can see from these points that the introduction of DSM lasers such as DBR and DFB lasers will enable us to use high-capacity optical-fiber communications at the minimum-loss wavelength of 1.5 – 1.6 μm. Furthermore, the mode partition noise generated by lasing mode-jump during high-frequency modulation (Okano *et al.,* 1980; Ogawa and Vodhanel, 1982) will be significantly reduced by use of the DSM lasers.

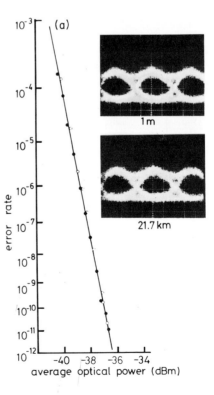

FIG. 36. Eye diagrams and transmission bit-error rates of signals transmitted through a single-mode fiber with a 1.55-μm-wavelength DFB-laser light source as a function of the average optical power at (a) 280 Mbits/sec; ●, 1 m; ○, 21.7 km (after Yamamoto *et al.,* 1982); (b) 445.8 Mbits/sec; x, 2 m; ●, 134.23 km (after Ichihashi *et al.,* 1983); and (c) 1.6 Gbits/sec; ○, 1 m; □ 40 km (after Nakagawa *et al.,* 1983a). *(Continued)*

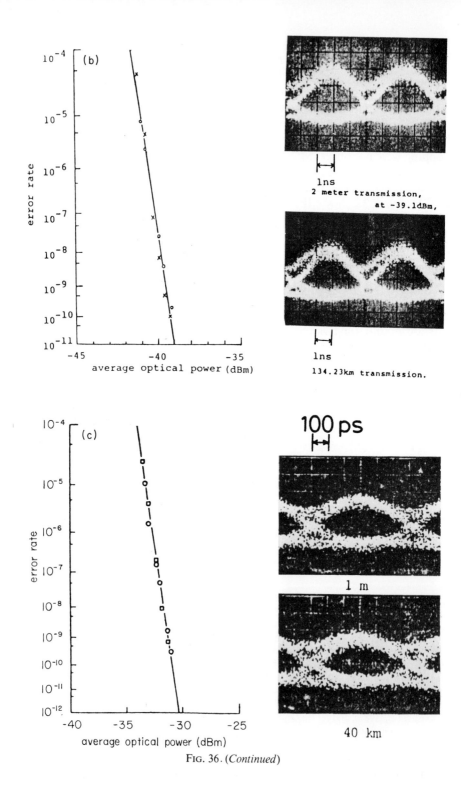

FIG. 36. (*Continued*)

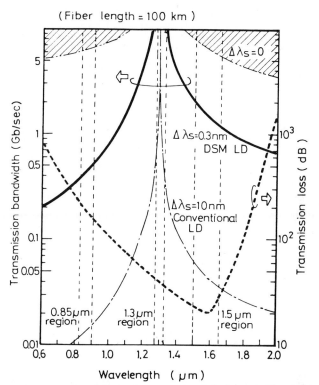

FIG. 37. Transmission bandwidth B of a conventional single-mode fiber 100 km long as a function of the wavelength of the light source. The thick dashed line indicates the transmission loss characteristics of the single-mode fiber. [From Suematsu *et al.* (1983). Copyright © 1983 IEEE.]

14. OTHER APPLICATIONS AND PROBLEMS FOR THE FUTURE

These DSM lasers could be applied to other fields such as optical video disks, fiber gyroscopes (Bergh *et al.,* 1982) and optical measurements. To optimize the DBR and DFB lasers, we have to know various physical parameters such as the details of the wavelength dependence of refractive index of $Ga_xIn_{1-x}As_yP_{1-y}$. Therefore, further study and development of DSM lasers are very essential in the near future. Monolithic integration of optical devices and optical circuits could also be accelerated by DBR and DFB integrated lasers developed for the DSM laser.

ACKNOWLEDGMENTS

The authors acknowledge the important data on DSM laser experiments supplied by Drs. H. Kawanishi presently at Kogakuin University, K. Utaka presently at Kokusai Denshin

Denwa Corporation, Y. Itaya presently at Nippon Telegraph and Telephone Public Corporation, and K. Furuya, M. Asada, and T. Tanbun-ek at the Tokyo Institute of Technology, and K. Kobayashi and Y. Abe who are presently at Nippon Electric Company, and the discussion of Professor Iga. They wish to acknowledge support in the form of a Scientific Research Grant-in-Aid from the Ministry of Education, Science, and Culture, Japan, NTT, and KDD.

REFERENCES

Abe, Y., Kishino, K., Suematsu, Y., and Arai, S. (1981). *Electron. Lett.* **17**, 945–947.

Abe, Y., Kishino, K., Tanbun-ek, T., Arai, S., Koyama, F., Matsumoto, K., Watanabe, T., and Suematsu, Y. (1982a). *Electron. Lett.* **18**, 410–411.

Abe, Y., Kishino, K., Suematsu, Y., and Arai, S. (1982b). *Top. Meet. Integr. Guided Wave Opt., Pacific Grove, Calif.* WB-2.

Aiki, K., Nakamura, M., and Umeda, J. (1976a). *IEEE J. Quantum Electron.* **QE-12**, 597–603.

Aiki, K., Nakamura, M., and Umeda, J. (1976b). *Appl. Phys. Lett.* **29**, 506–508.

Aiki, K., Nakamura, M., Kuroda, T., and Umeda, J. (1977). *Appl. Phys. Let.* **30**, 649–651.

Akiba, S., Utaka, K., Sakai, S., and Matsushima, Y. (1982). *Electron Lett.* **18**, 77–78.

Alferov, ZH. I., Andreyev, V. M., Gurevich, S. A., Kazarinov, R. F., Larionov, V. R., Mizerov, M. V., and Portnoy, E. L. (1975). *IEEE J. Quantum Electron.* **QE-11**, 449–451.

Anderson, D. B., August, R. R., and Coker, J. E. (1974). *Appl. Opt.* **13**, 2742–2744.

Arai, S., and Suematsu, Y. (1981). *Nat. Conv. Rec. IECE. Jpn.* 326.

Arnold, G., Peterman, K., Russer, P., and Berlec, F. J. (1978). *AEU* **34**, 129–136.

Asada, M., Adams, A. R., Stubkjaer, K., Suematsu, Y., Itaya, Y., and Arai, S. (1981). *IEEE J. Quantum Electron.* **QE-17**, 611–619.

Bergh, R. A., Cutler, C. C., Lefevre, H. C., Newton, S. A., Pavlath, G. A., and Shaw, H. J. (1982). *Top. Meet. Integr. Guided Wave Opt., Pacific Grove, Calif.* WB-2.

Burnham, R. D., Scifres, D. R., and Streifer, W. (1976). *Appl. Phys. Lett.* **29**, 287–289.

Burrus, C. A., Lee, T. P., and Dentai, A. G. (1981). *Electron. Lett.* **17**, 954–956.

Campbell, J. C., and Bellavance, D. W. (1977). *IEEE J. Quantum Electron.* **QE-13**, 253–255.

Casey, H. C., Jr., Somekh, S., and Ilegems, M. (1975). *Appl. Phys. Lett.* **27**, 142–144.

Chang, M. B., and Garmire, E. (1980). *Appl. Opt.* **19**, 2370–2374.

Chinn, S. R., and Kelly, P. L. (1974). *Opt. Comun.* **10**, 123–126.

Chinone, N., Aiki, K., and Ito, R. (1978). *Appl. Phys. Lett.* **33**, 990–992.

Choi, H. K., and Wang, S. (1982). *Appl. Phys. Lett.* **40**, 571–573.

Cohen, L. G., Lin, C., and French, W. G. (1979). *Electron. Lett.* **15**, 334–335.

Coldren, L. A., Miller, B. I., Iga, K., and Rentschler, J. A. (1981). *Appl. Phys. Lett.* **38**, 315–317.

DeWames, R. E., and Hall, W. F. (1973). *Appl. Phys. Lett.* **23**, 28–30.

Doi, A., Fukuzawa, T., Nakamura, M., Ito, R., and Aiki, K. (1979). *Appl. Phys. Lett.* **35**, 441–443.

Ebeling, K., Coldren, L. A., Miller, B. I., and Rentschler, J. A. (1983). *Appl. Phys. Lett.* **42**, 6–8.

Elachi, C., and Evans, G. (1975). *J. Opt. Soc. Am.* **65**, 404–412.

Fleming, M. W., and Mooradiani, A. (1981). *IEEE J. Quantum Electron.* **QE-17**, 4459.

Furuya, K., Suematsu, Y., and Hong, T. (1978). *Appl. Opt.* **17**, 1949–1952.

Furuya, K., Miyamoto, M., and Suematsu, Y. (1979). *Trans. IECE Jpn.* **E-62**, 305–310.

Haus, H. A., and Shank, C. V. (1976). *IEEE J. Quantum Electron.* **QE-12**, 532–539.

He, X. G., Gamo, K., Yuba, Y., and Namba, S. (1981). *Natl. Conv. Rec. Jpn. Appl. Phys., 42nd,* 7p-H-10. (In Jpn.)

Hirota, O., Suematsu, Y., and Kwok, K. (1981). *IEEE J. Quantum Electron.* **QE-17**, 611–619.

Hodgkinson, T. G., Wyatt, R., Smith, D. W., and Smith, D. R. (1982). *Top. Meet. Opt. Fiber Commun., Pheonix, Ariz. Pap.* PD6-1.

Hu, E. L., and Howard, R. E. (1980). *Appl. Phys. Lett.* **37**, 1022–1024.

Hurwitz, C. E., Rossi, J. A., Hsieh, J. J., and Wolfe, C. M. (1975). *Appl. Phys. Lett.* **27**, 241–243.

Ichihashi, Y., Nagai, H., Miya, T., and Miyajima, Y. (1983). *Proc. IOOOC '83, Tokyo* 29C5-2, p. 34 (postdeadline paper).

Iga, K., and Kawabata, K. (1975). *Jpn. J. Appl. Phys.* **14**, 427–428.

Iga, K., Kambayashi, T., Wakao, K., Kitahara, C., and Moriki, K. (1979). *IEEE Trans. Electron. Devices* **ED-26**, 1227–1230.

Ikegami, T. (1975). *IEE Conf. Publ.* No. 132, 111–114.

Ikegami, T. (1982). Personal communication.

Ikegami, T., Kuroiwa, K., Itaya, Y., Shinohara, S., Hagimoto, K., and Inagaki, N. (1982). *Proc. Eur. Conf. Opt. Fiber Commun., 8th, Cannes, Fr.*

Itaya, Y., Matsuoka, T., Kuroiwa, K., and Ikegami, T. (1984). *IEEE J. Quantum Electron.* **QE-20**, 230–235.

Kaminow, I. P., and Weber, H. P. (1971). *Appl. Phys. Lett.* **18**, 497–499.

Kasper, B. L., Linke, R. A., Campbell, J. C., Dentai, A. G., Vodhanel, R. S., Henry, P. S., Kaminow, I. P., and Ko, J.-S. (1983). *Eur. Conf. Opt. Fiber Commun. 9th, Geneva.*

Kawanishi, H., and Suematsu, Y. (1978). *Jpn. J. Appl. Phys.* **17**, 1599–1603.

Kawanishi, H., Suematsu, Y., and Kishino, K. (1977). *IEEE J. Quantum Electron.* **QE-12**, 64–65.

Kawanishi, H., Suematsu, Y., Itaya, Y., and Arai, S. (1978). *Jpn. J. Appl. Phys.* **17**, 1439–1440.

Kawanishi, H., Suematsu, Y., Utaka, K., Itaya, Y., and Arai, S. (1979). *IEEE J. Quantum Electron.* **QE-15**, 701–706.

Kawanishi, H., Hafich, M., Lenz, B., and Petersen, P. (1980). *Electron. Lett.* **16**, 738–740.

Kishino, K., Suematsu, Y., Utaka, K., and Kawanishi, H. (1978). *Jpn. J. Appl. Phys.* **17**, 751–752.

Kishino, K., Aoki, S., and Suematsu, Y. (1982). *IEEE J. Quantum Electron.* **QE-18**, 343–351.

Kitamura, M., Seki, M., Yamaguchi, M., Mito, I., Kobayashi, Ke., Kobayashi, K., and Matsuoka, T. (1983). *Electron. Lett.* **19**, 840–841.

Kobayashi, K., Utaka, K., Abe, Y., and Suematsu, Y. (1981). *Electron. Lett.* **17**, 366–368.

Kobayashi, K. (1982). Private communication.

Kobayashi, S., Yamada, J., Machida, S., and Kimura, T. (1980). *Electron. Lett.* **16**, 746–748.

Kogelnik, H., and Shank, C. V. (1971). *Appl. Phys. Lett.* **18**, 152–154.

Kogelnik, H., and Shank, C. V. (1972). *J. Appl. Phys.* **43**, 2327–2335.

Koyama, F., Arai, S., Suematsu, Y., and Kishino, K. (1981a). *Electron. Lett.* **17**, 938–940.

Koyama, F., Suematsu, Y., and Kishino, K. (1981b). *Natl. Conv. Rec. IECE Jpn., Kanagawa* s5-2. (In Jpn.)

Koyama, F., Suematsu, Y., Arai, S., and Tanbun-ek, T. (1983). *IEEE J. Quantum Electron.* **QE-19**, 1042–1051.

Kubota, K., Noda, J., and Mikami, O. (1980). *IEEE J. Quantum Electron.* **QE-16**, 754–760.

Kuroda, T., Yamanishi, S., Nakamura, M., and Umeda, J. (1978). *Appl. Phys. Lett.* **33**, 173–174.

Li, T. (1980). *Proc. IEEE* **68**, 1175–1180.

Lin, C., Burrus, C. A., Linke, R. A., Kaminow, I. P., Ko, J. S., and Miller, B. I. (1983). *Proc. IOOC'83, Tokyo,* pap. 29B5-5, p. 26 (postdeadline paper).

Linke, R. A., Kasper, B. L., Ko, J.-S., Kaminow, I. P., and Vodhanel, R. S. (1983). *Proc. IOOC '83, Tokyo* 29C5-2, p. 32 (postdeadline paper).

Matsumoto, N., and Kumabe, K. (1979). *Jpn. J. Appl. Phys.* **18**, 321–332.

Matsuoka, T., Nagai, H., Itaya, Y., Noguchi, Y., Suzuki, U., and Ikegami, T. (1982). *Electron. Lett.* **18**, 27–28.

Merz, J. L., and Logan, R. A. (1977). *Appl. Phys. Lett.* **30**, 530–533.

Mikami, O. (1981). *Jpn. J. Appl. Phys.* **20**, L488–L490.

Miya, T., Terunuma, Y., Hosaka, T., and Miyashita, T. (1979). *Electron. Lett.* **15**, 106–108.

Miya, T., Kawana, A., Terunuma, Y., Hosaka, T., and Ohmori, Y. (1981). *Trans. IECE Jpn.* **E-64**, 32–33. (In Jpn.)

Nakagawa, K., Hakamada, Y., and Suto, K. (1979). *Electron. Lett.* **15**, 747–748.

Nakagawa, K., Ohta, N., and Hagimoto, K. (1983). *Proc. IOCC '83, Tokyo* 27C2-1, p. 260.

Nakamura, M., and Yariv, A. (1974). *Opt. Commun.* **11**, 18–20.

Nakamura, M., Yariv, A., Yen, H. W., Smoken, S., and Garvin, H. L. (1973a). *Appl. Phys. Lett.* **22**, 515–516.

Nakamura, M., Yen, H. W., Yariv, A., Garmire, E., Somekh, S., and Garvin, H. L. (1973b). *Appl. Phys. Lett.* **23**, 224–225.

Nakamura, M., Aiki, K., Umeda, J., Yariv, A., Yen, H. W., and Morikawa, T. (1974). *Appl. Phys. Lett.* **25**, 478–488.

Nakamura, M., Aiki, K., Umeda, J., Katzir, A., Yariv, A., and Yen, H. W. (1975a). *IEEE J. Quantum Electron.* **QE-11**, 436–439.

Nakamura, M., Aiki, K., Umeda, J., and Yariv, A. (1975b). *Appl. Phys. Lett.* **27**, 403–405.

Namizaki, H. (1975). *IEEE J. Quantum Electron.* **QE-11**, 427–431.

Namizaki, H., Shams, M. K., and Wang, S. (1977). *Appl. Phys. Lett.* **31**, 122–124.

Ogawa, K., and Vodhanel, R. S. (1982). *Top. Meet. Opt. Fiber Commun., Pheonix, Az.* **THDD4.**

Okano, Y., Nakagawa, K., and Ito, T. (1980). *IEEE Trans. Comun.* **COM-28**, 238–243.

Okuda, M., and Kubo, K. (1975). *Opt. Commun.* **13**, 363–365.

Payne, D. N., and Gambling, W. A. (1975). *Electron. Lett.* **11**, 8–9.

Preston, K. R., Woolland, K. C., and Cameron, K. H. (1981). *Electron. Lett.* **17**, 931–933.

Reinhart, F. K., Logan, R. A., and Shank, C. V. (1975). *Appl. Phys. Lett.* **27**, 45–48.

Sakai, K., Utaka, K., Akibu, S., and Matsushima, Y. (1982). *IEEE J. Quantum Electron.* **QE-18**, 1272–1278.

Sakakibara, Y., Furuya, K., Utaka, K., and Suematsu, Y. (1980). *Electron. Lett.* **16**, 456–458.

Scifres, D. R., Burnham, R. D., and Streifer, W. (1974). *Appl. Phys. Lett.* **25**, 203–206.

Scifres, D. R., Burnham, R. D., and Streifer, W. (1975). *Appl. Phys. Lett.* **26**, 48–50.

Scifres, D. R., Burnham, R. D., and Streifer, W. (1976). *Appl. Phys. Lett.* **28**, 681–683.

Shams, M. K., and Wang, S. (1978). *Appl. Phys. Lett.* **33**, 179–173.

Shams, M. K., Namizaki, H., and Wang, S. (1978a). *Appl. Phys. Lett.* **32**, 179–181.

Shams, M. K., Namizaki, H., and Wang, S. (1978b). *Appl. Phys. Lett.* **32**, 314–316.

Shank, C. V., Schmidt, R. V., and Miller, B. I. (1974). *Appl. Phys. Lett.* **25**, 200–201.

Soda, H., Iga, K., Kitahara, C., and Suematsu, Y. (1979). *Jpn. J. Appl. Phys.* **18**, 3329–3330.

Stoll, H. M. (1979). *IEEE Trans. Circuits Syst.* **CAS-26**, 1065–1072.

Stoll, H. M., and Seib, D. H. (1974). *Appl. Opt.* **13**, 1981–1982.

Streifer, W., Burnham, R. D., and Scifres, D. R. (1975a). *IEEE J. Quantum Electron.* **QE-11**, 154–161.

Streifer, W., Scifres, D. R., and Burnham, R. D. (1975b). *IEEE J. Quantum Electron.* **QE-11**, 867–873.

Streifer, W., Scifres, D. R., and Burnham, R. D. (1976). *IEEE J. Quantum Electron.* **QE-12**, 74–78.

Streifer, W., Schifres, D. R., and Burnham, R. D. (1977). *IEEE J. Quantum Electron.* **QE-13**, 134–141.

Stubkjaer, K. E., Asada, M., Arai, S., and Suematsu, Y. (1981). *Jpn. J. Appl. Phys.* **20**, 1499–1505.

Subert, R. (1974). *J. Appl. Phys. Lett.* **45**, 209–215.

Suematsu, Y., Hakuta, M., Furuya, K., Chiba, K., and Hasumi, R. (1972). *Appl. Phys. Lett.* **21**, 291–293.

Suematsu, Y., Yamada, M., and Hayashi, K. (1975a). *Proc. IEEE* **63**, 208.

Suematsu, Y., Yamada, M., and Hayashi, K. (1975b). *IEEE J. Quantum Electron.* **QE-11**, 457–460.

Suematsu, Y., Arai, S., and Kishino, K. (1983). *IEEE J. Lightwave Technol.* **LT-1**, 161–176.

Suzuki, A., and Tada, K. (1980). *Proc. SPIE, Guided-Wave Opt. Surf. Acoust. Wave Devices, Syst. Appl., San Diego, Calif.* **239**, 10–18.

Takano, T., and Hamasaki, J. (1982). *IEEE J. Quantum Electron.* **QE-8**, 206–212.

Tanbun-ek, T., Arai, S., Koyama, F., Kishino, K., Yoshizawa, S., Watanabe, T., and Suematsu, Y. (1981). *Electron. Lett.* **17**, 967–968.

Tanbun-ek, T., Suzuki, S., Wang, S., Suematsu, Y., Koyama, F., and Arai, S. (1984). *IEEE J. Quantum Electron.* **QE-20**, 131–140.

Tohmori, Y., Suematsu, Y., Tsushima, H., and Arai, S. (1983). *Electron. Lett.* **19**, 656–657.

Tsang, W. T., and Wang, S. (1976). *Appl. Phys. Lett.* **28**, 596–598.

Tsang, W. T., Logan, R. A., and Johnson, L. F. (1979a). *Appl. Phys. Lett.* **34**, 752–755.

Tsang, W. T., Logan, R. A., and Johnson, L. F. (1979b). *J. Appl. Phys.* **50**, 5165–5167.

Tsang, W. T., Logan, R. A., Johnson, L. F., Hartman, R. L., and Koszi, L. A. (1979a). *IEEE J. Quantum Electron.* **QE-15**, 1091–1093.

Tsang, W. T., Olsson, N. A., Linke, R. A., and Logan, R. A. (1983). *Electron. Lett.* **19**, 415–416.

Tsukada, T. (1974). *J. Appl. Phys.* **45**, 4899–4906.

Umeda, Y., Yamamoto, M., and Unno, Y. (1977). *IEEE J. Quantum Electron.* **QE-13**, 646–651.

Utaka, K. (1981). Ph.D. Thesis, Tokyo Inst. Technol., Tokyo.

Utaka, K., Kobayashi, K., Kishino, K., and Suematsu, Y. (1980a). *Electron. Lett.* **16**, 455–456.

Utaka, K., Suematsu, Y., Kobayashi, K., and Kawanishi, H. (1980b). *Jpn. J. Appl. Phys.* **19**, L137–L140.

Utaka, K., Kobayashi, K., Koyama, F., Abe, Y., and Suematsu, Y. (1981a). *Electron. Lett.* **17**, 368–369.

Utaka, K., Kobayashi, K., and Suematsu, Y. (1981b). *IEEE J. Quantum Electron.* **QE-17**, 651–658.

Utaka, K., Akiba, S., Sakai, K., and Matsushima, Y. (1981c). *Electron. Lett.* **17**, 961–963.

Wang, S. (1973). *J. Appl. Phys. Lett.* **44**, 767–780.

Wang, S. (1974). *IEEE J. Quantum Electron.* **QE-10**, 413–427.

Wang, S. (1974–1975). *Wave Electron.* **1**, 31–59.

Wang, S. (1977). *IEEE J. Quantum Electron.* **QE-13**, 176–189.

White, K. I., and Nelson, B. P. (1979). *Electron. Lett.* **15**, 396–397.

Yamada, J., Kobayashi, S., Nagai, H., and Kimura, T. (1981). *IEEE J. Quantum Electron.* **QE-17**, 1006–1009.

Yamamoto, S., Utaka, K., Akiba, S., Sakai, K., Matsushima, Y., Sakaguchi, S., and Seki, N. (1982). *Electron. Lett.* **18**, 239–240.

Yamanishi, M., Ameda, N., Ishii, K., Kamura, T., Tsubouchi, K., and Mikoshiba, N. (1977). *IEEE J. Quantum Electron.* **QE-12**, 64–65.

Yariv, A. (1973). *IEEE J. Quantum Electron.* **QE-9**, 919–933.

Yariv, A., and Nakamura, M. (1977). *IEEE J. Quantum Electron.* **QE-13**, 233–253.

Yen, H. W., Nakamura, M., Garmire, E., Somekh, S., Yariv, A., and Garvin, H. L. (1973). *Opt. Commun.* **9**, 35–37.

Zory, P., and Comerford, L. D. (1975). *IEEE J. Quantum Electron.* **QE-11**, 451–457.

CHAPTER 5

The Cleaved-Couple-Cavity (C³) Laser

W. T. Tsang

AT&T BELL LABORATORIES
MURRAY HILL, NEW JERSEY

I. Introduction

For ultrahigh-capacity and very long-distance single-mode optical-fiber communications systems in the $1.3-1.7$-μm wavelength range, it is very desirable to have a semiconductor laser that emits *stably* at one single frequency even under high-bit rate direct modulation. This is particularly important in the wavelength range $1.5-1.7$ μm in which the present silica optical fibers with chromatic dispersion minimized at 1.3 μm have still lower losses. In this region, however, they are accompanied by a relatively larger chromatic dispersion of $15-18$ psec km^{-1} nm^{-1}. Unfortunately, conventional Fabry–Perot semiconductor lasers operate in multilongitudinal modes when directly modulated. As a result, the lowest-loss region at ~ 1.55 μm cannot be utilized owing to dispersion limitation. In addition, mode partition noise that orginates from mode partitioning in conventional laser diodes and is enhanced by chromatic dispersion further limits the bit-rate distance product severely. Although the zero-dispersion wavelength can be moved from 1.3 to 1.55 μm (or any other single wavelength above 1.3 μm) by changing the fiber design, this increases the attenuation at 1.55 μm such that little advantage is gained over operation at 1.3 μm. In addition, for very long-distance transmission (greater than 150 km) with dispersion-shifted fibers, the allowed spectral width still has to be less than ~ 15 Å. For a conventional laser diode operating at ~ 1.55 μm and having a cavity length of 250 μm, the longitudinal mode spacing is already about 15 Å. Thus, for such systems single-longitudinal-mode lasers are required. Furthermore, this method limits the usage of a wide wavelength range, even though two or more minimum dispersions can be realized. Consequently, the other alternative is to produce semiconductor lasers that operate stably in a single longitudinal mode even under high-bit-rate direct modulation. This has the important advantage that not only the lowest-loss 1.55-μm region can be utilized, but it also opens up the possibility of truly wideband transmission. Previously, single-longitudinal-mode operation has been obtained in a wide class of index-guided semiconductor lasers under cw operation. However, it has recently become apparent that the observation of single-longitudinal-mode output from an injection laser under steady-state excitation is insufficient to ensure that the laser will behave as a single-frequency device under actual modulation especially at high bit rates. In fact, in order to avoid a significant contribution to system error rates in a high-bit-rate communications system in which fiber dispersion is significant, the power ratio of the dominant longitudinal mode to the next-strongest mode must exceed $\sim 100:1$. This is because even though the lasers exhibit a nearly single-longitudinal-mode cw spectrum, mode-partition events frequently occur that can essentially extinguish the dominant mode for 1 to > 8 nsec, with the power loss appearing in a mode of very low average power. These events cause bit errors in communications systems. Consequently, considerable effort is being focused on the development of 1.3 – 1.55-μm InGaAsP-type injection lasers with better longitudinal mode control under modulation. Some examples include lasers with distributed feedback (DFB) (Sakai *et al.*, 1982; Matsuoka *et al.*, 1982), distributed Bragg reflectors (DBR) (Utaka *et al.*, 1980; Sakakibara *et al.*, 1980; Suematsu *et al.*, 1983), injection locking (Malyon and McComa, 1982), coupled cavities [either external (Preston *et al.*, 1981; Salathe, 1979) or integrated (Lang and Kobayashi, 1976; Coldren *et al.*, 1981)], double active layers (Tsang *et al.*, 1983), and very short cavities (Lee *et al.*, 1982). Of all these laser structures, transmission experiments at ~ 1.5 μm have been demonstrated with injection-locked diodes (Malyon and McComa, 1982), external cavity-controlled diodes (Preston *et al.*, 1981), DFB laser diodes (Ichihashi *et al.*, 1983), and very recently with cleaved-coupled-cavity (C^3) lasers (Tsang *et al.*, 1983a – c, e – i; Tsang and Olson, 1983; Linke *et al.*, 1983; Kasper *et al.*, 1983).

Coherent optical-fiber transmission systems that utilize frequency modulators (FM) or phase modulation (PM) comprise yet another area of intense interest (Yamamoto and Kimura, 1981; Saito *et al.*, 1980, 1981; Kobayashi *et al.*, 1982). Such systems are expected to push system performance toward even longer repeater spacings and higher information capacity than the presently employed pulse-coded-modulation – intensity-modulation (PCM – IM) systems. Extensive studies have been carried out by Kimura and colleagues (Yamamoto and Kimura, 1981; Saito *et al.*, 1980, 1981; Kobayashi *et al.*, 1982). Initial coherent-system experiments using direct frequency modulation of a conventional semiconductor laser with optical heterodyne detection, frequency shift keying (FSK) with a direct detection scheme, and phase shift keying (PSK) modulation with homodyne detection have been successful (Saito *et al.*, 1980, 1981). Even so, the development of FM optical systems has been relatively difficult due to the lack of suitable frequency-tunable lasers and a stable single-frequency local oscillator. In addition to the frequency tunability, the laser also needs to be mechanically compact and must emit stably at one single frequency. Previous approaches include the use of electro-optic frequency modulators [either integrated (Reinhart and Logan, 1975, 1980) or discrete devices in external-cavity semiconductor lasers (Tang *et al.*, 1976; Olsson and Tang, 1979)] and direct modulation of the injection current of conventional laser diodes. Envelope spectrum shifts of 0.4 Å and 20 Å are obtained either by the electro-optic effect (Reinhart and Logan, 1975) or by threshold gain variation through the modulation of the effective mirror reflectivity (Reinhart and Logan, 1980). In these demonstrations, only broad-area laser diodes were used. However, the most important drawback of this scheme is the difficult procedure involved in fabricating such complicated integrated structures. On the other hand, directly modulating the injection current of the conventional diode to obtain FM is simple, but the frequency shift Δf is very small. For example, the ratio of the frequency shift to the change in modulation current $\Delta f/\Delta I_m$ for an (AlGa)As-type channeled-substrate-planar (CSP) laser is ~ 100 MHz mA^{-1} in the modulation frequency region of 10 MHz to 1 GHz (Saito *et al.*, 1982; Dandridge and Goldberg, 1982). This is because the FM arises from index modulation due to carrier density modulation, and above threshold, the carrier density in most lasers should be pinned at the threshold value. This results in a relatively small value for $\Delta f/\Delta I_m$ and a corresponding large unintended intensity modulation. Yet for FSK systems, it is desirable to have a large Δf produced by a small ΔI_m without serious spurious intensity modulation. In addition, it is also desirable that $\Delta f/\Delta I_m$ be relatively independent of the modulation frequency, f_m.

In this chapter, a detailed presentation of the temporal–spectral behavior of the first, practical, electronically controllable, tunable, single-frequency semiconductor laser, the newly developed cleaved-coupled-cavity (C^3) laser (Tsang *et al.*, 1983a,d,j,k), and its various applications is given. These applications include ultrahigh-capacity and very long-distance PCM–IM lightwave transmission systems (Tsang *et al.*, 1983b; Linke *et al.*, 1983; Kasper *et al.*, 1983); multilevel, multichannel FSK (Tsang *et al.*, 1983c); high-speed and high-capacity optical switching and routing systems (Olsson and Tsang, 1983a); enhanced analog frequency modulation with negligible spurious intensity modulation (Tsang and Olsson, 1983; wideband FSK using spectral bistability (Olsson and Tsang, 1983b); gateable, mode-locked semiconductor lasers with an electronically controllable absorber (Tsang *et al.*, 1983e); optical logic operations (Tsang *et al.*, 1983f); self-aligned laser–detector modules for close tracking of laser output power (Tsang *et al.*, 1983g); Q-switching for subnanosecond optical pulse generation (Tsang *et al.*, 1983h; Lee *et al.*, 1983); optical bistable operation (Tsang *et al.*, 1983i; Dutta *et al.*, 1983); and analog-to-digital conversion.

The concept of coupled-cavity resonators for longitudinal mode selection was proposed as early as 1962 by Kogelnik and Patel (1962) and Kleinman and Kislink (1962) by working with gas lasers. An excellent review on mode selection in lasers (not including semiconductor lasers) was given by Smith (1972). Though this concept is available and it is also known that semiconductor laser diodes employing one Fabry–Perot cavity usually lase in multilongitudinal modes, no rigorous effort has been made to apply this coupled-cavity concept to semiconductor lasers for mode selection until recently. There are four reasons for this. First, there was little demand for truly single-frequency semiconductor laser diodes until the advent of very low-loss, single-mode fiber in the 1.55-μm region. Second, there is no practical fabrication technology for fabricating multicoupled cavity laser diodes whose total lengths are only in the order of 200 – 400 μm that is suitable for mass production. Third, there is the general belief that the length ratio of the two coupled cavities and the air gap separating them are crucial in achieving optimal mode selection. Finally, even if single-longitudinal-mode operation can be achieved in coupled-cavity semiconductor laser diodes, its stability with respect to changing biasing currents and temperatures will be very limited.

Nevertheless, techniques capable of implementing multiple coupled cavities in semiconductor diodes are available. Kosnocky and Connely (1968) created two coupled broad-area diodes by cracking a single diode into two halves after soldering it down onto a flexible metal spring. The purpose of their experiment was to study optical amplification. Hence, one of the cavities was intentionally destroyed by lapping the end mirror at an oblique angle. The same technique and experiment, again for optical amplification in which one section acts as the laser and the other as the amplifier, was repeated by Chang and Garmire (1980). However, instead of using broad-area lasers and spoiling the resonant cavity of the amplifier, they employed 50-μm oxide-stripe lasers and left the resonant cavity of the amplifier intact. No mode-selection study was performed, nor was any indication of single-frequency operation observed. The same concept of forming coupled cavities as that of Kosnocky and Connely (1968) was also employed by Allen *et al.* (1978). Though single-longitudinal-mode operation such as that observed in well-behaved conventional lasers was observed with their broken lasers under cw operation, no mention was made of pulsed operation. Furthermore, they went to the length of soldering together the electrode that was severed in the process of cleaving the device into a two-electrical-terminal device. As a result, no independent biasing of the two optically coupled diodes was possible. It is clear that they were unaware of the significance of refractive index adjustment to electrically alter the optical cavity length. The resulting two-terminal device could not be adjusted to optimize the intensity ratio of the selected mode relative to the undesired modes. More importantly, it precluded any possibility of an electrical feedback loop to stabilize the device against parameter drift during use. Furthermore, it also precluded the ability of electronic tuning between different selected single frequencies.

Another fabrication technique uses either wet or dry etching to form the reflecting mirror. This was first proposed and demonstrated by Merz *et al.* (1979). In their work, they utilized chemical etching to form the reflecting mirrors of the laser cavity and the coupled laser – detector module. Subsequently, Coldren and co-workers (1981) combined dry etching and chemical etching to form two-section lasers. This technique is capable of accurate control of the gap spacing separating the two sections. However, the mirrors formed by etching have not been of as a high quality as those formed by cleaving along crystallographic planes. (Furthermore, they usually are not perfectly parallel to the end facets that are formed by cleaving.)

II. Fabrication of C^3 Lasers

In the technique employed previously (Kosnocky and Connely, 1968; Allen *et al.*, 1978), a single diode chip is soldered down onto a flexible metal plate with In. A light scratch is then made at one edge. Bending the metal plate produces a cleavage along the scratch. This technique is thus applicable to one diode chip at a time. Furthermore, the cracked diodes must remain soldered to the metal plate and cannot be transferred to another more suitable heat-sink mounting and electrical package. This technique, therefore, is not suitable for mass production. In this part, a new technique for forming C^3 lasers by a simple variation in the usual cleaving procedures employed in the production process is described as proposed by Tsang (1983).

There are major and important differences in the present and previous techniques as will become evident in the following. Most importantly, the present technique is compatible with the mass-production processing procedures employed in the present production environment. Figure 1a shows schematically the final arrangement of the C^3 laser. The active stripes from the diodes are *precisely aligned* with respect to each other and separated from each other by a distance $d < 5$ μm. Unlike the monolithic two-section laser prepared by Coldren and co-workers (1981) by etching, all the facets are formed by cleaving along crystallographic planes and, hence, are perfectly mirror flat and parallel to each other for each cavity. Furthermore, complete electrical isolation between the two individual diodes is also obtained in the present case. The total length of the C^3 laser can be made as short as 100 μm with a typical length of 200–400 μm. Figure 1b shows a photograph of an actual C^3 laser cw-bonded to a Cu heat sink with independent electrical biasing to the two diodes.

In forming the C^3 laser, the standard laser wafer containing active stripes (such as buried heterostructure, buried crescent heterostructure, and or other types) is prepared in the same manner used in forming standard cleaved laser diodes. In the present case, however, it is essential that isolated, thick Au pads about the same size as the standard diode be electroplated onto the epilayer side (if cw-bonded epilayer side down) or the substrate side (if cw-bonded epilayer side up). The usual cleaving procedures are then applied at positions a and b shown in Fig. 2a to form cleaved bars. Because there are no thick Au pads, the bars are separated from each other at positions a. At positions b, however, the presence of the thick Au pads holds the two adjacent cleaved bars together. Note that the length aa typically is 200–400 μm

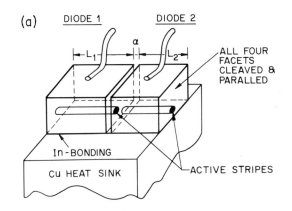

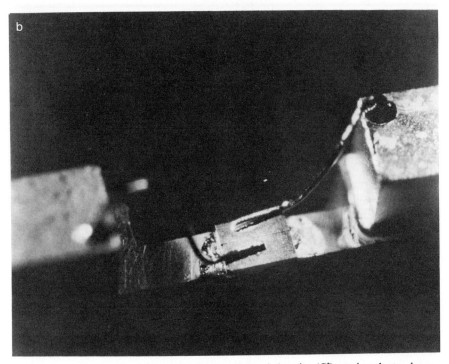

FIG. 1. (a) Schematic diagram of the cleaved-coupled-cavity (C^3) semiconductor laser. Drawing is not to scale. (b) A photograph of an actual C^3 laser.

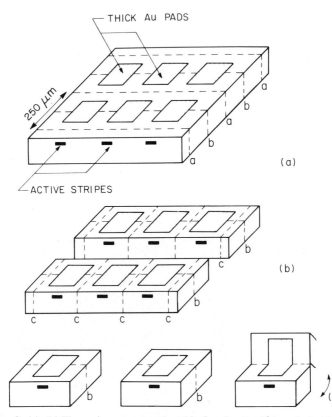

FIG. 2. (a)–(c) The various steps employed in forming the C³ laser in batch form.

long, and lengths *ab* and *ba* can be adjusted as desired. Individual pairs of diodes that are still held together by the Au pads are separated by sawing or deep scribing along positions *c* as shown in Fig. 2b. As shown by the diodes in Fig. 2c, cleaving along position *b* creates two precisely self-aligned, extremely closely coupled cavities that form a single C³ laser. The separation *d* is extremely small ($\lesssim 1$ μm). If a larger separation is desired, *d* can be conveniently increased by opening the two diodes apart and reclosing them by using the thick Au pad along the line of cleavage as the hinge. Finally, the two diodes that are hinged together by the Au pad are bonded simultaneously (epilayer side down) with In onto a Cu heat sink by employing standard cw bonding procedures. Separate electrical bondings are made to each diode on the substrate side. This forms the three-terminal device shown in Fig. 1. Figure 3 shows a photograph of the wafer in various stages of fabrication. The large wafer shows the isolated Au pads equivalent to Fig. 2a. The two separated bars (one substrate side up and one substrate down) show the cleaved

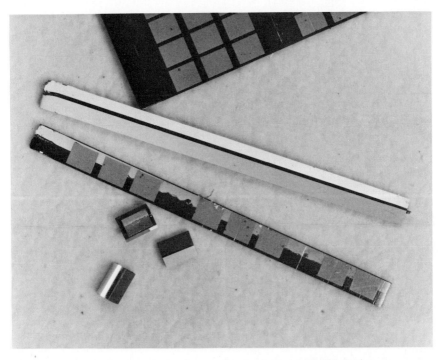

FIG. 3. Photograph showing a wafer at various stages of fabrication.

and hinged bars equivalent to Fig. 2b. The three individual lasers show the final stage of C³ lasers with the two hinged diodes opened up by different amounts, thus illustrating the simplicity and beauty of this technique. This technique differs from previous techniques (Kosnocky and Connely, 1968; Allen *et al.*, 1978) in a very significant way: On reclosing, the active stripe will automatically remain self-aligned. This is equivalent to the step described in Fig. 2c. Figure 4 shows an SEM photograph of a C³ laser on top of a penny. The cleaved gap is V-shaped and is not in the form of a parallel air etalon [as thought by Allen *et al.* (1978)]. Consequently, all the results obtained from theoretical analyses by assuming a parallel air gap can be considered qualitative especially with respect to the effects of the air gap.

It is evident that our technique is compatible with present processing technology employed for mass production of laser diodes. The C³ diodes can be freely transported and cw bonded to any desirable heat sink and packaged as usual.

Figure 5 shows the current–voltage $(I - V)$ characteristics of the original 250-μm GaInAsP-type buried crescent diode operating at 1.3-μm, the 136- and 120-μm diodes after the second cleavage and bonded to the Cu heat sink, and the electrical isolation between them. It is seen that the individual diode

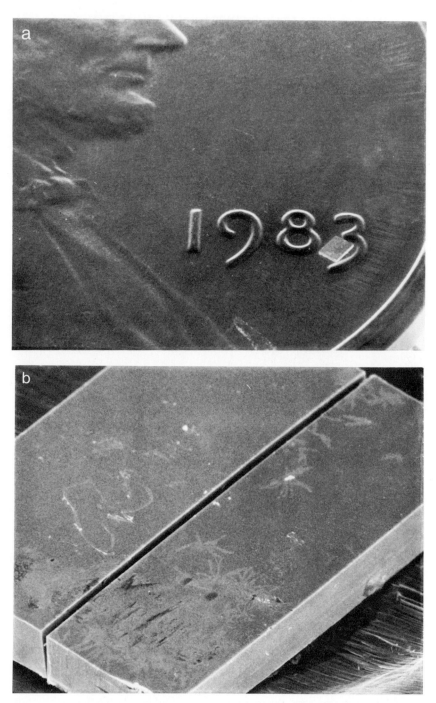

FIG. 4. (a) Scanning electron microscope photograph showing a C^3 laser on top of U.S. penny; (b) two cleaved diodes hinged together by an Au pad underneath. The cleaved-gap is V-shaped and not in the form of a parallel etalon.

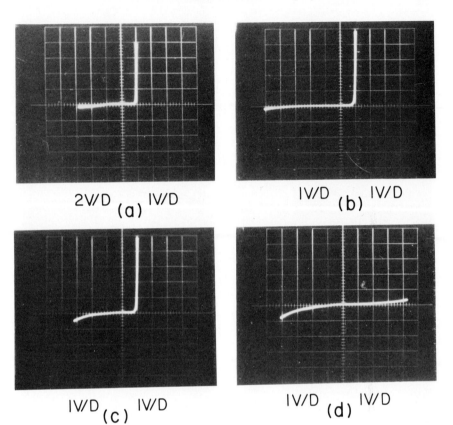

2V/D (a) IV/D

IV/D (b) IV/D

IV/D (c) IV/D

IV/D (d) IV/D

FIG. 5. $I-V$ characteristics of (a) the original 250-μm GaInAsP-type Fabry–Perot crescent diode, (b) the 136-μm and (c) 120-μm diodes after the second cleaving and bonding to the Cu heat sink, and (d) the electrical isolation between them.

$I-V$ characteristics are very well preserved and the electrical isolation between them is greater than 50 kΩ (corresponding to a back-to-back diode characteristic).

Figure 6 shows the pulsed light-current ($L-I$) characteristics measured from each side of a C^3 laser fabricated from a 1.3-μm GaInAsP-type buried crescent laser wafer. The output power intensity scales are the same for all curves. The present C^3 laser has a total cavity length of 255 μm. The left diode is 136 μm long, with a 5-μm separation (at the top) from the right diode (120 μm long). Unlike the monolithic two-section laser formed by etching, the present C^3 laser has a very low combined current threshold I_{thLR} of 46 mA and with only one laser pumped, the left and right lasers have thresholds $I_{\text{thL}} = 36$ mA and $I_{\text{thR}} = 30$ mA, respectively. We note that

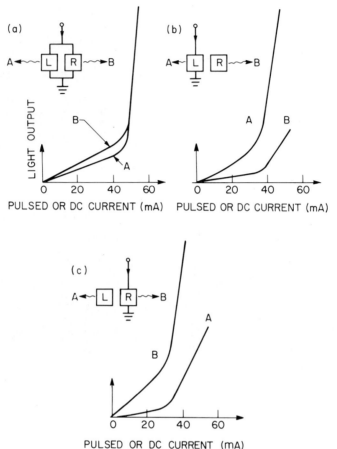

FIG. 6. Pulsed light-current characteristics measured from each side of a C^3 laser fabricated from a 1.3-μm wavelength GaInAsP-type crescent laser wafer. (a) Both diodes pumped together; (b) only the left diode pumped; (c) only the right diode pumped.

$I_{thL} + I_{thR} > I_{thLR}$, indicating that the two cavities are coupled (otherwise, the pumping of one diode should not influence the threshold of the other). When only one section of the C^3 laser is pumped, the optical beam generated by the electrically pumped diode is coupled and guided into the active stripe of the unpumped diode. The unabsorbed power was measured by curve B in Figs. 6b and c. The asymmetry of the two $L-I$ output characteristics in both the spontaneous emission level and differential quantum efficiency also serves as a good demonstration that nonuniform current injection along the active stripe of a conventional, single laser diode can be the cause of asymmetry in the two $L-I$ output characteristics. Similar $L-I$ characteristics are also obtained when the C³ laser is operated under cw conditions. The cw current thresholds are approximately the same as those under pulsed operation.

Recently, Bowers and co-workers (1983) introduced a very elegant technique for forming C³ lasers of very short cavity lengths. With the usual cleaving technique it is difficult to cleave a length that is less than the wafer thickness. Thus, to make a cavity length of 30 μm, the wafer would have to be extremely thin and fragile. One solution is to cut a groove through most of the wafer along the desired cleavage. Bowers *et al.* (1983) used laser-induced photochemical etching to cut such a deep, narrow groove in a bar of lasers. They used standard semiconductor laser fabrication techniques through to the step of cleaving the wafer into bars. The bars of lasers were then immersed in a 10% phosphoric acid solution and a focused laser beam ($\lambda = 5145$ Å) was swept along the bar at the desired distance from one cleaved end face. Both the Au contact layer and the InP substrate were photochemically etched.

Figure 7 shows the high-aspect-ratio groove produced by this process. The

50 μm

FIG. 7. SEM photo of a photochemically etched groove through electroplated Au and InP.

groove dimensions are 55 μm deep $\times$ 22 μm wide. The individual lasers on the bar are then separated from one another with a second photochemical etching step that etches through the 80-μm-thick substrates. The individual laser is formed by cleaving along the deep groove. The two separate diodes are held together by the Au pad on the epilayer side. This technique allows precise control of the cleaving process and the fabrication of very short-cavity C^3 lasers or C^3 lasers with large length ratios.

III. Single-Frequency Operation under High-Bit-Rate Modulation

Because of the cavity length and possibly the active stripe material and cross-sectional dimension differences of the two individual cavities, the laser mode spacings will be different and will be given by $\Delta\lambda_1 \sim \lambda_0^2/2N_{\mathrm{eff1}}L_1$ and $\Delta\lambda_2 \sim \lambda_0^2/N_{\mathrm{eff2}}L_2$, respectively. They are represented schematically in Fig. 8. As the two cavities are brought close to each other with their active stripes aligned with each other and as they become optically coupled, those modes from each cavity that coincide or nearly coincide spectrally will interfere constructively and become the enforced modes of the coupled-cavity resonator, while the others interfere destructively and become suppressed. The spectral spacing Λ of these enforced modes is significantly larger than either of the original individual mode spacings depending on the difference of $N_{\mathrm{eff1}}L_1$ and $N_{\mathrm{eff2}}L_2$ and is given approximately by

$$\Lambda = \Delta\lambda_1\,\Delta\lambda_2/|\Delta\lambda_1 - \Delta\lambda_2| = \lambda_0^2/2|N_{\mathrm{eff1}}L_1 - N_{\mathrm{eff2}}L_2|, \qquad (1)$$

for $N_{\mathrm{eff1}}L_1 \approx N_{\mathrm{eff2}}L_2$. This is shown schematically in Fig. 8c. Usually N_{eff1} and N_{eff2} are taken to be the same since both diodes are made from the same material and layer structure. Therefore, Λ is made large by having L_1 and L_2 slightly different. For situations in which L_1 and L_2 are very different, the coupled spacing is given by the mode spacing of the shorter cavity.

The spectral gain profiles for both diodes are taken as the same as shown in Fig. 8d. Thus, if an enforced mode falls near the gain maximum at the desired operating temperature as shown in Fig. 8, the normal gain roll-off is sufficient to suppress the adjacent enforced modes even during a fast ($\lesssim 1$-nsec) turn-on transient or under high-bit-rate direct modulation. Single-longitudinal-mode selection utilizing gain roll-off of the lasing medium and increased mode spacing of the resonator is also accomplished by making the single cavity very short; for example, $\lesssim 100$ μm for 1.3-μm and $\lesssim 50$ μm for 1.55-μm lasers (Lee et al., 1982).

In the work carried out by Tsang and co-workers (1983a – c,e,f; Tsang and Olsson, 1983), both 1.3- and 1.5-μm InGaAsP-type lasers investigated were large-optical-cavity, buried crescent (Tsang and Logan, 1978; Logan et al.,

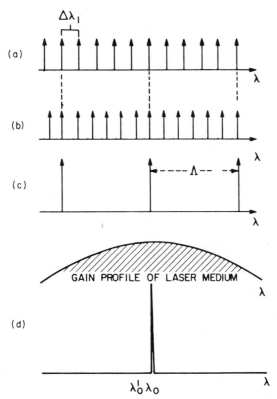

FIG. 8. Basic working principle of a C^3 laser for obtaining single-frequency operation. (a) Allowed Fabry–Perot modes for diode 1, (b) allowed Fabry–Perot modes for diode 2, (c) resultant modes for double active layer laser, and (d) resultant laser spectrum: $\delta\lambda = \lambda_0\,\delta N_{eff2}/N_{eff2}$, $\Lambda \approx \Delta\lambda_1\,\Delta\lambda_2/\Delta\lambda_1 - \Delta\lambda_2$.

1982) and ridge-waveguide (Kawaguchi and Kawakami, 1977; Kaminow *et al.*, 1983a) structures. The buried crescent laser has its active stripe completely surrounded by wide band-gap material. Hence, it should not have carrier out-diffusion as in the case of the ridge-waveguide laser. As a result, thresholds are typically 15–50 mA and 40–80 mA for buried crescent and ridge-waveguide lasers, respectively. Another important difference is that built-in, reverse-biased $p-n$ junctions are employed for current injection confinement in buried crescent lasers while oxide isolation is used in ridge-waveguide lasers. For the buried crescent lasers studied by Tsang *et al.* (1983a–k; Tsang and Olsson, 1983), current leakage becomes rather serious with increasing injection current levels. This leads to a bending of the $L-I$ curves and limits the maximum power emitted to a relatively low value of $\lesssim 5$ mW per mirror typically. Such a leakage problem is not observed in the ridge-waveguide lasers. In addition, the presence of the reverse-biased $p-n$ junction degrades the frequency response of the buried crescent lasers owing to the additional rc time constant.

In the following, the experimental results obtained by Tsang *et al.* (1983a,d,j,k) are described to show that very significant improvements in the temporal – spectral performances have been obtained with C³ lasers. To substantiate this, Fig. 9a shows the time-resolved transient mode spectra of a 1.3-μm buried crescent C³ laser under high-speed direct current excitation.

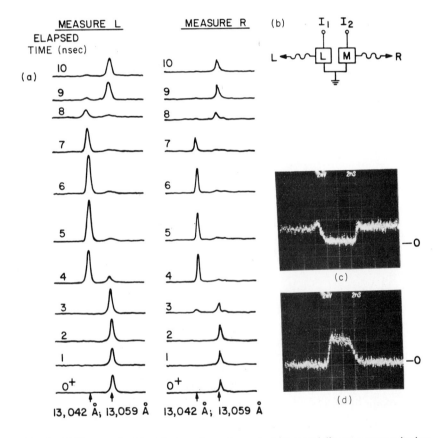

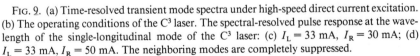

FIG. 9. (a) Time-resolved transient mode spectra under high-speed direct current excitation. (b) The operating conditions of the C³ laser. The spectral-resolved pulse response at the wavelength of the single-longitudinal mode of the C³ laser: (c) $I_L = 33$ mA, $I_R = 30$ mA; (d) $I_L = 33$ mA, $I_R = 50$ mA. The neighboring modes are completely suppressed.

Figure 9b shows the operating conditions. The rise time of the current pulse applied to the right diode (designated as modulator) is ~ 1 nsec, and that applied to the left diode (designated as laser) is a wide pulse or dc. The optical output was detected from the modulator side. The spectra shown in this figure were taken at various elapsed times after the onset of the light pulse that is emitted due to the application of the current pulse to the modulator. It is seen that the output power is concentrated into a single longitudinal mode right at the *onset* of the optical pulse and remains in the *same* mode thereafter. Such spectral stability under high-speed direct modulation of the present C^3 laser is in fact as good as those obtained with DFB or DBR lasers.

In this time-resolved spectral measurement, each resultant spectrum represents the averaged spectrum over a large number of pulses taken at the same instant of time on each pulse. As a result, the pulse-to-pulse spectral variations are not detected. To investigate such pulse-to-pulse spectral variations, the optical pulse response at the wavelength of the single longitudinal mode as selected with a spectrometer was displayed by using a sampling oscilloscope with the C^3 laser excited as before. Figures 9c–d show the photographs of such spectrally resolved pulse responses at two different modulator currents. The clean pulse response indicates that there are no pulse-to-pulse spectral variations in the C^3 lasers, while this in general is not true in regular crescent lasers. The neighboring longitudinal modes are completely suppressed. Thus, the spectrally resolved optical pulses shown here also represent the total power optical responses from the laser. Furthermore, no relaxation oscillations are observed.

To further demonstrate that the C^3 lasers continue to emit stably in single frequency even under high-bit-rate direct modulation in excess of 2 Gbit sec^{-1}, non-return-to-zero (NRZ) tests were run with the frequency controlling diode operating under cw conditions and the laser diode biased with a dc component and a pseudorandom word superimposed on it. Figure 10a shows the lasing spectrum of a 1.55-μm ridge-waveguide C^3 laser (Kaminow *et al.*, 1983b) under cw operation, while Figs. 10b–c show that under 2-Gbit sec^{-1} pulsed modulation. It is seen that the C^3 laser operates in a truly single longitudinal mode with no observable adjacent modes or line broadening even under high-bit-rate direct modulation.

Bit-error-rate measurements were also performed on these C^3 lasers to further ensure that they are truly single frequency and free of mode partitioning. In this experiment, the light from the front laser facet was collected by a 20 × microscope objective and focused onto the input slit of a spectrometer. The output slit of the spectrometer was set to pass only the dominant longitudinal mode. The wavelength-filtered output light was focused onto a high-speed photodetector and the resulting electrical signal was amplified and tested for errors by using a bit-error-rate test set. The resulting bit

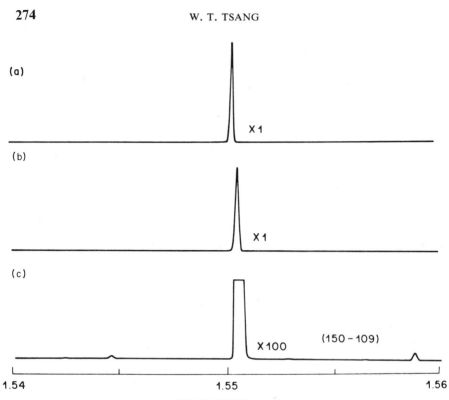

FIG. 10. Lasing spectra of a 1.55-μm ridge-waveguide C³ laser at 20°C (a) under cw opera-tion: $I_b = 13$ mA, $I_f = 59$ mA, (b) and (c) under 2 Gbit sec⁻¹ direct modulation of the laser section: $I_b = 13$ mA, $I_f = 43-75$ mA.

streams and eye patterns at 500 Mbit sec⁻¹, 1 Gbit sec⁻¹, 1.5 Gbit sec⁻¹, and 2.0 Gbit sec⁻¹ are shown in Fig. 11. Error-free bit rates were obtained through the spectrometer. By comparison, mode partition noise of a con-ventional buried crescent laser operating under the same conditions had error rates as high as $10^{-3}-10^{-2}$. The low error rate of C³ lasers under high-bit-rate direct modulation further confirms that they are operating stably in a truly single longitudinal mode with no mode-partitioning events. This indicates that C³ lasers should be suitable sources for high-bit-rate transmission through highly dispersive fibers. These results are of great inter-est since at 1.55 μm, the present silica-based fibers have the lowest loss, though they are accompanied with a large chromatic dispersion. Thus, in order to utilize this wavelength region for high-bit-rate and very long-dis-tance optical transmissions, an ultrapure, stable, single-frequency source at 1.55 μm is essential.

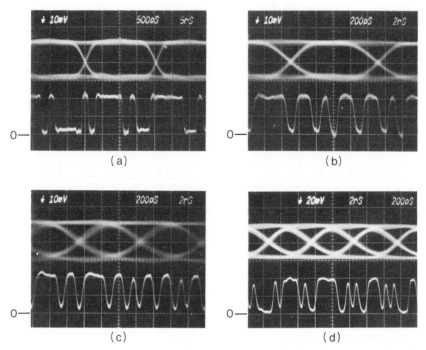

FIG. 11. Bit streams and eye patterns of a 1.55-μm ridge-waveguide C^3 laser at (a) 500 Mbit sec^{-1}, (b) 1 Gbit sec^{-1}, (c) 1.5 Gbit sec^{-1}, and (d) 2 Gbit sec^{-1}: $\lambda = 1.55$ μm, $T = 20°$C, $I_b = 13$ mA, $I_f = 43-75$ mA.

Figure 12a shows an example of the cw $L-I$ curves of a 1.5-μm buried crescent C^3 laser at room temperature under the operation conditions shown in the insert. For curves with $I_R < 20$ mA (the threshold of the right diode), the $L-I$ curves were smooth both below and above threshold. However, for curves with $I_R > 20$ mA, a modulation or ripple structure was observed in the $L-I$ curves below threshold. Sometimes the same structure is seen above threshold, but in those cases the ripples are more widely separated in current. Spectral measurements show that these structures are associated with the currents in which longitudinal mode switching occurs. A detailed study will be given in a later part.

Figure 12b shows the $L-I$ curve and the spectra at various current levels of another buried crescent C^3 laser. A wide range of stable, single-frequency operation was obtained even with highly nonideal original lasers. Mode discrimination ratios in excess of 500:1 were measured both under cw and high-speed modulation of 420 Mbit sec^{-1} as shown by the spectra given in Fig. 12c. The low power saturation was not due to the C^3 structure but was

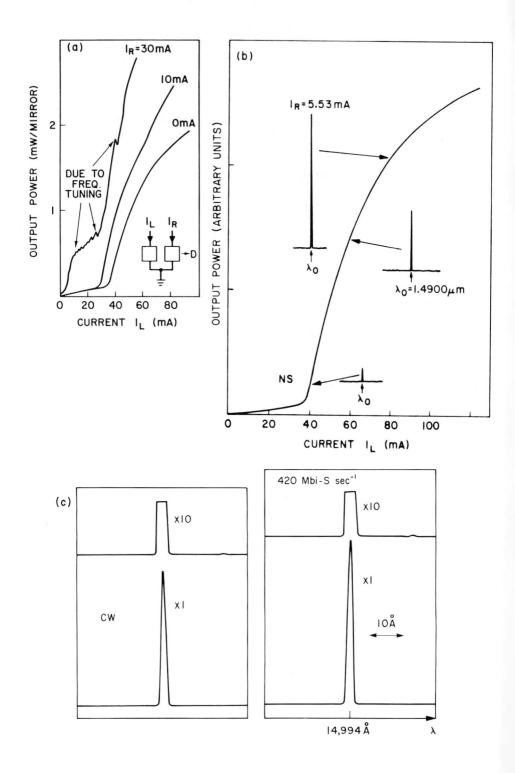

present in the original 1.5-μm crescent structure. It is believed to be due to serious current leakage across and/or around the reverse-biased current blocking junction. A typical threshold temperature-dependence coefficient T_0 of ~30 K was measured for both diodes pumped together. Again, this value is similar to that of the original crescent laser.

IV. Discrete Single-Frequency Tuning

In this part, a new mechanism of direct frequency modulation, the cavity-mode-enhanced frequency modulation (CMEFM), using a C^3 semiconductor laser demonstrated by Tsang *et al.* (1983a) is described. By using this mechanism, a very large-frequency excursion of 300 Å (typically ~150 Å) and a frequency tuning rate of 26 Å mA^{-1} (typically ~10 Å mA^{-1}) have been obtained with 1.3- and 1.5-μM C^3 lasers.

The basic working principle is illustrated schematically in Fig. 13. The propagating modes in each active stripe can have different effective refractive indexes N_{eff} even though they have the same geometric shape, size, and material composition. This is because N_{eff} is a function of the carrier density in the active stripe. This can be varied by adjusting the injection current below threshold when the junction voltage is not saturated. Thus, the mode spacings for active stripes 1 and 2 will be different and will given by $\Delta\lambda_1 \sim \lambda_0^2/2N_{eff1}L_1$ and $\Delta\lambda_2 \sim \lambda_0^2/2N_{eff2}L_2$, respectively. They are represented schematically by the solid lines in Fig. 13. As already discussed and demonstrated in Part III, stable single-longitudinal-mode operation is obtained as a result of longitudinal mode selection due to the coupled-cavity and gain roll-off of the active medium. This is again illustrated schematically in Fig. 13.

With this understanding and an examination of Eq. (1), we can then easily show how wide-range direct-frequency modulation with a large frequency-tuning rate can be achieved in C^3 lasers. Let laser 1 be biased with an injection current level I_1 above lasing threshold, so that it acts as a laser. Let laser 2 be biased with some current I_2 below threshold, so that it acts as an etalon. Under these conditions, the situation is described by the solid lines in Fig. 13. Now, if we increase I_2 to I_2' and keep I_1 constant, a change in the carrier density in active stripe 2 will correspondingly induce a decrease from N_{eff2} to N_{eff2}'. This results in a shift of the modes of laser 2 toward shorter wavelength as shown by the dashed lines in Fig. 13. As a result of such

---

FIG. 12. (a) Continuous wavelength $L-I$ curves at room temperature under the operation conditions shown in the insert. (b) Continuous wavelength $L-I$ and spectra of another C^3 laser showing the wide range of stable single-frequency operation. (c) Continuous wavelength and high-speed modulated (420 Mbit sec^{-1}) spectra for the same C^3 laser as in (b).

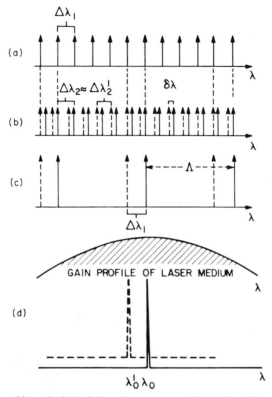

FIG. 13. Basic working principle of direct frequency modulation in a C^3 laser. In this operation, the left Fabrey–Perot diode is operated as the laser, while the right Fabry–Perot diode is operated as the modulator. The solid and dashed lines correspond to a pulsed current of I_2 and I_2' applied to the modulator, respectively. (a) Allowed modes for diode 1, (b) allowed modes for diode 2, (c) resultant modes for C^3 laser, and (d) resultant laser spectrum: $\delta\lambda = \lambda. \delta N_{eff2}/N_{eff2}$, $\Lambda \approx \Delta\lambda_1 \Delta\lambda_2/\Delta\lambda_1 - \Delta\lambda_2$.

changes, the modes from laser 1 and etalon 2 that originally coincide become misaligned and the adjacent mode on the shorter wavelength side becomes coincident and is enforced instead. This results in a shift by one mode spacing of the laser toward the shorter wavelength. Since the change in N_{eff2} necessary to shift the next adjacent modes into alignment is small, only a small change in I_2 is sufficient.

The amount of the frequency shift is greatly enhanced by the necessity to jump at least one discrete mode spacing of laser 1 (~ 15 Å for a 136 μm cavity at 1.3-μm, ~ 20 Å at 1.5-μm). Further increases in I_2 will continue to shift the lasing mode to the next one. This new mechanism, which we call cavity-mode-enhanced frequency modulation (CMEFM), results in a very large frequency-tuning rate and a very wide frequency-tuning range (at least half of the spectral width of the gain profile, i.e., $\gtrsim$ Å in one direction). If the roles of the laser diode and the frequency tuning diode are reversed, tuning will proceed in the opposite direction (i.e., toward longer wavelengths). This is because, in this case, the modes of cavity 1 are being shifted with respect to cavity 2 (see Fig. 13). In either direction, the number of modes that can be shifted depends on the number of modes within Λ, which, in turn, depends on the relative lengths L_1 and L_2. Note that the range can be further increased by temperature control. In addition, the C³ laser is also always operating in a highly stable, single longitudinal mode even under high-speed direct amplitude or frequency modulation. The former has been shown in the last part, and the latter will be shown in the following. Both 1.3- and 1.5-μm buried crescent C³ lasers have been studied in this discrete single-frequency tuning experiment.

Figure 14 shows the various spectra obtained with different current levels applied to the modulator (right) diode of a 1.3-μm buried crescent C³ laser of 135 μm (left diode) and 120 μm (right diode). The insert shows the detailed operation conditions. The laser (left) diode was biased with a 48-mA current pulse 500 nsec long. This corresponds to 1.3 times the threshold of this diode when no current was applied to the right diode. The current threshold of the

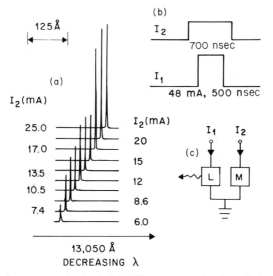

FIG. 14. (a) Various spectra obtained with different current levels applied to the modulator. (b) and (c) The detailed operation conditions. The C³ laser operated in single-longitudinal mode under all conditions.

right diode alone was 30 mA. The total threshold with both diodes pumped together was 46 mA. It is seen that a frequency shift of 150 Å has been achieved. Such a frequency shift is to be compared with the 0.4-Å shift obtained in intracavity electro-optically modulated lasers (Reinhart and Logan, 1975) or ~ 20 Å in the spectrum envelope shift (Reinhart and Logan, 1980) and the approximately 1-GHz shift in direct modulation of single laser diodes (Saito et $al.$, 1982; Dandridge and Goldberg, 1982). Furthermore, a frequency tuning rate of 10 Å mA^{-1} was obtained. Even larger values can be obtained by having shorter cavities. Such values are again to be compared with ~ 100 MHz mA^{-1} in direct modulation of single laser diodes (Saito et $al.$, 1982; Dandridge and Goldberg, 1982) and 0.02 Å V^{-1} in intracavity electro-optic modulation (Reinhart and Logan, 1975). The fact that the C^3 laser is operated in a clean single longitudinal mode is also very important.

Figure 15a shows the time-resolved transient mode spectra under high-speed direct frequency modulation achieved under the conditions described by Fig. 15b. The spectra were measured from the left and right sides of the C^3 laser, respectively. A dc current of 29.2 mA was applied to the laser (left) diode and a dc current of 20.3 mA with a superimposed pulse current of 10 mA (~ 1 nsec rise time) was applied to the modulator (right) diode. The spectra shown were taken at various elapsed times at 1-nsec intervals starting 4 nsec before the onset of the frequency-switch due to the application of the current pulse to the modulator. Complete frequency switching (from 13,058 to 13,042 Å) was accomplished in ~ nsec, which was limited by the rise time of the current pulse used. The laser operated $stably$ in single longitudinal mode. To investigate the pulse-to-pulse spectral variations, the optical pulse responses at the single longitudinal modes before frequency switching (13,058 Å) and after frequency switching (13,042 Å) as selected by the spectrometer were displayed with a sampling oscilloscope as shown in Figs. 15c–d. The clean flat pulse responses indicate that there was no pulse-to-pulse spectral variation.

This discrete frequency-tuning experiment was also performed with 1.5-μm buried crescent lasers (Tsang et $al.$, 1983a). Figure 16a shows a typical example of the various spectra obtained with different current levels applied to the modulator diode for a 1.5-μm C^3 buried crescent laser. A frequency excursion of ~ 150 Å and a tuning rate averaged over many steps of ~ 13 Å mA^{-1} was typically obtained. Again, highly single-longitudinal-mode operation was obtained at the same time. In Fig. 16b we show a 1.5-μm C^3 buried crescent laser that allows a frequency shift as large as 300 Å in 13 discrete steps and an averaged frequency tuning rate as large as 26 Å mA^{-1}. Figure 16c shows the increment in current ΔI_L needed for the C^3 laser to shift by one longitudinal mode spacing as a function of lasing wavelength. It is

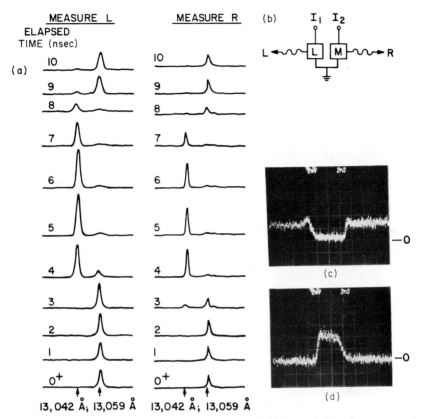

FIG. 15. (a) Time-resolved transient mode spectra under high-speed direct frequency modulation achieved under conditions described by (b): $I_1 = I_{dc} = 29.2$ mA, $I_2 = I_{dc} + I_{pulse}(5$ nsec$) = 20.3$ mA $+ 10$ mA. The rise time of the current pulse applied to the modulator was approximately 1 nsec. Spectrally resolved optial pulse responses at the single-longitudinal modes (c) before frequency switching (13,058 Å) and (d) after frequency switching (13,042 Å).

seen that ΔI_L increased parabolically with increasing lasing wavelength. This is consistent with the fact that as the modulator diode approaches threshold, the carrier density (or junction voltage) begins to saturate.

With such large frequency shifts we can, in fact, use the C^3 lasers as optical sources for multilevel multichannel FSK transmission systems (Tsang *et al.,* 1983c) by employing a single optical fiber. Each individual wavelength can be utilized as one level. The spectral separation of each level can be controlled by using the proper cavity lengths for the C^3 laser. They can also be used as sources in wavelength division multiplexing (WDM) systems.

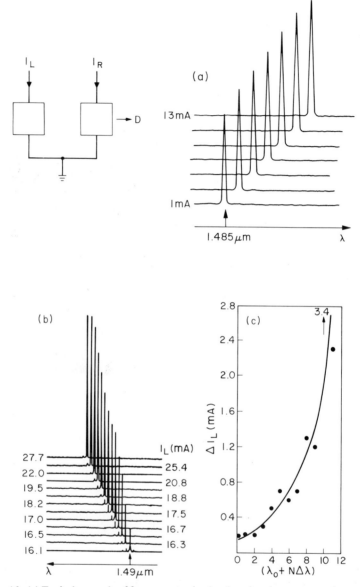

FIG. 16. (a) Typical example of frequency tuning by changing the current to the modulator diode of a 1.5-μm wavelength C^3 laser: $I_R = 38$ mA. (b) Spectra from a 1.5-μm wavelength C^3 laser with the largest frequency tuning range of 300 Å and a tuning rate of 26 Å mA^{-1}: $I_R = 22$ mA. (c) The increment ΔI_L in modulator current needed for the C^3 laser wavelength to shift by one longitudinal mode spacing plotted as a function of lasing wavelength: $\lambda_1 = 1.49$ μm, $\Delta\lambda = 23$ Å.

V. Analog Frequency Modulation

As discussed in Part I, coherent optical-fiber transmission systems are expected to significantly improve system performance toward longer repeater spacings and higher information capacity. For such coherent transmission applications, it is desirable to have a large Δf by a small modulation current ΔI_m without serious spurious intensity modulation. In addition, it is also desirable that the frequency deviation per unit modulation current amplitude $\Delta f/\Delta I_m$ be relatively independent of the modulation frequency f_m.

Part 4 described a new mechanism of frequency modulation, CMEFM, using a C^3 semiconductor laser. In this mode of operation, the laser output is electronically tunable among as many as 15 discrete single frequencies, each separated by ~ 20 Å. This makes the C^3 lasers suitable as light sources for ultrahigh-capacity WDM systems and multilevel multichannel FSK systems. In this part, a new and different mode of frequency modulation using C^3 lasers is described.

To better understand this mode of operation, a brief summary of the operational characteristics of C^3 lasers will be helpful. The C^3 laser has two optically coupled but electrically isolated diodes. The laser diode is always operated above threshold. The other diode, the modulator, can either be below or above threshold, corresponding to two modes of operation of the C^3 laser. When the C^3 laser operates in a tuning mode, the modulator is below threshold, and the lasing wavelength can be controlled by varying the current to the modulator diode. The tuning in effect is a result of the shift in the modes of the modulator with respect to that of the laser section resulting from the index change that is controlled by the modulator current via the carrier density. Because of the optical coupling of the two diodes, such a slight shift of the modulator modes results in a shift (~ 20 Å) in the enforced mode of the coupled cavity to the adjacent mode of the laser diode. This discrete longitudinal mode tuning is what Tsang *et al.* (1983a) reported as the CMEFM. Figure 17 shows a recording of the light output power from the modulator diode versus the modulator current for seven values of the laser current for a buried crescent C^3 laser. The bumps in the $L-I_m$ curve result from the tuning of one longitudinal mode to the next in the C^3 laser as the modulator current is varied. The peak of each bump corresponds to the optimum modulator current for each selected longitudinal mode. Such structures on the $L-I$ curves were also present in Fig. 12a.

Conventional analog (as opposed to discrete mode tuning) FM can also be accomplished as described earlier by operating one diode above threshold (producing the necessary output power) and biasing the other diode with an appropriate dc current that is below threshold and superimposed with a

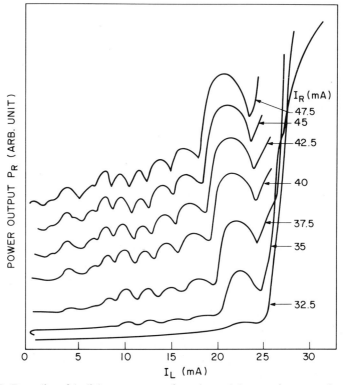

FIG. 17. Recording of the light output power from the modulator section versus the modulator current for seven values of the laser currents measured from 1.3-μm wavelength buried crescent C³ laser.

small modulating current I_m as proposed and demonstrated by Tsang and Olsson (1983). In comparison with direct FM of conventional semiconductor lasers, this scheme offers the following advantages:

(1)   The necessary output power can be produced independently by biasing the laser section above threshold, while FM is accomplished by modulating the modulator section below threshold.

(2)   Since the modulator is below threshold, effective carrier density modulation is achieved with a small modulating current. This produces very significantly larger values for $\Delta f / \Delta I_m$.

(3)   This scheme is applicable to all laser structures including buried heterostructure (BH) lasers. In fact, since the injected carriers are well confined in the active stripes in BH lasers, they should be more effectively utilized in inducing a corresponding index change in BH–C³ lasers than in

C^3 lasers formed from laser structures with lateral carrier spreading. Hence, BH-C^3 lasers should have larger values for $\Delta f/\Delta I_m$ than other C^3 lasers.

(4)   Since the modulator section is below threshold and only a small modulating current is needed, there is negligible spurious intensity modulation.

(5)   The laser output from the C^3 laser is highly single longitudinally moded.

Figure 18 shows the experimental configuration for measuring the frequency modulation characteristics using an unbalanced stabilized Michelson interferometer. A typical recording of $\Delta f/\Delta I_m$ at $f_m = 100$ KHz as a function of the dc bias current to the modulator section I_m is shown in Fig. 19a for a buried crescent InGaAsP-type C^3 laser operating at 1.5 μm. The periodic structure has the same origin as the bumps in the $L-I_m$ curves shown in Fig. 17 (not obtained from the same laser). The abrupt transitions correspond to the currents at which longitudinal mode switching occurs. They also correspond to the locations of the cusps on the $L-I_m$ curves shown in Fig. 17. The zero crossovers of the sloped portions are the locations in which the longitudinal mode discrimination ratios are the highest (i.e., the two cavities are exactly tuned to each other as shown in Fig. 19b). They also correspond to the peaks of the bumps in the $L-I_m$ curves. Therefore, even though the C^3 laser operates in the same single longitudinal mode within each bump, it is purest (i.e., its longitudinal mode discrimination ratio is highest) at the peak. At this point, variation in the modulator current will not produce frequency fluctuation even in the gigahertz range. The frequency emitted is extremely stable. Therefore, the ability to tune the C^3 laser exactly to these points by electrical means is very important for error-free, high-bit-

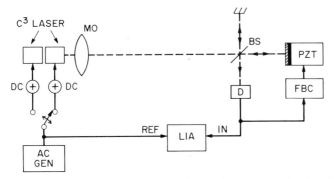

FIG. 18. Experimental configuration for measuring the frequency modulation characteristics using an unbalanced Michelson interferometer: (MO) microscope objective, (PZT) piezoelectric transducers, (D) detector, FBC feedback circuitry, (LIA) lock-in amplifier, (BS) beam splitter, and (ac GEN) signal generator.

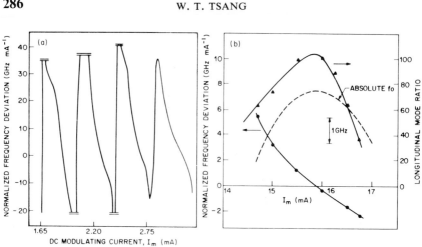

FIG. 19. (a) Typical recording of frequency deviation per unit modulation current amplitude $\Delta f/\Delta I_m$ at $f_m = 100$ kHz as a function of the dc bias current to the modulator section I_m. (b) Measured $\Delta f/\Delta I_m$ and longitudinal mode discrimination ratio and the optical center frequency f_0 as a function of I_m.

rate, very long-distance transmission systems. It also allows for frequency stabilization against environmental fluctuations and laser aging by using feedback loops. This will be discussed in detail in the next part. The longitudinal mode discrimination ratio is defined as the ratio of the intensity of the dominant mode to the next most prominent mode. Since $\Delta F/\Delta I_m$ is actually a measure of the differential slope of the center frequency f_0 versus the dc bias current I_m to the modulator, the f_0-I_m curve in Fig. 19b is obtained by integrating the $\Delta f/\Delta I_m$ versus I_m curve. It says that as the two cavities become detuned, the absolute frequency of the lasing longitudinal mode increases very slightly in the gigahertz range. This can be understood tentatively as follows. As the two coupled cavities become detuned owing to changing I_m, the round-trip optical loss increases. This increases the lasing threshold of the C^3 laser resulting in increased carrier density at threshold. Higher carrier density at threshold lowers the effective refractive index and, hence, decreases the lasing frequency as the two cavities are detuned in agreement with the observation shown in Fig. 19b. This detuning also degrades the longitudinal mode discrimination ratio. Meanwhile, it also decreases the output power producing the bumplike structures on the $L-I_m$ curves. It also becomes plausible that as the two cavities gradually become detuned, the magnitude of $\Delta f/\Delta I_m$ increases and is zero when the two cavities are exactly tuned as observed in Fig. 19a. A value for $\Delta f/\Delta I_m$ in excess of 30 GHz mA^{-1} has been obtained with longitudinal mode ratios larger than 50:1.

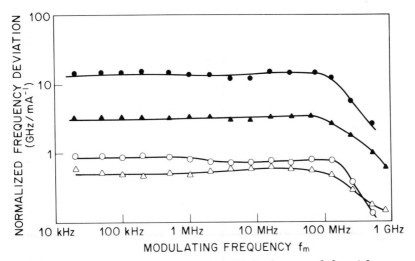

FIG. 20. Values for $\Delta f/\Delta I_m$ as a function of modulating frequency f_m for $\sim 1.5\text{-}\mu m$ wavelength buried crescent and ridge-waveguide C^3 InGaAsP-type lasers under two schemes of modulation. The filled circles represent the buried crescent C^3 laser under modulator modulation; the filled triangles represent the ridge waveguide C^3 laser under modulator modulation. The open circles represent the buried crescent C^3 laser under laser modulation; the open triangles represent the ridge waveguide C^3 laser under laser modulation.

Figure 20 shows $\Delta f/\Delta I_m$ as a function of f_m for a $\sim 1.5\text{-}\mu m$ wavelength buried crescent and a ridge-waveguide C^3 InGaAsP-type laser under two schemes of modulation. In one case, the modulator section biased below threshold was modulated. In the other case, the laser section biased above threshold was modulated for comparison purposes. As discussed earlier, very significantly larger values for $\Delta f/\Delta I_m$ were obtained by modulating the modulator section than were obtained by modulating the laser section. In fact, the present $\Delta f/\Delta I_m$ values are significantly larger than any previously reported values obtained by directly modulating conventional semiconductor lasers. Furthermore, the present FM responses are also much more uniform. The roll-off beyond 150 MHz is due to the carrier lifetime limitation in the modulator section. As discussed earlier and in contrast to previously reported results obtained from modulating conventional semiconductor lasers, the buried crescent C^3 laser demonstrates a significantly larger $\Delta f/\Delta I_m$ value than the ridge-waveguide C^3 laser. In fact, as shown in Fig. 19a, $\Delta f/\Delta I_m$ values as large as 30 GHz mA^{-1} for $f_m < 150$ MHz were obtained for the crescent buried C^3 laser by modulating the modulator section.

Figure 21 shows the frequency modulation (FM converted to AM) obtained by modulating the modulator and the laser with the same modulating current amplitude of 0.2 mA peak to peak. Modulating the modulator sec-

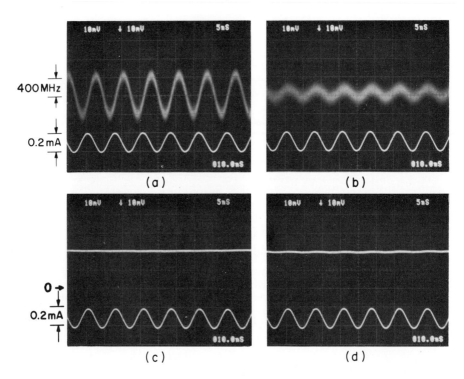

FIG. 21. Frequency modulation (FM converted AM) obtained by modulating (a) the modulator and (b) the laser with the same modulating current amplitude. The upper traces of (c) and (d) are the unintended intensity modulations corresponding to (a) and (b), respectively. The lower traces are the modulating currents.

tion produced significantly larger frequency deviation, as already shown in Fig. 20. The associated unintended intensity modulation under both conditions was obtained by blocking one arm in the Michelson interferometer. The upper traces of Fig. 21c–d are the intensity modulations corresponding to Fig. 21a–b, respectively. There was negligible intensity modulation when FM was accomplished by modulating the modulator section.

VI. Spectral Stabilization with Feedback Loop

Optical transmitters to be installed in the field must operate unattended for extended periods of time and require a high degree of stability. This is even more true for single-frequency emitters and WDM systems in which, in addition to amplitude stability, a high degree of spectral stability is necessary for satisfactory operation of the transmitter. It is therefore desirable to incorporate in the transmitter design some stabilization scheme that can compensate for environmental fluctuations and changes due to normal aging of the laser diode. In this part, a simple spectral stabilization scheme that will maintain the C^3 laser in single-longitudinal-mode operation and effectively compensate for changes due to aging and environmental fluctuations proposed and demonstrated by Olsson and Tsang (1983c) will be described.

When the C^3 laser operates in a tuning mode, the lasing wavelength can be controlled by varying the current to the modulator section. (The modulator is below threshold.) The stabilization technique described here is intended for a C^3 laser operating in the tuning mode (i.e., it will be especially useful for WDM transmitters in which close tracking of the laser wavelength is necessary. The stabilization technique of Olsson and Tsang is similar to previously reported schemes for stabilization of external-cavity semiconductor lasers

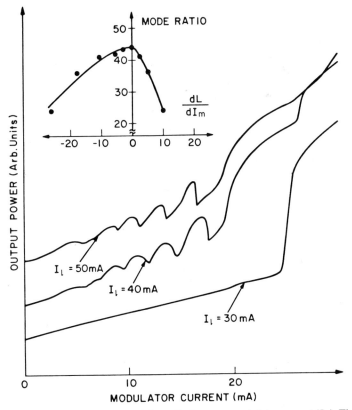

FIG. 22. Output power from the modulator diode versus modulator current (I_m). The different curves correspond to a laser current I_L of 30, 40, and 50 mA. Insert shows measured mode discrimination ratio versus dL/dI_m.

and interferometers (Preston, 1982; Olsson *et al.,* 1980) and relies on the existence of a nonlinearity in the output power of the C^3 laser versus the modulator current.

Figure 22 shows a recording of the light output power from the modulator section versus the modulator current for three values of the laser current. The bumps in the $L-I$ curve result from the tuning of the C^3 laser when the modulator current is varied as discussed in Part V. As will be shown later, the peak of each bump corresponds to the optimum modulator current for each

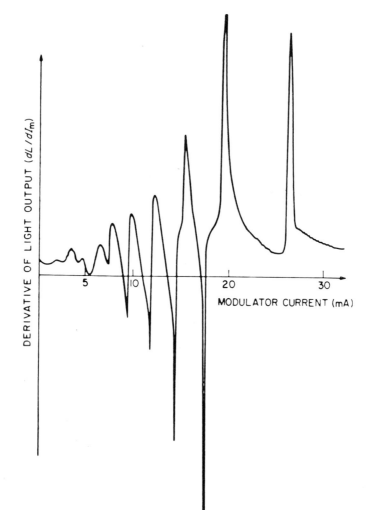

FIG. 23. Measured dL/dI_m versus modulator current measured for $I_1 = 40$ mA.

selected longitudinal mode. At the optimum point, the derivative of the light output with respect to modulator current dL/dI_m is zero, and it has opposite signs on each side of the zero point. Therefore a simple proportional regulator or feedback loop can be constructed using dL/dI_m as the error signal. Figure 23 shows the measured dL/dI_m as a function of modulator current for a laser current of 40 mA.

The experimental setup is shown in Fig. 24. A small ac current ($\sim 50\ \mu A$) at 22 KHz is applied to the modulator section of the C^3 laser. A measure of dL/dI_m is obtained by detecting the 22-kHz component of the output power with a lock-in amplifier (LIA). After amplification, the output from the LIA is fed to the modulator and the feedback loop is closed. The loop time constant is 1 nsec. As confirmed earlier and repeated here, the peak of the bump in the $L-I_m$ curve actually corresponds to the optimum modulator current. We measured the longitudinal mode discrimination ratio as a function of dL/dI_m.

The measurement was done at the 5-mA bump (see Fig. 22), and as seen from insert in Fig. 22, the best mode ratio coincides with the zero derivative. To evaluate the performance of the feedback system, we measured the range of single-mode operation both with respect to temperature and drive current through the laser section. The result is shown in Fig. 25 in which the respective operating ranges are shown. The measurement was done for three different modes at $I_m = 5$, 8, and 11 mA. An approximately 100% improve-

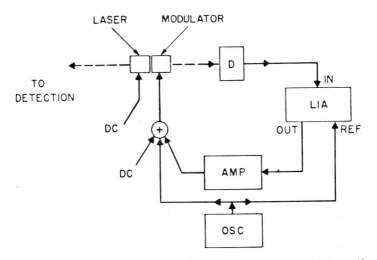

FIG. 24. Experimental setup: (D) detector, (LIA) lock-in amplifier, (AMP) amplifier, and (OSC) oscillator.

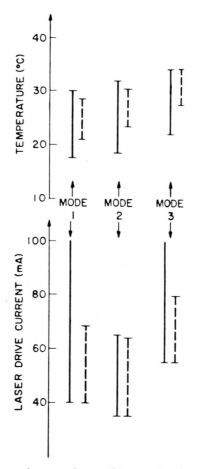

FIG. 25. Single-mode operating range for the C³ laser. Solid bars represent the ranges with feedback, and dotted bars represent the ranges without feedback.

ment is obtained in the range over which the C³ laser operates in a single longitudinal mode, both with respect to temperature and laser current. The improvement with respect to laser current is probably even higher; we did not operate the laser beyond 100 mA to avoid possible damage to the laser. As stated earlier, when 1.3 – 1.6-μm stripe lasers with more ideal Fermi level clamping properties become available, ultrastable C³ lasers operating in a single longitudinal mode over wide temperature and current ranges can be expected. The present laser is employed as an illustrative example of feedback control.

VII. Operational Characteristics of C³ Lasers

To recapitulate, the C³ laser has two optically coupled but electrically isolated diodes. Note that the assignment of the laser diode and the modulator diode is arbitrary. Similar behaviors are obtained when the roles are interchanged. When the modulator reaches threshold the Fermi level should more or less pin at the threshold level, and no further increase in carrier density or tuning is possible. Thus, one way to operate the C³ laser as a stable single-frequency source is to have both the laser and modulator sections above threshold. The other way is to operate the C³ laser in the tuning regime but with the dc current to the modulator diode biased at a point where the emitted light is the purest (i.e., the longitudinal mode discrimination ratio is the highest). This latter mode of operation requires stricter biasing control than the former, as will be shown in the following.

Figure 26 shows the various domains of single-longitudinal-mode opera-

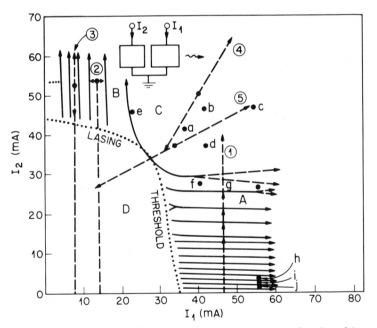

FIG. 26. Various domains of single longitudinal-mode operation as a function of the current applied to each section in a 1.55-μm ridge waveguide C³ laser. Regions A and B are the tuning regions. Region C corresponds to operation in one single longitudinal mode in which both diodes are above threshold. Region D represents the case in which both diodes are below threshold. The dashed numbered lines represent the various operation lines discussed in the text. Longitudinal mode ratios: a, 400 : 1; b, c, d, 260 : 1; e, 100 : 1; f, 230 : 1; g, 180 : 1; h, 166 : 1; i, 160 : 1; j, 150 : 1.

tion as a function of the currents applied to each section obtained for a 1.55-μm ridge-waveguide C^3 laser (Tsang *et al.,* 1983d). Regions A and B are the tuning regions corresponding to one diode biased above its lasing threshold and the other diode biased below threshold at all times. Each narrow stripe corresponds to one single-longitudinal-mode operation. Switching from one longitudinal mode to an adjacent one occurs abruptly at the boundary line. Region C, which is square in shape, corresponds to the situation in which both diodes are above threshold. This entire region corresponds to one single-longitudinal-mode operation. Thus, it is seen that the present experimental result is in excellent agreement with the above understanding. A spectral bistable region is also observed in a narrow region between regions A and C. This is due to spectral hysteresis with respect to modulator current. Region D is that for which both diodes are below threshold.

Thus, for *discrete* single-frequency tuning, one diode is biased above threshold at a constant current, and the current to the other diode is varied as shown by operation line 1. For gigahertz-range analog direct frequency modulation of the *same* single longitudinal mode, only a small modulating current is superimposed on the dc current applied to the modulator so that it is operating within one of the narrow stripes in region A or B as shown in operation line 2. For stable, single-frequency operation, there are two ways to operate the C^3 laser. First, we can operate in regions A or B by adjusting the optimal dc current applied to the modulator diode so that it is operating within one of the narrow stripes at a point where the light emitted is purest (i.e., the longitudinal mode discrimination ratio is highest or the longitudinal mode partition events are lowest). Information is then pulse-code modulated by modulating the injection current above the threshold of the laser diode as shown by operation line 3. The 119-km, 420-Mbit sec^{-1} transmission experiment to be described later was performed under this mode of operation. The second way involves operating both sections above threshold so that they are within the square region C. Longitudinal mode discrimination ratios have been measured at different points within region C and found to be approximately the same ($\sim 250:1$), and while in regions A and B, they are $\sim 160:1$ for this particular C^3 laser. Ratios as large as $\sim 2000:1$ have been obtained with other lasers operating at 1.3 and 1.5 μm (Tsang *et al.,* 1983j,k; Kaminow *et al.,* 1983b).

Similar data were also obtained from a buried crescent C^3 laser as shown in Fig. 27 (only the single-frequency region C is shown). The curved boundaries are believed to result from the increased current leakage of the reverse-biased current-blocking $p-n$ junction with increasing injection current level. For single-frequency operation in region C, the question remains as to what the optimal way of pulse-code-modulating the C^3 laser is in order to achieve wide dynamic range and good extinction ratios. A poor extinction ratio results in

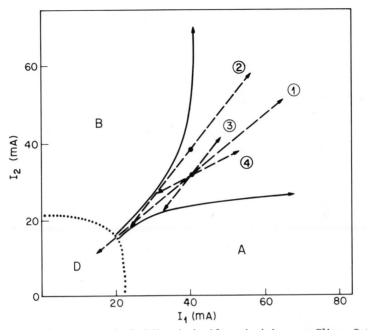

FIG. 27. Data similar to those in Fig. 26 but obtained from a buried crescent C^3 laser. Only the single-frequency region C is shown in detail. The various dashed lines show the dynamic ranges of direct modulation under various operation conditions as described in the text.

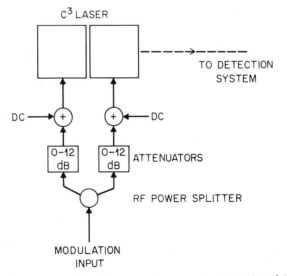

FIG. 28. Circuit arrangement for dc biasing and pulse-code modulating of the two C^3 laser diodes.

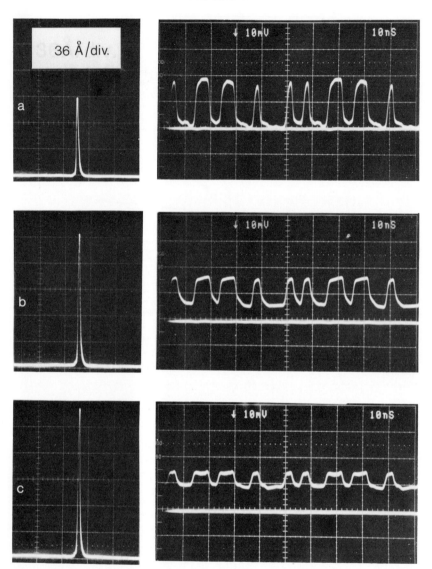

FIG. 29. Spectra and corresponding optical bit trains at 300 Mbit sec^{-1} for the various operation conditions shown in Fig. 27: (a) operation condition 1, (b) operation condition 2, (c) operation condition 3 and 4.

system penalty. In fact, the preliminary C^3 lasers employed in the 119-km transmission experiment had extinction ratios of $1:3-1:6$, resulting in a penalty of $6-3$ dB.

Figure 27 shows four different operation lines in region C obtained by

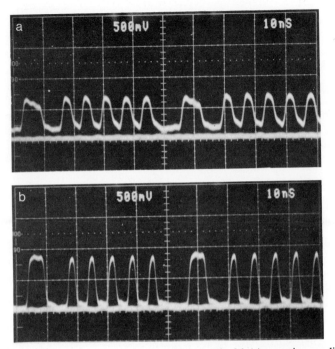

FIG. 30. Optical bit trains for operation conditions in Fig. 26: (a) operation condition 4, (b) operation condition 5.

independently dc biasing the two sections of the C³ lasers and pulse modulating both at 300 Mbit sec⁻¹ with a 64-bit word with different magnitudes by using the circuit arrangement shown in Fig. 28. The solid dot on each operational line represents the dc biasing point. The length of the line represents the dynamic range of the pulse modulation within which a longitudinal-mode intensity discrimination ratio of 150 : 11 was obtained. The different slopes are produced by having different combinations of relative pulse modulation magnitudes applied to the two sections. The corresponding optical bit trains at 300 Mbit sec⁻¹ and spectra are shown in Fig. 29. Although in all cases the spectra are highly single longitudinally moded, the extinction ratios differ significantly. Complete modulation was obtained for operation condition 1. Only in this case is the lower extreme of the modulation reaching below threshold, while in all other cases the lower limit stays above threshold, since it is limited by the boundary lines of longitudinal mode switching. Operation line 1 is the optimal line that passes through the apex of region C into the nonlasing region without trespassing switching regions. Hence, with this operational line, single-frequency operation with wide dynamic range and 100% modulation depth is obtained.

In the case of the C³ ridge-waveguide lasers studied, because of the better device characteristics of the base lasers, correspondingly higher-performance C^3 lasers were obtained. Figure 30 shows the optical bit trains for operation lines 4 and 5 in Fig. 26. It is seen that the ability to modulate deep below threshold without mode switching again results in 100% modulation.

The thermomechanical integrity of C^3 lasers has also been demonstrated (Olsson et al., 1983a) by investigating the effect of repeated thermal cycling on the dynamic spectral properties of a C^3 laser. Thermal cycling between 20 and 100°C is shown to give no measurable change in operating wavelength, mode rejection ratio, or extinction ratio.

VIII. 1.55-μm, Single-Frequency Transmission Experiments using C^3 Lasers

In the preceding parts, we have shown that C^3 lasers operate stably in single longitudinal mode with no observable mode-partition problem even under direct bit-rate modulation in excess of 2 Gbits sec^{-1}. In this part, the results of unrepeatered transmission with 1.55-μm single-frequency C^3 lasers and using optical fiber that has 2080 psec nm^{-1} chromatic dispersion are described.

As discussed in Part I, the present silica optical fibers with dispersion minimized at 1.3 μm have still lower losses in the 1.5-μm wavelength window. However, they are accompanied by a relatively large chromatic dispersion of 15–18 psec km^{-1} nm^{-1}. This leads to significant system performance penalites at high bit rates. Though the zero-dispersion wavelength can be moved from 1.3 to 1.5 μm by changing the fiber design, at present this increases the attenuation at 1.5 μm so that little advantage is gained over operation at 1.3 μm (Yamada et al., 1982; Ainslie et al., 1981).

Several transmission experiments at ~ 1.5 μm have been reported employing single-frequency lasers obtained from injection-locked diodes (Malyon and McComa, 1982), external cavity controlled diodes (Cameron et al., 1982), and DFB laser diodes (Iwashita et al., 1982).

1. 420-MBIT SEC^{-1}, 119-KM TRANSMISSION EXPERIMENT

In this experiment, C^3 lasers formed from both 1.5-μm crescent (Tsang et al., 1983b) and 1.55-μm ridge-waveguide GaInAsP-type laser wafers were employed. When operated as a single-frequency source for the optical transmitter, the back diode was dc biased below threshold for optimal frequency control, while the front control diode (exit beam coupled to the fiber) was dc biased with an NRZ signal superimposed.

The important laser characteristics of the 1.5-μm C^3 crescent and 1.55-μm ridge-waveguide lasers are given in Table I. Both lasers emit in a single longitudinal mode even under direct modulation as high as 1 Gbit sec^{-1}. The

TABLE I

SINGLE-FREQUENCY C³ LASER CHARACTERISTICS

Parameter	Crescent laser	Ridge-waveguide laser
Wavelength	1.5 μm	1.546 μm
Linewidth	$\lesssim 1.5$ Å	$\lesssim 1.5$ Å
Threshold	15–30 mA	30–50 mA
Far-field	$\theta\perp \sim 50°, \theta_{\parallel} \sim 40°$	$\theta_{\P} \sim 40°, \theta_{\parallel} \sim 20°$
T_0	~ 30 K	~ 35 K
Peak power launched	1.5 mW	1.8 mW

longitudinal mode discrimination ratio is in excess of 150 : 1 and values as high as 600 : 1 have been obtained. The linewidths are less than 1.5 Å. Figure 31 shows the lasing spectra of a C³ crescent and a C³ ridge-waveguide laser modulated as described earlier at 420 Mbit sec^{-1} with an NRZ $2^{15} - 1$ pseudorandom word. The fine structure in the spectrum of the C³ crescent laser is due to the presence of the first-order transverse mode in addition to the fundamental transverse mode. Typical threshold currents are 15–30 mA for the C³ crescent lasers and 30–50 mA for the C³ ridge-waveguide lasers at 20°C.

During the transmission experiment, the laser was mounted in a thermoelectrically controlled module and was operated at a variety of temperatures in the 11–30°C range. In general, it was found that the buried crescent C³ lasers (and also regular crescent lasers) had a roll-off beyond 200 MHz in the modulation response. This is believed to be caused by a nonoptimized reverse-biased $p-n$ junction employed for current confinement. As a result of this, the output power in general tends to saturate at ~ 3 mW per mirror as shown in previous parts.

In contrast, the C³ ridge-waveguide lasers possess excellent modulation response even beyond 2 GHz and do not appear to have any serious power saturation problem. Because of the differences in lasing wavelengths, extinction ratio for maintaining stable single-frequency operation, and $L-I$ characteristics, a transmission of 108 km was achieved with a Ge avalanche photodiode receiver for a variety of C³ crescent lasers operating at 1.486, 1.500, and 1.512 μm, and a transmission of 119 km was achieved with an InGaAs$-p-i-n$ receiver for a C³ ridge-waveguide laser operating at 1.546 μm. Detected signal power rather than chromatic dispersion limited the transmission distance in all these experiments.

The fiber for the C³ crescent laser experiment consisted of six sections each 18 km long, which made up the cable that was previously deployed in a deep-sea trail (Runge, 1983). An additional 11 km of fiber on spools was

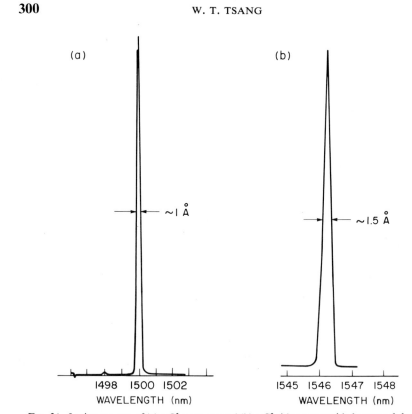

FIG. 31. Lasing spectra of (a) a C^3 crescent and (b) a C^3 ridge-waveguide laser modulated as described in the text at 420 Mbit sec^{-1} with an NRZ $2^{15}-1$ pseudorandom word.

added for the C^3 ridge-waveguide laser experiment as shown in Fig. 32. The fiber was designed for minimum dispersion at 1.315 μm, (Pearson *et al.,* 1982) and had 13-psec km^{-1} nm^{-1} chromatic dispersion at 1.5 μm and 17.5 psec km^{-1} nm^{-1} chromatic dispersion at 1.546 μm. Because of the single-frequency nature of the C^3 lasers, this fiber was useful in the 1.5–1.6-μm region. The loss of the entire 108-km section at 1.5 μm was 31.6 dB, while that of the entire 119 km at 1.546 μm was 32.2 dB. In both cases, the loss included single-mode connectors and flame-fusion splices. Connectors were used here for convenience, but in an actual system, all six connectors and six of the twelve loop-back splices would be eliminated. The loss associated with these connectors and extra splices corresponded to an effective margin of 3 dB for this experiment.

For the buried crescent C^3 laser transmission experiment, the receiver included a Ge APD and an integrating one-transistor GaAs FET preamplifier. For the C^3 ridge-waveguide transmission experiment, the receiver in-

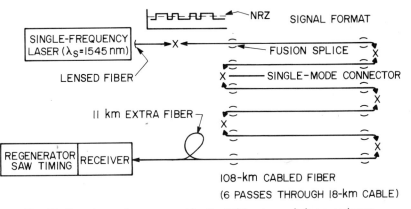

FIG. 32. Experimental arrangement for the 119-km transmission experiment.

cluded an InGaAs–p–i–n and a monolithic preamplifier of transimpedance type using Si bipolar transistors. The receiver efficiency for both was ~70%, which included fiber coupling, surface reflections, and detector quantum efficiency.

Transmission measurements were made using an NRZ format and a $2^{15}-1$ pseudorandom word. The details of the experiment will be published separately. Figure 33a shows the resulting error-rate performance for transmission paths of 108 km and 3 km with the C^3 crescent laser as the single-frequency source. Similarly, Figure 33b shows the resulting error-rate performance for transmission paths of 119, 108, and 2 km with the C^3 ridge-waveguide laser as the source at 27°C. In both experiments, since the data for all these measurements fall on the same line, there is no penalty due to mode-partition noise or other dispersion effects. As expected and shown in Figure 34 for the two experiments, the eye patterns at the output of the linear channel after 108 km in both experiments show no degradation from those after 2 km or from the back face of the laser. For the C^3 crescent laser transmission experiment, a 6-dB penalty occurred at the receiver because the laser was operated at an extinction ratio of 0.6. An additional 1-dB penalty is thought to be due to a combination of excess noise associated with the extinction ratio and intersymbol interference from imperfect equalization of the receiver. For the C^3 ridge-waveguide transmission experiment, a 2.6-dB penalty occurred because the laser was operated at an extinction ratio of 0.29. The detected NRZ signal powers for a bet error rate of 10^{-9} were -40.0 dB m and -35.6 dB m with the Ge APD and InGaAs–p–i–n receivers, respectively.

The present results confirm that C^3 lasers formed from different base laser structures all operate in a stable single longitudinal mode, and no dispersion

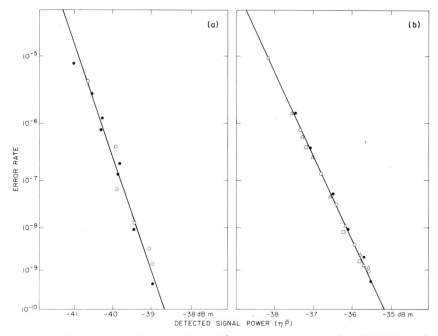

FIG. 33. (a) Error rate performance for the C^3 crescent laser transmission; 420 Mbit sec^{-1}, $\lambda_1 = 1500$ nm, $r = 0.6$, $m = 13$ psec km^{-1} nm^{-1}. Open circles represent 3-km fiber; filled circles represent 108-km fiber. (b) Ridge-wavelength laser transmission; 420 Mbit sec^{-1}, $\lambda_1 = 1546$ nm, $r = 0.29$, $m = 17.5$ psec km^{-1} nm^{-1}. Filled circle, 2-km fiber; open triangle, 108-km fiber; open circle, 113-km fiber; open square, 119-km fiber.

penalty was observed in any of these experiments. The transmission distances in these experiments were all limited by the detected signal power rather than by chromatic dispersion. Greater repeater spacings and speeds, therefore, can be achieved with C^3 lasers that operate at higher peak power and better extinction ratios. Such improvements can be achieved by having better quality basic laser diodes from which C^3 lasers can be formed.

2. 420-Mbit Sec^{-1}, 161.5-Km Transmission Experiment

After the demonstration of the record transmission experiment of 119 km by Tsang *et al.* (1983b), Ichihashi *et al.* (1983) also demonstrated another transmission experiment with a 1.55-μm DFB laser at 445.8 Mbit sec^{-1} over 134 km. Very recently, using a single-frequency C^3 ridge-waveguide laser at 155 μm and a receiver incorporating an InGaAs/InGaAsP/InP APD (Campbell *et al.*, 1983) with a GaAs FET preamplifier, Kasper *et al.* (1983) achieved transmission with a bit error rate of 5×10^{-10} at Mbit sec^{-1} over

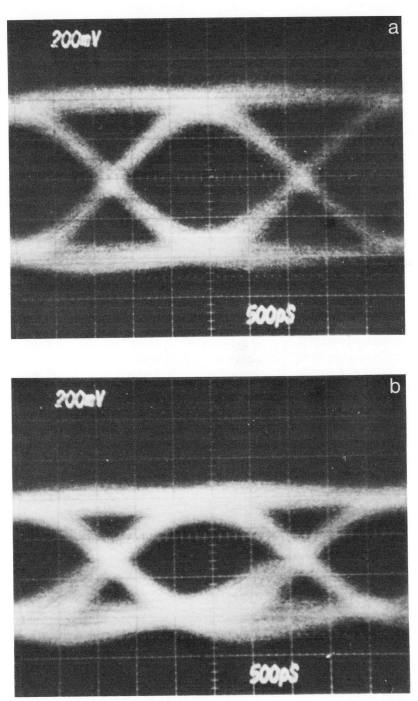

Fig. 34. Eye patterns at the output of the linear channel (a) after 2 km for C³ ridge-waveguide laser and (b) after 119 km in both experiments.

161.5 km of single-mode fiber. Although the fiber had a dispersion of 17 psec km^{-1} nm^{-1} at this wavelength, they observed no penalty due to the laser.

The block diagram of the overall transmission system is shown in Figure 35. The laser stud temperature was held at 20.1°C, and the rear cavity bias was held constant at 12.1 mA. The front cavity current was switched between 43.9 and 94.9 mA to generate a pseudorandom NRZ bit stream. The laser output was coupled with 40% efficiency to a lensed single-mode fiber jumper. The average power launched into the fiber was +0.9 dB m.

The overall loss of the 161.54 km of single-mode fiber, including 32 epoxy splices and 4 biconic connectors, was 42.8 dB. The average fiber loss before splicing was 0.239 dB km^{-1} at 1.55 μm, and the average splice loss was 0.11 dB. The dispersion minimum was at 1.3 μm.

The fiber output was coupled to a 46-μm-diameter, back-illuminated, mesa InGaAs/InGaAsP/InP APD by using a 0.29-pitch GRIN lens. At unity gain (40-V bias) the APD dark current was 35 nA, and the quantum efficiency was 68%. The detector was followed by a high-impedance 0.5-μm-gate GaAs FET preamplifier. The total input capacitance was 1.0 pF, and the load resistance was 20 kΩ. The front-end output was connected to a chain of commercial amplifiers with an rc equalizer to correct for input integration and a third-order Bessel–Thompson low-pass filter to limit noise.

With this preamplifier, the optimum APD gain was found to be $\langle M \rangle = 9.1$ with a bias voltage of 105.0 V and a dark current of 90 nA. The receiver sensitivity for a bit error rate of 10^{-9} at 420 Mbits sec^{-1} was $P = -42.6$ dB m.

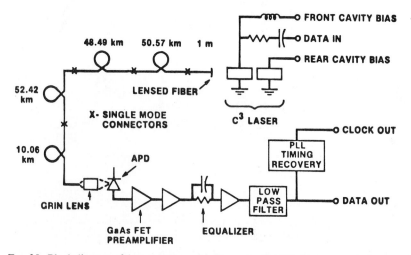

FIG. 35. Block diagram of the overall transmission system for 161.5-km transmission experiment.

An error rate of 5×10^{-10} was obtained through 161.5 km of fiber with a received power level of -41.7 dB m. The transmitted and received eye patterns are shown in Fig. 36. Error-rate curves for 161.5 and 10 km of fiber are given in Fig. 37. Average NRZ power levels for the experiment are given in the accompanying tabulation.

Average launched power	$+0.9$ dB m	Observed penalty with 161.5-km fiber	0.7 dB
Fiber and connector loss	-42.8 dB	Required power for 10^{-9} BER	-41.9 dB m
Average received power ($\bar{P}$)	-41.7 dB m	Remaining margin	0.2 dB
Receiver sensitivity for 10^{-9} BER ($\bar{P}$)	-42.6 dB m		

As shown in Figure 36a, the laser extinction ratio is greater than 40 : 1. In previous transmission experiments with C^3 lasers, the extinction ratio had to be 3.4 : 1 (Tsang *et al.,* 1983b) in order to maintain single-mode operation. It has been observed that reflections from a fiber can prevent C^3 lasers from oscillating in a single mode (Vodhanel, 1983). The first 50 km of fiber in this experiment had particularly low reflections from splices owing to the use of a very well index-matched epoxy. This may be the reason that the very good laser extinction ratio was possible.

Since the error-rate curve for 161.5 km of fiber in Figure 37 does not exhibit any bending, there is no evidence of mode-partition noise. However, the approximately 0.7-dB difference in the curves for 10 km and 161.5 km does indicate a dispersion penalty that may be due to the finite laser linewidth. The C^3 laser is very stable as a single-frequency source and is not affected by small changes in bias currents.

3. 1-Gbit Sec^{-1}, 101-Km Transmission Experiment

A 1-Gbit sec^{-1} lightwave transmission experiment using the C^3 mode-stabilized laser at 1.55 μm in conjunction with 101 km of conventional single-mode fiber with a dispersion minimum at 1.3 μm was also successfully demonstrated by Linke *et al.* (1983).

In this experiment, a 1.55-μm ridge-waveguide laser was mounted in a high-speed laboratory fixture, and its light was coupled with an efficiency of 40% into a lensed single-mode fiber pigtail. The laser stud temperature was held constant at 20°C by using a thermoelectric cooler and controller. The output laser diode was switched between two current levels (100 and 60 mA), corresponding to the *mark* and the *space* signaling levels of a pseudorandom NRZ bit stream with a repetition period of $2^{15} - 1$ bits. The current in the rear diode was held constant at 11 mA. These levels gave an average output per facet of $+5.5$ dB m and a rather poor extinction ratio e of

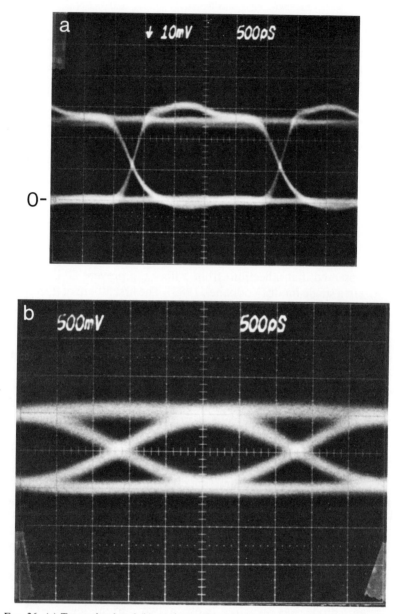

Fig. 36. (a) Transmitted and (b) received (after 161.5 km) eye patterns obtained through 161.5 km of fiber. Extinction ratio > 40 : 1.

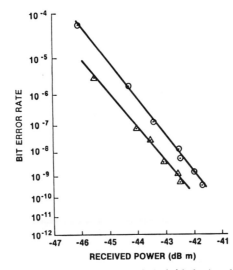

FIG. 37. Error-rate curves for 161.5 km (open circles with dots) and 10 km (open triangles with dots) of fiber at 420 Mbit sec^{-1}.

3.7:1 that gave rise to a theoretical power penalty of 2.4 dB. Larger extinction ratios could be obtained at lower prebias points but only at the expense of unacceptable error rates due to partition noise. The main-to-side-mode power ratio for the modulated laser was better than 1500:1, and the modulated line width (FWHM) was about 1 Å. Figure 38 shows the spectrum of the laser obtained through the lensed pigtail.

The fiber consisted of 100.9 km of single-mode fiber (Lazay and Pearson, 1982) made at AT&T Bell Laboratories. Its loss was 27.4 dB at 1.55 μm, including 2.9 dB of loss from 20 epoxy splices and 3 biconic connector pairs. The fiber has a dispersion minimum at 1.3 μm and a dispersion of 17.5 psec km^{-1} nm^{-1} at 1.55 μm.

The receiver front end consisted of a back-illuminated mesa $p-i-n$ diode and a packaged 0.5-μm gate GaAs FET. The FET was connected as an integrating amplifier with a front-end load resistance of 20 kΩ and a total input capacitance of 1.2 pF. The receiver pulse response did not have a perfect raised cosine characteristic owing to the use of a commercially available three-pole Butterworth filter for channel shaping; the eye was only 75% open and a 1.2-dB intersymbol interference penalty was introduced. Even with this penalty, the measured receiver sensitivity at 1 Gbit sec^{-1} and 1.55 μm was $n\overline{P} = -32.7$ dB m for an error rate of 10^{-9}. The external quantum efficiency η of the detector is 67%, which gives a received power sensitivity $\overline{P}$ of -31.0 dB m.

An error rate of 2×10^{-10} was obtained after transmission through

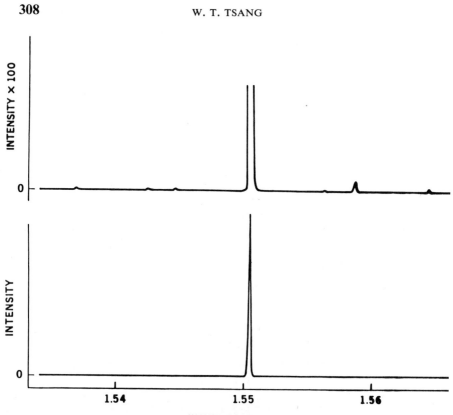

Fɪɢ. 38. Spectrum of the 1.55 –μm C³ ridge-waveguide laser under 1 Gbit sec⁻¹ pseudorandom NRZ modulation. The main-to-side mode power ratio is better than 1500:1 and the modulated line width (FWHM) ~ 1 Å.

101 km with a received power level of $\bar{P}$ of −26.2 dB m. The transmitted and received eye patterns are shown in Fig. 39. Error-rate curves obtained by decoupling the fiber from the receiver are given in Figure 40. Points are also included for transmission over an 84-km section of fiber. Power levels at a bit error rate of 10^{-9} are summarized in the accompanying tabulation.

Average launched power	+ 1.6 dB m	Received average power $\bar{P}$	−27.1 dB m
Fiber and connector loss	27.4 dB	Receiver sensitivity for $e = 3.7\ \bar{P}$	−28.2 dB m
Loss added for a bit error rate of 10^{-9}	−1.3 dB	Remaining unassigned penalty	1.1 dB

The actual extinction ratio penalty is obtained from the curves in Fig. 40 for transmission over 3 m of fiber under good (>20:1) and bad (3.7:1)

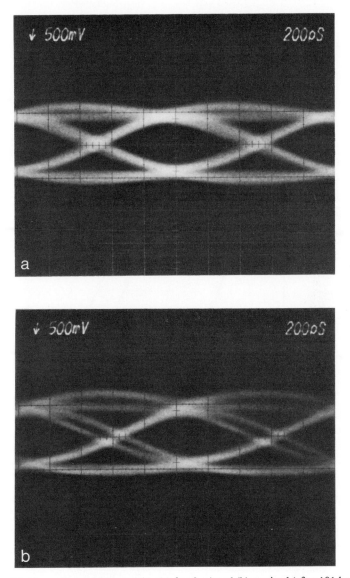

FIG. 39. Eye patterns. (a) Transmitted (after 3 m) and (b) received (after 101 km).

extinction ratio conditions. The measured sensitivity difference of 2.8 dB agrees well with a calculated penalty of 2.4 dB. An additional power penalty of about 1 dB, which is nearly independent of error rate, is observed for both 84- and 101-km transmission. Pulse broadening resulting from the finite

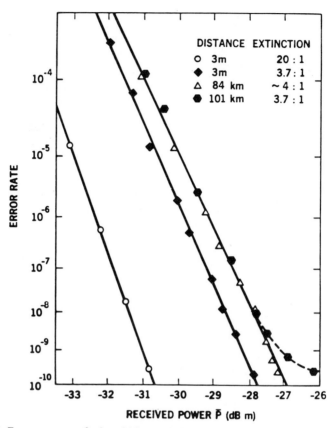

FIG. 40. Error-rate curves for 3 m, 84 km, and 101 km obtained by decoupling the fiber from the receiver.

(1-Å) linewidth can account for only about $\frac{1}{4}$ of this penalty. A bend in the error-rate curve for 101-km transmission indicates the presence of an error rate floor probably due to residual partition noise. This noise floor was not observed in the 84 km experiment, which used the same laser, but appeared at 101 km possibly as a result of the need to drive the laser slightly harder.

4. 1-GBIT SEC⁻¹, 120-KM TRANSMISSION EXPERIMENT

Linke *et al.* made use of the high-speed performance of a new APD (Campbell *et al.*, 1983) as well as the demonstrated high-speed modulation ability and single-frequency nature of a 1.55-μm ridge-waveguide C³ laser. The present 120-km result represents the longest unrepeatered span reported at speeds above 500 Mbit sec⁻¹ as well as the largest product of distance times bit rate (120 km Gbit sec^{-1}) for any speed. This product is a measure of the information-carrying capacity of a transmission system.

Figure 41 shows the component configuration for the experiment described here. The front-facet output was 5.6 and 1.5 mW for the mark and space bits, respectively, of a $2^{15}-1$ NRZ bit stream. The transmitted and received eye patterns are shown in Figure 42. The main-to-side-mode power ratio (determined from the output of the back of the laser during modulation) was 5000 : 1. Coupling of the laser light into the standard single-mode fiber was accomplished with 40% efficiency by means of a microlensed fiber pigtail. The fiber had a dispersion of 17.5 psec km^{-1} nm^{-1} at 1.55 μm and a loss of 0.24 dB km^{-1} before splicing. Epoxy splices with an average loss of 0.1 dB per splice were used. Single-mode connectors allowed for flexibility in the experimental configuration, and a variable attenuator made possible the measurement of the error-rate curves shown in Fig. 43.

The fiber was coupled with a GRIN lens to the new APD that incorporated an intermediate-band-gap InGaAsP grading layer between an InGaAs absorption layer and the InP multiplication layer (Campbell *et al.*, 1983). This intermediate layer improved the response speed of the APD by a factor of about 20. At unity gain, the APD dark current was 24 nA, and the quantum efficiency of the detector (which had no antireflection coating) was 68%. The receiver sensitivity $\bar{P}$ was -37.0 dB m at 1 Gbit sec^{-1}.

Error-rate curves for three different conditions are given in Fig. 43. Curve a shows the performance of the receiver at 1 Gbit sec^{-1} and 1.55 μm for a laser on-to-off ratio of 20 : 1 and a short fiber pigtail. A second baseline curve

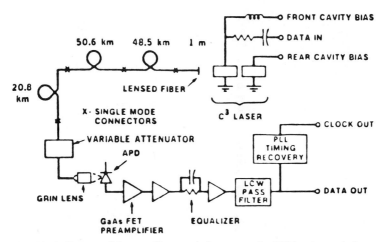

FIG. 41. Block diagram of the overall transmission system for 120-km transmission experiment.

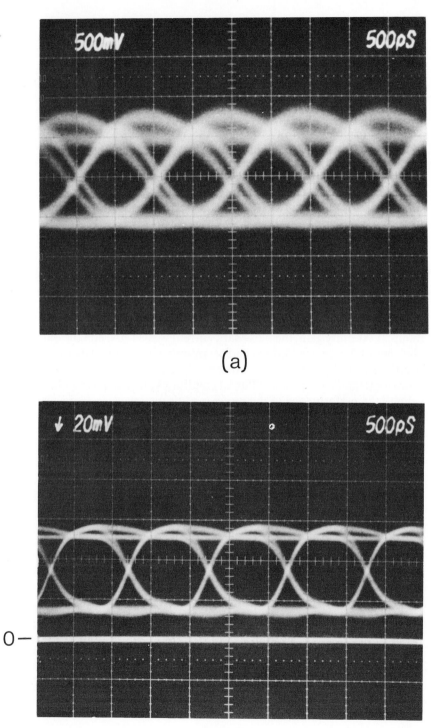

(a)

(b)

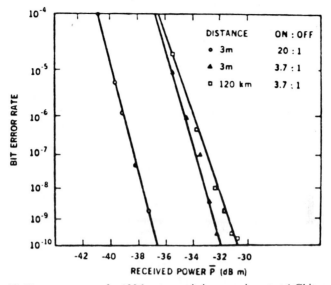

FIG. 43. Error-rate curves for 120-km transmission experiment at 1 Gbit sec⁻¹.

with the short fiber (curve b) was obtained for the on-to-off ratio that was found to be necessary in the experiment (3.7 : 1). It can be seen from the figure that the lower ratio results in a power penalty of 4.5 dB at a bit error rate of 10^{-9}. Curve c shows the actual system performance after transmission through 120 km. Another penalty of 1.1 dB at a bit error rate of 10^{-9} is seen as a result of the 0.2-nsec Å^{-1} dispersion of the fiber and the 1.4-Å linewidth of the chirped laser line under modulation. A minimum error rate of 2×10^{-10} was obtained with no evidence for an error-rate floor. This improvement is largely because of the increased sensitivity of the APD.

5. 432-MBIT SEC⁻¹, THREE-CHANNEL WAVELENGTH DIVISION MULTIPLEXING (WDM) IN SINGLE-MODE FIBER

Single-mode fiber is gaining importance in telecommunications as it allows high-bit-rate transmission with long repeater spacing in both the 1.3- and 1.5-μm regions. The large information capacity can further be exploited by WDM as either an initial installation or a system upgrade at a later date. A 11.27 – 1.34-μm two-channel WDM experiment at 432 Mbit sec⁻¹ has previously been reported (Kaiser, 1983). Recent advances in single-frequency

---

FIG. 42. Eye patterns (a) transmitted after pigtail and (b) after 120 km.

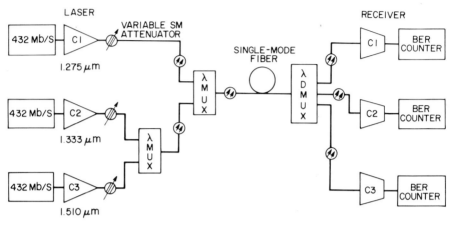

FIG. 44. Schematic diagram of the three-channel WDM experiment.

lasers make it attractive to utilize the wide band of very low-loss region near 1.5 μm. A 32-km, three-channel WDM experiment that was an extension of the two-channel experiment by adding a 1.5-μm frequency-stabilized laser and a modified multiplexer (MUX) and demultiplexer (DMUX) pair was demonstrated by Cheung *et al.* (1983).

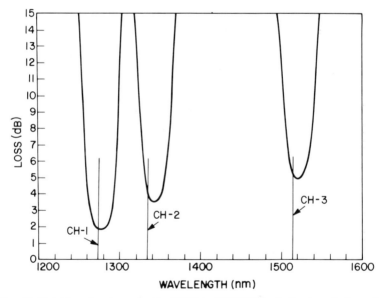

FIG. 45. Total insertion losses for the MUX and DMUX, including two single-mode connectors for the 1.27-μm channel and three for the other two channels.

A schematic of the experimental system is shown in Fig. 44. A channeled-substrate and etched-mesa-type InGaAsP/InP buried heterostructure (BH) laser operating at 1.274 and 1.333 μm, respectively, were used for the first two channels. As the third channel, a frequency-stabilized, buried crescent C^3 laser operating at 1.512 μm was used. By precisely controlling the laser stud temperature, the dc bias current, and the peak-to-peak modulation current, the laser operated in a single longitudinal mode at 432 Mbit sec^{-1} NRZ with a side-mode suppression ratio better than 1 : 90.

The multiplexer consisted of a pair of two-channel interference filter multiplexers (Lipson and Harvey, 1983) connected in tandem, first combining the 1.33- and 1.51-μm channels with a 1.41-μm long wave pass (LWP) filter, and then the 1.27-μm channel with a 1.31-μm LWP filter. The demultiplexer was a grating device with three appropriately positioned output fibers. Fig.

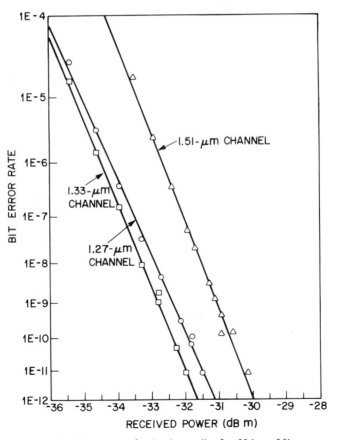

FIG. 46. Bit error rate for the three rails after 32 km of fiber.

45 shows the total insertion losses for the MUX and DMUX, including two single-mode connectors for the 1.27-μm channel and three for the other two channels. The loss values at the three operating wavelengths were 1.8, 4.5, and 4.9 dB. Optical crosstalk was 38–40 dB between the two 1.3-μm channels and 50 dB between a 1.3- and the 1.5-μm channel.

The fiber used in this experiment was a 125-μm od, 8.3-μm-core depressed-cladding-index fiber with a total loss of 17.5, 17.7, and 11.3 dB for a span of 32 km at the three laser wavelengths. The specific dispersion for the three rails were -3.4, 1.8, and 13.2 psec km^{-1} nm^{-1} for the fiber whose zero-dispersion wavelengths were located at 1.311 μm.

A 432-Mbit sec^{-1} regenerator was used at the receiver side, which consisted of an equalized receiver, a discrete linear-channel amplifier followed by a surface acoustic wave (SAW) filter timing recovery and decision circuit. The receiver included a pigtailed InGaAs $p-i-n$ and a bipolar transimpedance amplifier. The bit error rates for the three rails are shown in Fig. 46. The power penalty at a bit error rate 10^{-9} for the 1.51-μm channel was less than 0.1 dB, indicating the absence of fiber dispersion penalty. The receiver sensitivity of this channel was lower than that of the other two because of the slower rise/fall times of the particular laser used. The power penalties for the 1.27- and 1.33-μm channels (due entirely to fiber dispersion) were 0.5 and 0.2 dB, respectively, a result consistent with previous experiments (Cheung, 1983). Since the operating wavelengths of the three lasers were within the 20-nm width of the MUX/DMUX, mode-partition noise due to the loss boundary of the MUX/DMUX (Cheung, 1983) was insignificant.

IX. Multilevel Multichannel Optical Frequency Shift Keying (FSK)

All optical-fiber communications systems developed so far are based on amplitude modulation (AM) and direct detection of transmitted optical energy. Recently, interest has also been placed on optical frequency modulation (FM) and demodulation of coherent laser waves. As shown in Part IV, step-frequency tuning as large as 15 discrete frequencies can be achieved with a C^3 laser operating in the tuning mode. In addition, it is also shown that the C^3 laser operates in a clean, single longitudinal mode even under high-speed direct modulation. It has also been shown that complete frequency switching was accomplished in $\lesssim$ nsec as limited by the rise time of the current pulse used. With such large frequency shifts, Tsang et al. (1983c) demonstrated the use of a C^3 laser as the optical source for a four-level, two-channel FSK transmission system employing a single optical fiber.

Figure 47a shows the electrical pulses for channels A (2 mA) and B (4 mA) alone and multiplexed (A + B) as applied to the modulator diode. (The pulse

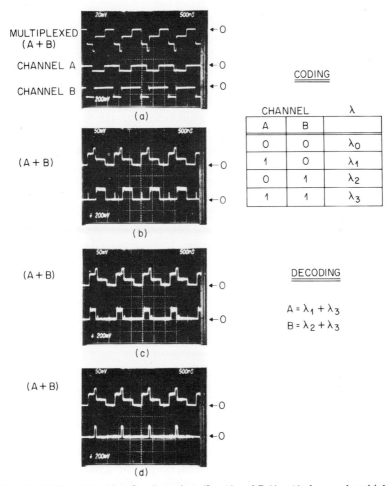

CODING

CHANNEL		λ
A	B	
0	0	λ_0
1	0	λ_1
0	1	λ_2
1	1	λ_3

DECODING

$A = \lambda_1 + \lambda_3$
$B = \lambda_2 + \lambda_3$

FIG. 47. (a) Electrical pulses for channels A (2 mA) and B (4 mA) alone and multiplexed (A + B) as applied to the modulator diode. The pulse amplitudes are increasing downward. (b)–(d) The same multiplexed electrical pulses (A + B) are shown in each as the top trace but with pulse amplitude increasing upward. The bottom trace in each figure shows the optical pulses obtained at (b) 13,107 Å, (c) 13,123 Å, and (d) 13,139 Å. Pulse intensity increases upward. The charts on the right show the coding and decoding of the four-level, two-channel optical FSK.

amplitudes are increasing downward in Fig. 47.) A dc current of 40 mA was applied to the laser diode. In Fig. 47b–d, the same multiplexed electrical pulses (A + B) are shown in each of the figures but with pulse amplitudes increasing upward. Since the lasing wavelength is a function of the current level applied to the modulator (Figs. 14 and 15), the three different current levels resulting from multiplexing channels A and B will yield lasing modes at three different wavelengths as shown by the bottom curves in Fig. 47b–d. The table on the right shows the coding and decoding of such a four-level, two-channel optical FSK system. With the C^3 laser shown in Fig. 14, an eight-level, three-channel FSK can also be constructed in a similar manner. Unlike the two-level, one-channel FSK system demonstrated by Saito *et al.* (1981, 1982), the present frequency shift is so large (16 Å versus < 1 GHz) that a direct detection scheme instead of heterodyne detection can be employed. This eliminates the use of an ultrastable local oscillator. Other time-resolved and spectrally resolved pulse response measurements have also shown that such C^3 lasers operate in highly stable longitudinal modes even under high-speed direct frequency modulation. This indicates that in actual FSK systems, high bit rates can be achieved. In a separate experiment in which the laser was frequency shifted between two wavelengths that were 15 Å apart, bit rates of up to 500 Mbit sec^{-1} were achieved with bit error rates of less than 10^{-9}.

X. Optical Switching and Routing Systems

Communications networks require some means for switching information signals between different routes and channels. With the increasing use of optical-fiber communications systems, there has evolved a need for optical switching systems. Indirectly, this can be achieved by first converting the optical signals to electrical signals and then performing the switching and routing. It is evident that switching systems that can manipulate the optical signals directly (such as integrated optic devices and the device presented here) have important advantages.

Olsson and Tsang (1983a) proposed and demonstrated a new type of optical switching and routing system based on the frequency-tunable C^3 semiconductor laser described earlier. In this switching system, the route or output port that an optical signal will follow is determined by the wavelength of the optical signal. By switching the lasing wavelength of the C^3 laser transmitter, different routes or output ports can be addressed. The key parameters of a switching system, the access time and number of addressable ports, are determined by how fast the C^3 laser can be wavelength switched and how many discrete frequencies are available from the C^3 laser. Switching times as short as 1 nsec have been observed and up to 15 discrete frequencies are available from the C^3 laser as demonstrated in the preceding part; indicating that high-performance switching systems can be built with this new scheme.

A general $N \times N$ switch network is outlined schematically in Fig. 48. It consists of N C^3 lasers that can be tuned to N discrete wavelengths $\lambda_1 - \lambda_N$.

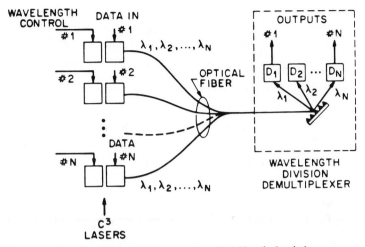

FIG. 48. Schematic diagram on an $N \times N$ optical switch.

The optical outputs from the lasers are combined and enter a wavelength division demultiplexer (WDD). In the WDD, the different wavelengths are separated, and each wavelength is detected with a separate detector. If desirable, the separated wavelengths can be launched into optical fibers for further transmission before detection. In this application, the C³ laser is operated in the tuning regime (i.e., the laser section is operated above threshold, and the optical output is taken from this side; the modulator is below threshold, and the lasing wavelength is controlled by varying the current to this section). The C³ laser step tunes between the longitudinal modes of the laser section, which are typically 20 Å apart for a 100-μm laser operating at 1.3 μm.

An experimental realization of a 1 × 4 switching system using a C³ laser operating at 1.3 μm has been demonstrated by Olsson and Tsang (1983a). A schematic of the experimental arrangement is shown in Fig. 49. The laser current I_L consists of a 46-mA dc bias and a 300-Mbit sec^{-1}, 64-bit, NRZ word with a 20-mA peak-to-peak amplitude. The wavelength control current applied to the modulator section of the C³ laser is synchronized with the data signal and is a four-level step waveform. (See top trace in Fig. 50.) The step amplitudes must be adjusted such that each of the current levels corresponds to a different wavelength. As a result, the 64-bit data word will be transmitted as four subwords, each subword transmitted at a different wavelength. By using a diffraction grating, the different wavelengths can be separated and detected. The top trace in Fig. 50 shows the wavelength control signal and the lower traces are the optical outputs at $\lambda = 1.313, 1.315, 1.317,$

threshold, the switching time is determined by the carrier lifetime. It is therefore advantageous to operate the modulator at a high current level (and, hence, high carrier density) and shorter carrier lifetime. For the experiments presented here in which only four channels were used, the modulator current can be kept at a fairly high value (11 – 19 mA) for all four channels, and the switching speeds are very fast (see Fig. 51b). When the number of channels is increased, however, some of them must be accessed at lower modulator currents and the access time will be longer. Some initial experiments show that access times up to 7 – 8 nsec can be expected when the modulator current is only a few milliamperes. The inherent crosstalk in this new optical switch is limited by the mode discrimination ratio at the C^3 laser. With the mode discrimination ratio defined as the ratio of the power in dominant mode to the power in the next most prominent mode, ratios of 100 : 1 are routinely obtained.

XI. FSK Utilizing Spectral Bistability

In Part IX, the use of a C^3 laser for multilevel multichannel FSK by Tsang *et al.* (1983c) was discussed. Olsson and Tsang (1983b) also proposed and demonstrated a new kind of FSK transmitter based on a spectrally bistable C^3 laser. They found that for some buried crescent C^3 lasers and for some ranges of modulator and laser currents (both above threshold), the C^3 laser exhibits spectral bistability and hysteresis, as shown in Part VII. That is, for a given set of current levels, the C^3 laser is operating in a pure single longitudinal mode, but which of two possible modes is lasing depends on how the operating point was approached.

This bistability can be predicted from an analysis of the nonlinear rate

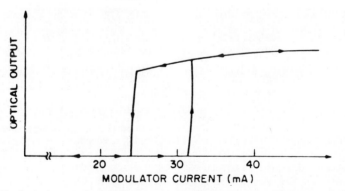

FIG. 52. Optical output at 1.5122 μm versus modulator current. The arrows indicate the direction transversed by the hysteresis loop. An identical but inverted hysteresis loop is obtained at 1.5102 μm.

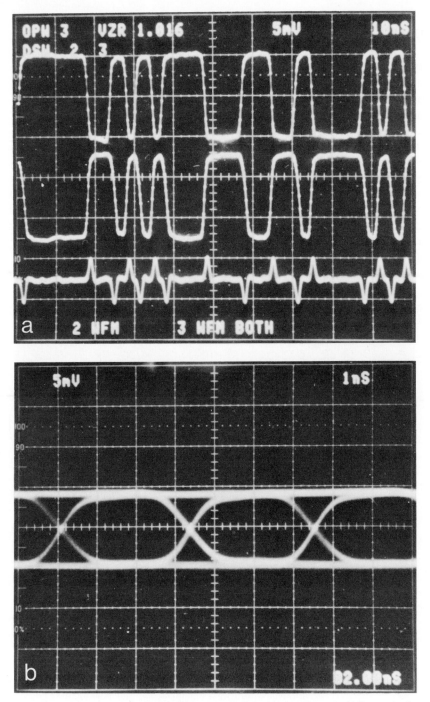

FIG. 53. (a) Optical outputs at 1.5122 μm (top tace) and 1.5102 μm (middle trace) when modulated at 300 Mbit sec^{-1}. The bottom trace is the modulating signal (20 mA div^{-1}). The time scale is 10 nsec div^{-1}. (b) Eye diagram for the optical output at 1.5 μm and 1 nsec div^{-1}.

equations and the coupling of the index of refraction to the carrier density in the two diodes. Figure 52 shows the output power in one of the bistable modes versus modulator current. The arrows indicate the direction in which the hysteresis loop was traversed. The laser current was held constant at 50 mA ($2.5I_{th}$). The threshold current of the modulator section was 25 mA. The bistable region extends from approximately 35 mA to at least 70 mA (maximum tested) with respect to the laser current. The experiments presented here pertain to an InGaAsp-type buried crescent laser operating at 1.5 μm. To operate this laser as a FSK transmitter one simply bises the modulator current somewhere inside the hysteresis loop; the laser current is adjusted to give the desired output power. By applying a short negative- or positive-current pulse to the modulator, the laser can be switched between the two wavelengths.

As will be shown later, these switching pulses can be very short, and once

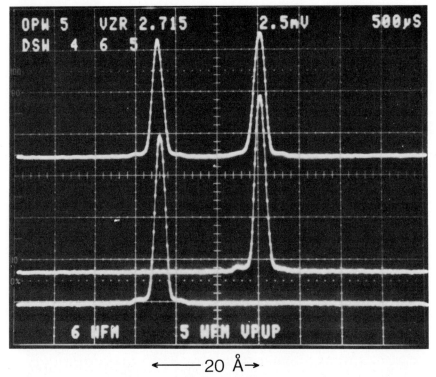

$\longleftarrow$ 20 Å $\rightarrow$

Wavelength

FIG. 54. Top trace: optical spectrum modulated at 300 Mbit sec^{-1}. Lower traces: spectra of the two bistable states. The peak separation is 22 Å.

the laser is brought into one state it will stay in that state until a switching pulse of opposite polarity is applied. Also, note that the amplitude of the switch pulses is not critical. The only requirement is that they be large enough to bring the modulator outside the hysteresis loop. This relaxes the requirement on the design of reliable optical-fiber systems. In Fig. 53a, the optical outputs at $\lambda = 1.5122$ and $1.5102\ \mu$m when the laser is frequency shifted at 300 Mbit sec^{-1} are shown. The bottom trace in Fig. 53a shows the switching pulses applied to the modulator. The switch pulses have a 1-nsec FWHM and 10-mA peak amplitude. Varying the amplitude between 6.5 and 13.5 mA (maximum available) has no noticeable effect on the observed waveforms. The eye diagram in Fig. 53b was obtained from the optical signal at $1.5122\ \mu$m and shows a detection-limited switching time of ≈ 1 nsec. The total output power varies less than 5% when the laser is switched between the two wavelengths. The spectral output from the bistable C³ laser is shown in Fig. 54. The top trace is the spectrum in which the laser is frequency shifted at 300 Mbit sec^{-1} and the two lower traces are the two stable states without modulation. The mode discrimination power ratio is approximately 250:1 for both states.

XII. Gateable Mode-Locked Semiconductor Lasers with an Electrically Controllable Absorber

It has been well established that picosecond and subpicosecond optical pulses can be obtained by active or passive mode locking of semiconductor laser diodes in external cavities (P. T. Ho *et al.,* 1978; H. Ho *et al.,* 1980; Holbrook *et al.,* 1980; Ippen *et al.,* 1980; van der Ziel *et al.,* 1981). With active mode locking, a pulse width (FWHM) of 5.3 psec has been obtained by van der Ziel (1981), while with passive mode-locking, a pulse width as short as 0.65 psec was obtained (van der Ziel *et al.,* 1981). Such ultrashort optical pulses are very attractive for many applications. However, in some important applications, like optical communications, it is a serious drawback that the optical pulses are always emitted in a continuous train. Gating of individual pulses has so far not been demonstrated. Yet the ability to gate on or off any particular optical pulse in the train is of great importance as it will allow for the first-time coding of information with such ultrafast pulses.

The mechanisms involved in active and passive mode locking are different. In active mode locking, optical pulse sharpening results from gain modulation due to the injection of an rf current at the frequency corresponding to the round-trip transit time of the optical pulse in the external cavity. In passive mode locking, optical pulse sharpening results from optical absorption of the leading edge of the pulse by a localized region of saturable absorption. Previously, the needed saturable absorber was simply the dark-line defects formed as a result of degradation of the diode (Ippen *et al.,* 1980). Since the concentration of such defects increases rapidly with time, this results in unstable characteristics of the mode-locked laser and a very limited lifetime at the same time. Recently, a technique to produce the necessary amount of saturable absortion in a well controlled manner by proton bombardment of the exit face of the diode was demonstrated (van der Ziel *et al.,* 1981). Since the ranges of penetration and proton dosage are externally variable, the resulting mode-locked diode characteristics can be optimized for the first time. This results in 0.65-psec pulses. Meanwhile, reliable operation is obtained as a result of the stable, nonlinear element introduced by proton bombardment. However, in all of these schemes, the nonlinear elements (saturable absorbers) are built-in and fixed; they cannot be varied.

Recently, Tsang *et al.* (1983e) introduced two new concepts. The first enables us to gate on or off any particular optical pulse in the train and hence allows coding of information. The second enables us for the first time to electrically adjust the amount of optical absoprtion and, hence, electrically control the mode-locked laser characteristics.

The experiment of gating the ultrashort pulses of the actively mode-locked laser will be described first. The experimental arrangement is shown in Fig. 55. Unlike the conventional mode-locking experiment, the regular semiconductor laser diode used in the external cavity is replaced with a three-terminal laser. Since the main purpose is to demonstrate the gating experiment, no effort was made to produce extremely short pulses. Hence, no antireflection (AR) coating was applied to the interior laser–air facet. Further, no effort was made to precisely determine the pulse width of the ultrashort pulses by the second-harmonic autocorrelation technique. The external cavity consisted of a 0.50-NA microscope objective that collimated the light and a 1200 lines mm^{-1} grating. The optical cavity length was ~ 70 cm, corresponding to

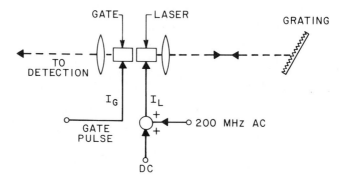

FIG. 55. Schematic diagram of an electronically gateable, actively mode-locked C^3 laser in an external cavity.

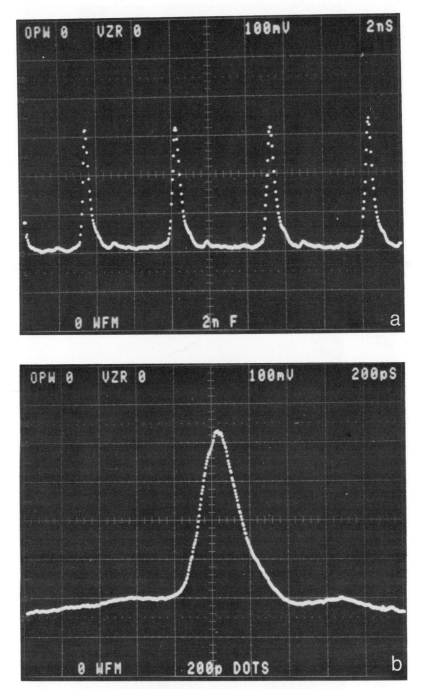

FIG. 56. (a) Continuous train of ultrashort optical pulses produced by active mode locking. (b) Detection-limited (200-psec) optical pulse emitted.

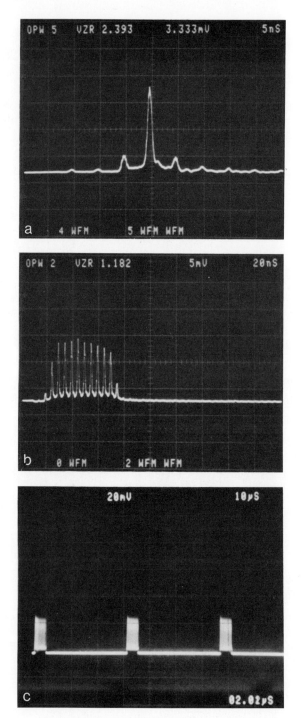

FIG. 57. Gated output from the mode-locked semiconductor laser for (a) a single optical pulse gated on, (b) and (c) for multiple optical pulses gated on with two different gating pulses.

a round trip transit time of ~ 5 nsec. The use of the grating ensures single longitudinal mode operation of the mode-locked laser and allows for longitudinal mode selection. In this experiment, active mode locking was employed. Thus, diode 2 was operated as the mode-locked laser with a 200-MHz rf (sine wave) superimposed on a dc current, while diode 1 was operated as the gating diode. Since the active stripes in the two diodes of the C³ laser are self-aligned, excellent optical coupling of the mode-locked and gating diodes was automatically achieved. On the other hand, there was complete electrical isolation between them. With no current injection to the gating diode, the large optical absorption (both diodes have the same active medium) blocks the output of the mode-locked diode and no mode-locked optical pulses are emitted. With dc current pumping to the gating diode, it becomes optically transparent to the mode-locked pulses. Consequently, a continuous train of ultrashort pulses produced by active mode locking is emitted as shown in Fig. 56a. Figure 56b shows the detection-limited (200-psec) optical pulses emitted. Since an external grating was used, the optical pulses are single frequency. Even though no second-harmonic auotcorrelation was made to determine the precise pulse width, the emitted pulses were confirmed to have been produced by active mode locking in the external cavity; that is, no optical pulse was emitted when the optical path in the external cavity was blocked. By pulsing the gating diode with different pulse widths, the desired number of mode-locked pulses can be conveniently obtained as shown in Figure 57. Figure 57a shows the case in which a single optical pulse was gated on. Figures 57b and c are that in which mulitple

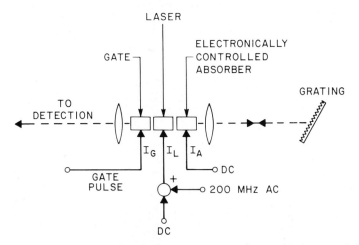

FIG. 58. Schematic diagram of an actively mode-locked C³ laser with three coupled diodes for electronic gating and optical absorption control.

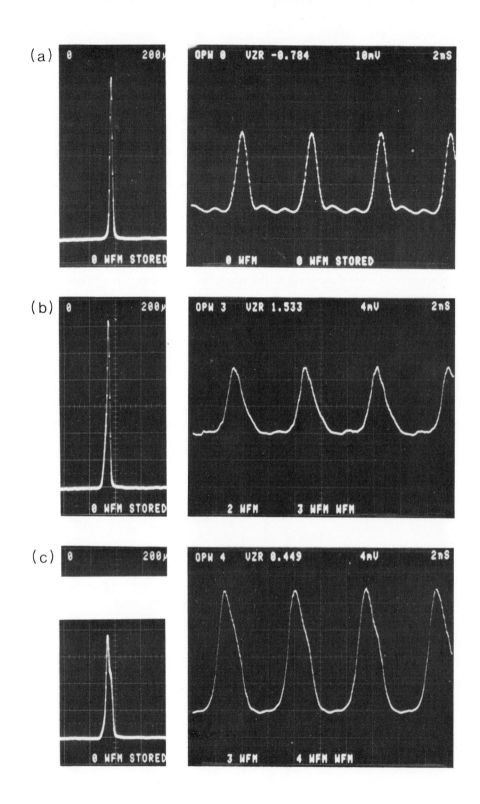

optical pulses are gated on with two different gating pulses. The residual pulses at the beginning and end result from the slow rise and fall time of the gating current pulse used.

The scheme using two coupled diodes shown in Fig. 55 can be further extended to three coupled diodes as shown in Fig. 58. In this scheme, an additional diode (diode 3) is coupled to the gating and laser diodes. Since this diode is also of the same active medium, the optical absorption can be varied by varying the current applied to it. As a result, it constitutes an *in situ* electronically controllable absorber. Consequently, with this arrangement, we have a gateable mode-locked semiconductor laser with an electronically controllable optical absorber. Gating results similar to those shown in Fig. 57 were also obtained.

Figure 59 shows the effect of different amounts of optical absorption (controlled by the dc current level injected into the absorber diode) on the optical pulses and spectra. It is evident that decreasing the amount of optical absorption (increasing the dc current to the absorber diode) increases both the optical pulse width and the spectal width. In fact, for currents $\gtrsim 15$ mA, the optical pulses shown in the figure are no longer detection limited. For currents $\lesssim 14$ mA, however, they were detection limited. As a result, we were not able to determine whether or not continuing increases in optical absorption with a current below 14 mA would result in a continual decrease in pulse width without using the autocorrelation measurement technique. The present results, however, did demonstrate the first electronically controllable optical absorber in a mode-locked semiconductor laser.

XIII. *Q*-Switching a C³ Laser with an Integrated Intracavity Modulator

A continuous train of light pulses of very short duration (< 100 psec) at high repetition rates (a few hundred megahertz to a few gigahertz) can be generated by semiconductor lasers constructed with an intracavity saturable absorber (Basov, 1968). Previously, these pulse-train generators were fabricated as tandem double-section, stripe-geometry GaAs-type lasers (Lee and Roldan, 1970). One section was biased above lasing threshold, and the other, serving as a saturable absorber, was biased below threshold. However, the achievement of PCM for a *train* of light pulses would require a separate external electro-optical modulator. Furthermore, when the conventional

---

FIG. 59. Effect on optical pulses (right) and spectra (left) of different amounts of optical absorption as controlled by the dc current I_A applied to diode 3. Scale is 15 Å div⁻¹. (a) $I_A = 14$ mA; (b) $I_A = 15.5$ mA; (c) $I_A = 17$ mA.

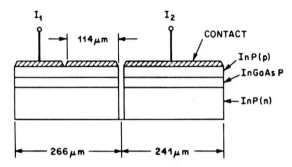

FIG. 60. Schematic diagram of a Q-switching C^3 laser with one cavity partially pumped.

lasers are made to produce deeply modulated pulsations, the optical spectrum broadens and multilongitudinal-mode output results.

Recently, Lee *et al.* (1983) used a C^3 laser that was subdivided into two parts with the electrode on one of the two diodes (Fig. 60) to produce an absorber section. This section is only partially pumped by applying current to one of the divided electrodes and thus serves as a saturable absorber. The other cavity (with full contact) serves as an amplifier and is fully pumped. Oscillation in the form of repetitive light pulses begins when light that is sufficient to bleach its optical loss is coupled into the absorber cavity from the amplifier cavity. In addition, since the mode gain is highest when the fields in both cavities interfere constructively, adjustment of the bias currents to conditions for constructive interference between the cavities can lead to single-frequency operation.

The experimental Q-switching C^3 laser was made from a V-groove buried crescent laser that operated in a single transverse mode at 1.3 μm. For the particular example described here, the lasing threshold of the fully pumped amplifier section alone occurred at 22 mA and for the partially pumped absorber section at 40 mA.

To determine the dc bias condition for sustained pulsation, the laser light output was detected by a fast InGaAs $p-i-n$ photodiode whose output was monitored by a microwave spectrum analyzer as the bias currents were adjusted. The presence of strong repetitive light pulses was indicated by the appearance of a noise–peak trace on the analyzer screen; the central frequency of this trace could be tuned by the dc bias of either the amplifier or the

FIG. 61. (a) Waveform of the repetitive light pulses produced by the Q-switching C^3 laser with dc bias and a small microwave signal for injection locking. (b) Eye diagram of the light output when the Q-switched light pulses are modulated by an NRZ pseudorandom word at a 322-Mbit sec^{-1} data rate.

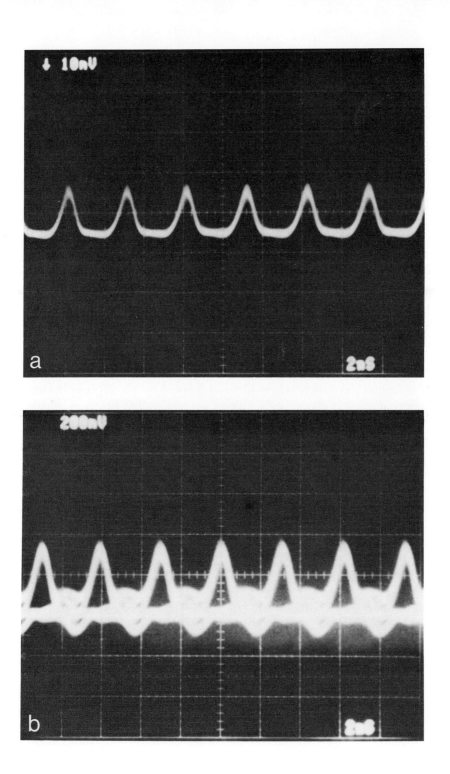

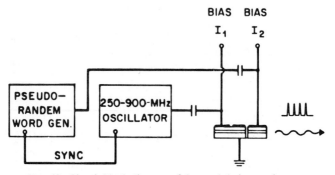

FIG. 62. Circuit block diagram of the modulation equipment.

absorber laser cavity. With a microwave injection signal of about -20 dB m, the pulse repetition frequency could be tuned from 320 to 450 MHz; with a large injection signal, oscillation frequencies as high as about 1.5 GHz were observed. Because the bandwidth of our real-time oscilloscope was approximately 400 MHz, most of our experiments were performed at about 350 MHz. Figure 61a shows the stable light output that resulted from properly adjusted bias currents ($I_1 = 42.5 - 46.5$ mA; $I_2 = 22 - 36$ mA); the modulation depth was about 100% with an average output power of approximately 1 mW.

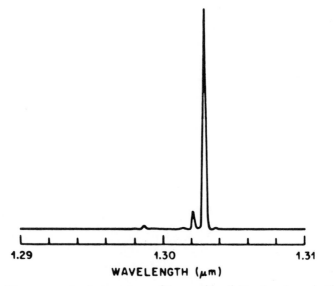

FIG. 63. Time-averaged output spectrum of the repetitive light pulses shown in Fig. 61a.

These Q-switched light pulses were modulated by a pseudorandom word generator at a bit rate that was synchronized to the pulse repetition rate, as shown in the schematic diagram of Fig. 62. Figure 61b shows the clearly opened eye diagram obtained with an NRZ pseudorandom word. The normal expectation is that, along with a modulating current pulse covering many cycles of the self-pulsation (corresponding to a string of ones), the output would be stable; however, with a short modulating pulse (corresponding to successive ones and zeros), the self-pulsation might be unstable. The observed eye diagram indicates that any such instabilities were insignificant.

The time-averaged light output spectrum from a three-section laser biased to produce stable repetitive Q-switched pulses (Fig. 61a) is shown in Fig. 63. As expected, a nearly single longitudinal mode is seen. As noted before, this is quite different from the multimode spectrum usually produced by a conventional single-cavity laser undergoing Q-switched self-pulsation.

XIV. An Amplifier–Modulator Integrated with a C³ Laser

Lee *et al.* (1984) demonstrated an external amplifier–modulator integrated with a C³ injection laser on the same 1.3-μm InGaAsP/InP-type laser chip. The maximum available gain was 20 dB at an output power of 1 mW. High on–off extinction ratios and side-mode suppression greater than 20 dB were achieved at about 1-Gbit sec^{-1} (NRZ) modulation rate.

The construction of the integrated amplifier–modulator C³ laser is rather similar to that of the C³ laser itself, except that the device consists of three cleaved sections instead of two. The original lasers, 500 μm long and emitting near 1.3 μm, were cleaved from a channelled-substrate (V-groove) buried heterostructure wafer. Figure 64 shows a schematic diagram of the integrated devices. Briefly, the two shorter sections (143 μm and 121 μm) were operated as a cw C³ laser, and the longer section (248 μm) was used as the

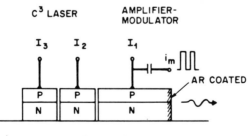

FIG. 64. Schematic diagram of the amplifier–modulator integrated with a C³ laser.

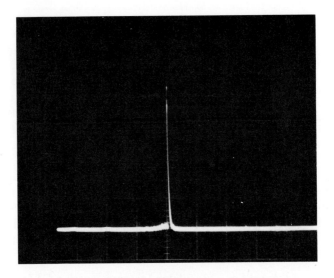

(a)

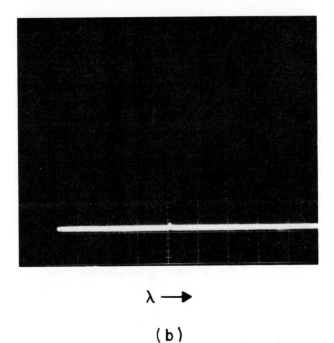

$\lambda \longrightarrow$

(b)

FIG. 65. Spectrum of the amplifier output (a) when the amplifier is on, i.e., $I_1 = 44.6$ mA (the vertical scale is 20 mV cm^{-1} and (b) when the amplifier is off, i.e., $I_1 = 0$ (the vertical scale is 5 mV cm^{-1}).

amplifier–modulator. (This section should be as long as practical to maximize absorption.) The output facet of the amplifier–modulator was antireflection (AR) coated with a single layer of sputtered silicon nitride, which is necessary to prevent the amplified signal from being reflected back into the laser cavity. With an accurate AR coating, the frequency stability of the C³ laser can be maintained even though the gain of the amplifier is changed by the modulating current. In this experiment, the reflectivity of the coated output facet was estimated to be less than 1%.

In order to achieve the best possible extinction ratio, the signal input to the amplifier must be small so that the output signal from the amplifier is low when the amplifier is off. The input signal then is amplified by the maximum available gain when the amplifier is on.

In operation, the C³ laser was pumped by two dc currents, I_2 and I_3. The center section, with current I_2, was used as a tuner to select the oscillation frequency; the outer section, with current I_3, served as the primary laser. The lasing threshold current I_{3th} was 16 mA at 15°C, while the current I_2 ranged from 4 to 8 mA. The single-frequency output from the tuner facet was much

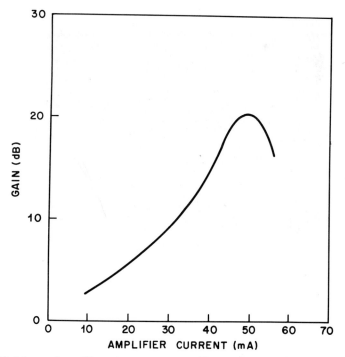

FIG. 66. Measured amplifier gain versus current. The maximum available gain is 20.2 dB. $P_{in} = -19.6$ dB m.

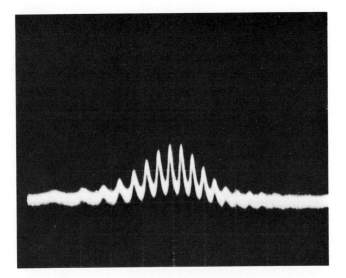

(a)

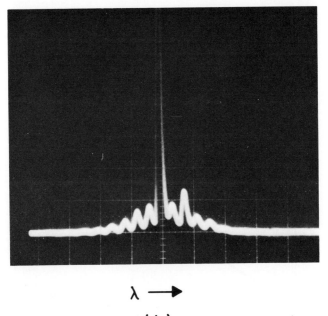

λ ⟶

(b)

FIG. 67. (a) Amplified spontaneous emission spectrum at $I_1 = 49.2$ mA; (b) gain saturation occurs at $I_1 = 53.5$ mA, when the amplitude of a side mode begins to grow.

lower than that from the laser facet, which satisfies the amplifier low-input-power condition that is required to achieve a high extinction ratio under modulation conditions. Figure 65a shows the output spectrum when the amplifier current is turned on at $I_1 = 44.6$ mA dc, and Fig. 65b shows the output spectrum when the amplifier is turned off (i.e., $I_1 = 0$). Single-frequency operation was also maintained under modulation, and a high extinction ratio (20 dB) was achieved.

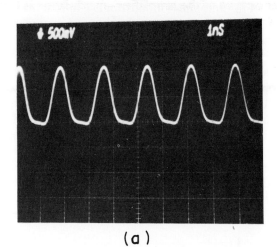

(a)

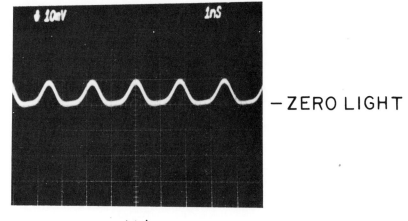 — ZERO LIGHT

(b)

FIG. 68. (a) Modulating signal pulse from the word generator; (b) detected amplifier output light pulses.

The gain versus current characteristics of the amplifier were determined by measuring the ratio of the amplifier output at a given current to that at zero current. The result is plotted in Fig. 66. The maximum available gain is 20.2 dB at $I_1 = 50$ mA, and 1-dB gain suppression occurs at $I_1 = 54$ mA. The output power at the maximum gain is about 1 mW.

Note that, with zero bias current ($I_1 = 0$), the amplifier also can serve as an optical detector. An input power to the amplifier of 10.8 μW is estimated from the detected photocurrent of 11.4 μA in the detector (amplifier section). Since the power output of the amplifier–modulator is 1 mW at maximum gain, the resulting amplifier gain is 20 dB, which is in good agreement with the directly measured value.

The amplified spontaneous emission spectrum of the amplifier section (with the laser turned off) is shown in Fig. 67a. The best side-mode suppression under modulation is obtained when the lasing frequency is tuned to coincide with the mode at the peak of the spontaneous emission spectrum. The amplifier gain saturation (at the laser wavelength) occurs when a side mode begins to grow in amplitude as the current is increased, as shown in the output spectrum in Fig. 67b.

For high-speed modulation experiments, a Tau–Tron 1-Gbit sec^{-1} word generator was used as a signal generator. The amplifier section was dc biased at the mid-gain quiescent point, $I_1 = 38$ mA, and a modulating signal with a

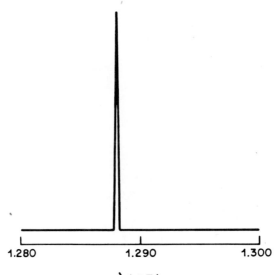

1.280   1.290   1.300

$\lambda \, (\mu m)$

FIG. 69. Time-averaged output spectrum showing a single frequency at 920 Mbit sec^{-1} modulation rate. The side-mode suppression is better than 100 : 1.

1-V peak-to-peak amplitude was superimposed. The output light pulses from the amplifier were fed through a grating spectrometer (Spex 1700) and detected by a fast InGaAs $p-i-n$ photodiode. The photodiode current was displayed on a 1-GHz real-time oscilloscope (Tektronics 7104). The upper trace in Fig. 68 shows the electrical driving pulses (a 16-bit 101010... NRZ word pattern) at a 920 Mbit sec^{-1} rate, and the lower trace shows the detected light pulses with a 100% modulation depth. However, at a 1.1-Gbit sec^{-1} modulation rate, the light pulse shape becomes somewhat sinusoidal and the modulation depth begins to diminish. Figure 69 shows the time-averaged single-frequency output spectrum at 920 Mbit sec^{-1} modulation rate. The side-mode suppression, determined primarily by the dc-operated C³ laser itself, was better than 100 : 1. The operation of the modulator had no effect on the spectrum.

XV. Optical Logic Operations

The present digital computers are based on electronic components whose inputs and outputs are, at any point in time, in one of two possible states. These two states (usually designated as 0 and 1 can be two voltage levels, two current levels, or forward and reverse magnetic field directions. The fundamental building blocks of the two-state circuitry are logic gates and memory (bistable) devices. A logic gate is a device that uses a fixed set of rules to transform a set of logical variable inputs into a single logical variable output. A memory device is one that can assume either of two states and will remain in its current state until an external excitation causes it to change state.

Recently, a very significant interest has been shown in optical bistable devices (Smith, 1972; Bowden *et al.*, 1980). Such bistable devices are capable of performing a wide range of optical signal processing functions such as optical limiting, optical switching, optical memory, and optical logic operations.

A new scheme of performing optical logic operations that is not based on optical bistability was proposed and demonstrated (Tsang *et al.*, 1983f). This scheme is based on a C³ semiconductor laser. The lasers used in this demonstration are InGaAsP-type, 1.3-μm buried crescent lasers, although any gain- or index-guided, standard Fabry–Perot cavity, stripe-geometry semiconductor laser diodes can be used.

As already shown earlier, the wavelength of the output laser beam can be controlled by adjusting the currents to the two separate diodes. A very large lasing frequency excursion of 150–300 Å and a frequency tuning rate of 10–26 Å mA^{-1} have been obtained. Further, the laser always operates in a highly stable single longitudinal mode even under high bit-rate modulation.

Based on the above unqiue characteristics of the C³ laser, a complete set of basic logic operations [AND, OR, INVERT (or NOT), and EXCLUSIVE OR,] can be performed. The schematic diagrams given in Fig. 70 show two possible arrangements for the present demonstration. In Fig. 70a, a single output was detected by setting the grating at the angle for deflection of the desired wavelength into the detector. In Fig. 70b, multiple outputs each at a different wavelength are detected simultaneously using an array of detectors. In both cases, an additional output can be obtained by detecting the entire output of all wavelengths from the other side of the laser as shown in the figure.

The two photographs shown in Fig. 71 illustrate the logic operations of OR, AND, and EXCLUSIVE OR gates. Traces 1 and 2 in both photographs are the current pulses I_A and I_B applied to diodes A and B, respectively (see

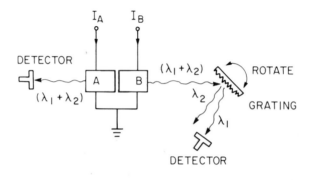

(a)

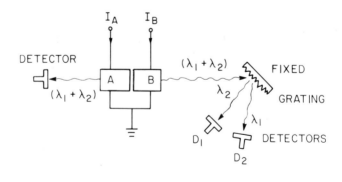

(b)

FIG. 70. Schematic diagrams showing two possible arrangements for the present demonstration of optical logic operations by using C^3 lasers. (a) Single output detection and (b) multiple output detection.

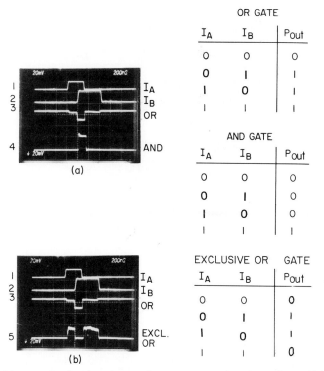

OR GATE		
I_A	I_B	P_{out}
0	0	0
0	1	1
1	0	1
1	1	1

AND GATE		
I_A	I_B	P_{out}
0	0	0
0	1	0
1	0	0
1	1	1

EXCLUSIVE OR	GATE	
I_A	I_B	P_{out}
0	0	0
0	1	1
1	0	1
1	1	0

FIG. 71. Photographs showing the two input current pulses $I_A = 17$ mA (dc) + 15 mA (pulsed) and $I_B = 17$ mA (dc) + 20 mA (pulsed) and the various optical output pulses at (a) 1.3036 and (b) 1.3116 μm, illustrating the OR, AND, and EXCLUSIVE OR logic gates. The corresponding truth tables are given on the right.

Fig. 70). In both cases, a dc bias was applied such that $I_A = I_{dc} + I_{pulse} = 17$ mA + 15 mA, and $I_B = I_{dc} + I_{pulse} = 17$ mA + 20 mA. Trace 3 in both photographs represents the total output power detected (power scale increases downwards) by the detector on the left side (see Fig. 70). This operation is an OR gate as shown by the truth table on the right. The overlap of the two current pulses I_A and I_B causes the increased output power over the duration of the overlap. Trace 4 shows the output pulse measured at 1.3036 μm. It is seen that a pulse is present only when both I_A and I_B are present simultaneously. This operation corresponds to an AND gate as shown by the truth table on the right. Trace 5 shows the presence of an optical pulse as measured at 1.3116 μm only when pulses I_A and I_B are present alone. Such an operation, according to the truth table shown on the right, is an EXCLUSIVE OR gate. It is seen that these logic operations are made possible because the wavelength of the optical output depends only on

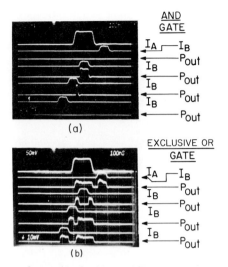

(a)

(b)

FIG. 72. Various states of operation for (a) an AND gate detected at 1.3006 μm and (b) an EXCLUSIVE OR gate detected at 1.3088 μm. The input pulse I_B was gradually moved across the input pulse I_A in time scale. The operating conditions were $I_A = 10$ mA (dc) + 22 mA (pulsed) and $I_B = 14.5$ mA (dc) + 12 mA (pulsed).

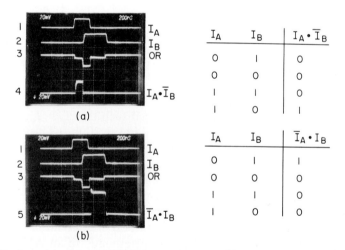

(a)

(b)

FIG. 73. Operation of the INVERTER gate. The operation conditions were $I_A = 18.7$ mA (dc) + 17 mA (pulsed) and $I_B = 17$ mA (dc) + 17 mA (pulsed) at (a) 1.3112 and (b) 1.3048 μm. Trace 3 is the OR gate as in Fig. 71. The logic operations correspond to $(I_A \cdot \bar{I}_B)$ and $(\bar{I}_A \cdot I_B)$ and their truth tables are given on the right. The unary operations $\bar{I}_B$ and $\bar{I}_A$ can be obtained simply by having $I_A = 35.7$ mA (dc) in the top photograph and $I_B = 34$ mA (dc) in the bottom photograph.

the injection current level and that the C³ laser always operates in a clean single longitudinal mode as shown in previous parts.

Figure 72 shows the various states of operation for an AND gate detected at 1.3006 μm and an EXCLUSIVE OR gate detected at 1.3088 μm as the input pulse I_B is gradually moved across the input pulse I_A in the time scale. The operating conditions are $I_A = 10$ mA (dc) + 22 mA (pulsed) and $I_B = 14.5$ mA (dc) + 12 mA (pulsed).

Figure 73 shows the operation of the INVERTER gate. The operation conditions are $I_A = 18.7$ mA (dc) + 17 mA (pulsed) and $I_B = 17$ mA (dc) + 17 mA (pulsed) and are given by traces 1 and 2, respectively. In both photographs trace 3 represents the OR gate as in Fig. 71. Under the present operating conditions, trace 4 (detected at 1.3112 μm) represents the logic operation $(I_A \cdot \bar{I}_B)$, where $\bar{I}_B$ stands for NOT I_B, according to the truth table on the right. Under the same conditions (but with the output detected at 1.3048 μm) trace 5 represents the logic operation $(\bar{I}_A \cdot I_B)$, where $\bar{I}_A$ stands for NOT I_A, according to the truth table on the right side. The unary operation NOT I_B can be obtained simply by having $I_A = 3.57$ mA (dc) in the top photograph, while the unary operation NOT A can similarly be obtained by having $I_B = 34$ mA (dc) in the bottom photograph.

In this demonstration, long current pulses (50–300 nsec) were used for convenience. However, we have found by carrying out time-resolved spec-

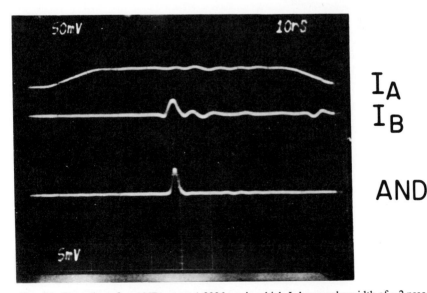

FIG. 74. Operation of an AND gate at 1.3006 μm in which I_B has a pulse width of ~ 2 nsec. The biasing conditions were the same as those used in Fig. 72.

tral measurement that the frequency switching processing as the current pulse is turned on is extremely fast ($<$ 1 nsec; limited by the rise time of the current pulse used) and highly stable in single-longitudinal-mode operation. This indicates that extremely high-bit-rate information processing can be achieved by using the present newly developed C^3 semiconductor lasers. Figure 74 shows the operation of an AND gate in which I_B has a pulse width of about 2 nsec. The biasing conditions were the same as those used in Fig. 72. From these examples, it is seen that with the same two inputs applied, several *different* logic operations can be obtained at different wavelengths simultaneously if the detection scheme shown in Fig. 70b is employed. The variation of lasing wavelength with temperature can be remedied by having, for example, heat-sink temperature stabilization or having photodetectors with sufficiently large areas as in Fig. 70.

XVI. A Self-Aligned Laser–Detector Module

Semiconductor lasers have been widely used as light sources in optical transmitters for optical fiber communications systems. Figure 75a shows a cutaway schematic view of a typical assembled package (Shumate *et al.*, 1978). Light from the front face of the laser diode is coupled into the optical fiber for transmission. Light from the back face of the laser diode falls onto a small-area ($\sim$ 1-mm-diameter) photodiode. This photodiode signal is used as

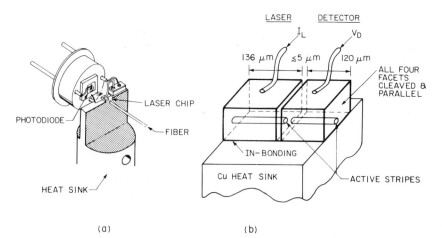

(a)   (b)

FIG. 75. (a) Cutaway schematic view of a typical laser–photodiode assembled package used for feedback control of the injection current applied to the laser diode in order to maintain a constant average power launched into the front-face optical fiber. (b) Schematic diagram of the self-aligned cleaved-coupled laser-detector (CCLD) unit. In this experiment a 1.3-μm GaIn-AsP-type buried crescent heterostructure laser was used. Drawing is not to scale.

a feedback to control the injection current into the laser diode in order that the average power launched into the front-face fiber be maintained constant at all times and under all environmental fluctuations. Thus, the stability of the coupling between the laser and detector is very important in ensuring the proper performance of the optical transmitter. However, with time, an erroneous photodiode signal can be fed back to the laser diode owing to, for example, changes in the mechanical alignment of the photodiode with the laser, asymmetric aging of the $L - I$ characteristics from each face of the laser diode, beam wandering of the laser diode, and changes in the photodiode itself. As a result, the functional lifetime of the optical transmitter can be significantly reduced even though the laser diode is still operative. This behavior has been recently confirmed experimentally. Another characteristic of this scheme is that the photodiode is typically $\sim 3 - 4$ mm away from the backface of the laser diode, so that the light-collection efficiency is in general $\sim 50\%$ for the typical photodiode area used.

Several solutions have been proposed to remedy this problem. One is the use of a front-fiber light tap as the source of the feedback control signal. However, this requires the laser to be operated at a much higher output power (typically about twice) than that used in the backface monitoring described earlier. Such increased output power operation can shorten the lifetime of the laser and, for gain-guided lasers, will push the operating point into the more nonlinear regime and also cause the nonlinear regime to occur sooner (at lower power levels) as the laser ages. Another possible scheme is to utilize an integrated laser–detector formed by etching as demonstrated by Merz et al. (1979). However, such lasers formed by etched mirrors in general have poor performances compared to those with cleaved facets. Further, the laser–detector power coupling efficiency demonstrated is only about 10%. Because one mirror of the laser is formed by cleaving and the other formed by etching, it is likely that such a laser will also be more susceptible to having asymmetric output power develop with aging. Tsang et al. (1983g) proposed and demonstrated a new approach: the cleaved-couple laser–detector (CCLD) shown schematically in Fig. 75b. In this arrangement, both the laser and the detector are cleaved from the same buried crescent heterostructure laser wafer. Further, the active layers of the laser and detector are precisely *self-aligned* and are optically *very closely coupled* with a separation distance $\lesssim 5$ μm. To form a laser–detector system, we simply forward-bias one of the diodes (in this experiment, the 136-μm section) and operate it as a laser. The other section is unbiased or reverse-biased and operates as the photodiode.

Some of the advantages of this new scheme are obvious.

(1) Because of the inherent precise self-alignment and small separation of the two active GaInAsP stripes, the optical beam emitted from the laser backface is extremely efficiently coupled to the photodiode.

(2)   Because of the small separation and the use of built-in index wave-guiding in both the laser and detector diodes, erroneous photodiode signal feedback due to beam wandering can essentially be eliminated.

(3)   Since the stripes in both the laser and detector are of the same material, the photodiode is automatically sensitive to the light generated by the laser diode.

(4)   Since the light coupled into the active stripe of the detector is guided by the bulit-in index waveguide and the absorbing waveguide can be made longer for complete absorption (in this case $\sim 120\,\mu$m), very efficient photon–carrier conversion can be achieved. In fact, one can consider this as a *waveguided photodetector,* in contrast to the conventional broad-area photodiodes that, in general, have an absorbing layer thickness of only $1-2\,\mu$m.

(5)   Since both the laser and the detector are bonded rigidly together, they should be mechanically very stable and compact.

(6)   From the processing point of view, all the procedures are the same as those employed for laser diodes.

(7)   Since the photodiode is used only for monitoring the average power launched into the fiber, high-speed response and amplification are not essential for this application.

Figure 76a shows the light output power versus current as detected by a Ge broad-area photodiode from the exit face of the laser diode and the detected voltage across a 50-Ω resistor from the cleaved-coupled waveguide photodetector. The insert shows the arrangement of the detection set-up. It is seen that the tracking of the photodiode voltage with the output power of the laser is excellent. In Fig. 76b, the photodiode voltage is plotted as a function of the laser output power. A linear response curve was obtained. A linear response is particularly important for the photodiode to function as a backface feedback control. The photodiode was operated in the zero-biased mode. Figure 76c shows the optical output pulse as detected with a broad-area Ge detector from the exit side of the laser (top) and the cleaved-coupled photodiode detected voltage pulse (bottom) under zero bias. For a conventional broad-area back-illuminated InGaAs photodiode, the responsivity is about 0.6 mA mW^{-1} (without antireflection coating). From this measurement, an estimate yields a responsivity of ~ 0.5 mA mW^{-1}, which includes any possible coupling losses at the cleaved separation between the laser and the detector. Such a value indicates that an extremely efficient coupling of the laser and the detector was achieved in the present CCLD scheme. This is especially so when we consider the fact that the laser beam energy falls on the absorption edge of the absorbing medium of the photodiode, instead of inside as in the case of InGaAs photodiodes, since the same GaInAsP material was used for both the laser and the detector.

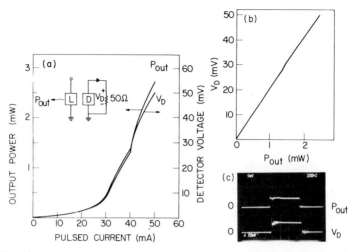

FIG. 76. Light output power versus current as detected by a broad area Ge photodiode from the exit face of the laser diode and the detected voltage across a 50-Ω resistor from the cleaved-coupled waveguide photodiode. The insert shows the arrangement of the detection set-up. (b) The photodiode voltage is plotted as a function of the laser output power. (c) The optical output pulse as measured by the broad area Ge photodiode (top) and the cleaved-coupled waveguide photodiode (bottom).

XVII. Theoretical Analysis of C^3 Lasers

6. INTRODUCTION

In this part, a theoretical analysis carried out by Henry and Kazarinov (1983) of the steady-state operation of coupled-cavity lasers at threshold and above threshold will be given. When compared with the experimental results obtained earlier, their theoretical results can be considered semiquantitative. But most importantly is the fact that their approach is straightforward and yields the various key features about coupled-cavity lasers. Coupled-cavity lasers have also been treated by Coldren *et al.* (1982), Ebling and Coldren (1983), and Marcuse and Lee (1983). Henry and Kazarinov's (1983) starting equation for the optical fields is basically the same as that of Coldren and Koch (1983).

In describing the two cavities, they specify how the refractive indexes and gains change with carrier density and how the carrier density is coupled to the currents and to optical power. The exact relationships among these quantities will have a strong influence on the resultant curves calculated, though the qualitative picture will be the same. Thus, the calculated results should be considered semiquantitative rather than quantitative. The linking of refractive index and gain change is crucial to understanding coupled-cavity laser

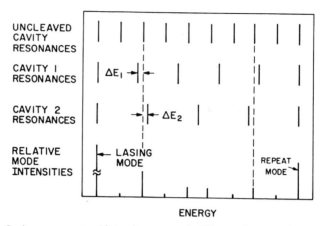

FIG. 77. Cavity resonances and intensity spectra of a C³ laser using an analyses by Henry and Kazarinov (1983).

operation. Indeed, the greatest effect of a small change in carrier density is to shift the cavity resonances via refractive index changes; the small gain change that also occurs is much less significant.

The analysis shows that there are two selection mechanisms. The first mechanism works best when one cavity acts as a resonant reflector that modulates the end losses of the lasing cavity. The reflection from the resonator is periodic in energy and has maxima at energies between the modulator cavity resonances. This is a relatively weak mode selection mechanism providing gain–loss differences on the order of 11 cm^{-1} in 15-mil-long devices.

The second mechanism is related to diffraction losses in the gap between the two cavities. The modes of highest net gain avoid the gap losses by interfering in such a way as to reduce transmission into a lossy gap. This reduction is maximized when the gap width d is an integral number of half wavelengths and when the cavities are both resonant. The condition of simultaneous cavity resonances leads to sharp mode discrimination with gain–loss differences on the order of $5–10$ cm^{-1} in 15-mil-long devices. Both mechanisms derive their mode selectivity from the registration of the two sets of cavity resonances. This is illustrated in Fig. 77 for a case in which the gap loss mechanism is dominant.

7. Equations of Steady-State Operation

The fields in the two cavities of a coupled-cavity laser are indicatd in the sketch in Fig. 78. The magnitudes of the complex transmission and reflection coefficients T_2 and R_2 that completely characterize the gap are also plotted in Fig. 78. The field E_1' at the gap results from reflection of E_1 and

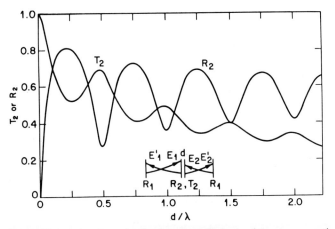

FIG. 78. Field transmission T_2 and reflection R_2 coefficients of the gap separating cleaved sections versus the gap separation $d\lambda$. Here λ is the lasing wavelength and h is the active layer thickness of the laser.

transmission of E_2,

$$E_1' = R_2 E_1 + T_2 E_2. \tag{2}$$

On the other hand, the round trip through cavity 1 results in a relationship between E_1 and E_1'

$$E_1 = R_1 \exp[(g_1 - \alpha)L_1 + 2ik_1L_1]E_1', \tag{3}$$

where g_1 is the gain, α the loss, L_1 the length, and k_1 the propagation vector given by $k_0 n_1$, where $k_0 = 2\pi/\lambda$, and n_1 is the refractive index of cavity 1. By eliminating E_1' from Eqs. (2) and (3) and dividing the resulting equation by $\exp[(g_1 - \alpha)L_1 + 2ik_1L_1]R_1R_2$, we find

$$\left\{ \frac{\exp[-(g_1 - \alpha)L_1 - 2ik_1L_1]}{R_1R_2} - 1 \right\} E_1 = \frac{T_2}{R2} E_2. \tag{4}$$

Similar considerations for cavity 2 lead to

$$\left\{ \frac{\exp[-(g_2 - \alpha)L_2 - 2ik_2L_2]}{R_1R_2} - 1 \right\} E_2 = \frac{T_2}{R_2} E_1. \tag{5}$$

The homogeneous equations (4) and (5) have solutions only if the secular equation

$$\left\{ \frac{\exp[-(g_1 - \alpha)L_1 - 2ik_1L_1]}{R_1R_2} - 1 \right\}\left\{ \frac{\exp[-(g_2 - \alpha)L_2 - 2ik_2L_2]}{R_1R_2} - 1 \right\}$$

$$= \frac{T_2^2}{R_2^2} \tag{6}$$

is satisfied. The key parameter describing the coupling between the two cavities is

$$T_2/R_2 = T \exp(i\phi). \qquad (7)$$

Setting each term in brackets equal to zero in Eq. (6) determines the lasing condition for each cavity if the cavities were uncoupled.

The values of R_2 and T_2 are readily calculated using the standard treatment of a two-surface Fabry–Perot resonator, once the transmission losses due to diffraction in the air gap have been estimated. These diffraction losses have been calculated by Coldren *et al.* (1982), who neglected diffraction spreading parallel to the junction plane. This is not accurate since the beam divergence in this direction is also significant especially in BH-type lasers. They also used a Gaussian model to estimate diffraction spreading perpendicular to the junction plane. Nevertheless, Henry and Kazarinov (1983) incorporated the same assumption in their calculation of diffraction losses in the air gap.

Using the calculations of T_2 and R_2, the magnitudes of which are plotted in Fig. 78, the magnitude of the coupling coefficient T is plotted in Fig. 79 and its phase ϕ in Fig. 80 as a function of d/λ.

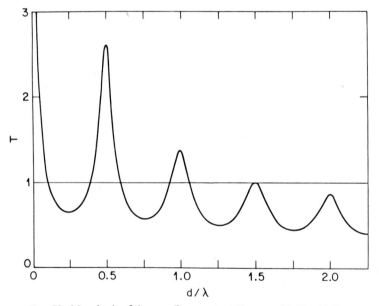

FIG. 79. Magnitude of the coupling constant T versus d/λ. $T = T_2/R_2$.

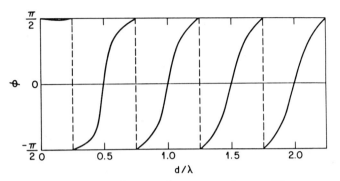

FIG. 80. Phase of the coupling constant ϕ versus d/λ.

8. GENERAL ANALYSIS

In the three-terminal independently derived coupled-cavity lasers, the currents I_1 and I_2 to each cavity are under experimental control, and g_1 and g_2 are not completely independent. For a single-cavity laser operating above threshold, stimulated emission pins the value of carrier density required for gain at approximately equal to the loss for the lasing mode. In the coupled-cavity laser, there are two carrier densities supplied by two different currents. In their theoretical analysis, Henry and Kazarinov (1983) assumed that stimulated emission only provides a relationship between the two carrier concentrations, but does not pin them. For example, they argue as follows. If the current to the first cavity I_1 with the laser above threshold is increased, the additional optical power generated in the first cavity will increase stimulated emission in the second cavity. For a fixed I_2, g_2 will decrease, and g_1 must increase by a definite amount in order to compensate for this decrease and maintain the lasing condition. Hence, neither g_1 nor g_2 are pinned, but there is a relationship between them. This also applies to the carrier densities and to the refractive indexes n_1 and n_2 in the two cavities. The relationship between g_1 and g_2 depends on how strong this mutual optical pumping effect is. Since the lasing wavelength in a semiconductor diode laser is always at the lower energy side, this mutual pumping effect may be low. If so, g_1 and g_2 may not be strongly related to each other. This will have a significant influence on the final results of the analysis especially with regard to the ranges of I_1 and I_2 that will have the same single frequency. However, the assumption made by Henry and Kazarinov (1983) that g_1 and g_2 are intimately related to each can be considered as a general case for the purpose of analysis.

To simply describe this effect, they further assume both gain and refractive index change *linearly* with carrier density in the region of laser operation. The phases in each cavity ($-2k_1L_1$ and $-2k_2L_2$) vary linearly [see Eqs.

(4–6)] with changes in gain and photon energy. The variation of the phase $-2k_1L_1$ is given by

$$-\Delta(2k_1L_1) = \beta \, \Delta g_1 L_1 - 4\pi n_e \, \Delta E L_1, \tag{8}$$

where the first term originates from the change in refractive index with gain, and the second term arises from the change in phase with energy at a fixed carrier density. The parameter β is the ratio of the changes in the real and imaginary parts of the refractive index with changes in carrier density and n_e the effective index associated with the group velocity v_g. The value of β has been determined to be about 5 in AlGaAs-type lasers (Henry, 1982). The value of this parameter is not accurately known for quaternary lasers. However, refractive index changes in reaching threshold are similar in the two materials (Manning et al., 1983), so we will assume $\beta = 5$ for the numerical examples in this chapter.

The mode gains are determined by secular equation (6). This complex equation can be regarded as two nonlinear equations involving three unknowns, g_1, g_2, and E. One unknown can be taken as the independent variable and the equations can be solved for the other two unknowns. The general procedure for solving nonlinear equations with one or more unknowns employs Newton's method, which finds the solution in the neighborhood of a point. To solve Eq. 6 Henry and Kazarinov used a standard program, zone J of the AT&T Bell Laboratories port library. If the values of g_1, g_2, and E are approximately known for a mode at one point, all other points for this mode can be found by varying one of these parameters and solving for the other two.

As a starting point they chose the point of symmetric operation in which $(g_1 - \alpha)L_1 = (g_2 - \alpha)L_2$. A general point can be expressed by adding Δg_1, Δg_2, and ΔE to g_1, g_2, and E of the initial point. In terms of these new variables, the secular equation (6) becomes

$$\{q \exp[-\Delta g_1 L_1 + i(\beta \, \Delta g_1 - 4\pi n_e \, \Delta E)L_1]\}$$
$$\{q \exp[-\Delta g_2 L_2 + i(\beta \, \Delta g_2 - 4\pi n_e \, \Delta E)L_2]\} = T^2 \exp(2i\phi), \tag{9}$$

where $q = x \exp(-\delta_1/\beta)$ is chosen so that at the initial point, $\Delta E = 0$ and the magnitudes and phases of the exponential factors in Eq. (9) are x and δ_1, respectively. By varying Δg_1 and solving Eq. (9), mode 0 in Fig. 81 can be traced out.

In going from mode 0 to mode 1, the phases $-2ik_1L_1$ and $-2ik_2L_2$ change by approximately 2π. These phase changes are primarily due to a change in energy ΔE. This energy change can be estimated by using Eq. (8) with $\Delta g_1 = 0$. This estimate of ΔE is accurate enough to switch one mode to the next and trace out all the modes shown in Fig. 81 and in subsequent figures. Mode switching occurs at the crossings near $\delta = \pm 40°$.

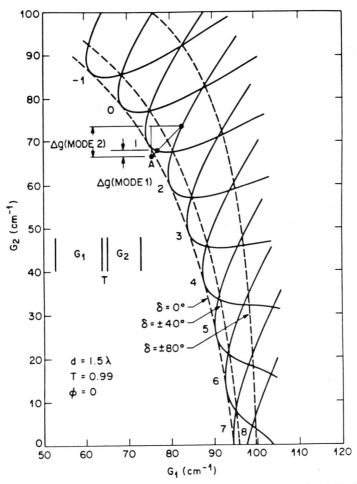

FIG. 81. Gain curves in the (g_1, g_2) plane for $d = 1.5\lambda$, $I = 0.99$, and $\phi = 0$. In this case, maximum selectivity occurs at $\delta = 0$ and mode hopping occurs near $\delta \approx \pm 40°$, where $\delta = \delta_1 - \delta_2$ is a measure of the energy separation of the two cavity resonances. The relative intensities of each mode at operating point A are inversely proportional to Δg, where Δg is the side of a 45° triangle down from point A to the mode line.

9. MODE SELECTIVITY

For a single-cavity laser, each mode i has a definite value of loss–gain $\alpha - g_1$ that determines its intensity. In a coupled-cavity laser, each mode is described by a curve in the (g_1, g_2) plane. Consider modes 1 and 2 of Fig. 81 in which the laser is operating at point A, the ratio of mode intensities ι_1 and

ι_1 is

$$\iota_2/\iota_1 = \Delta g \text{ (mode 1)}/\Delta g \text{ (mode 2)}, \qquad (10)$$

where the Δg's are the sides of the $45°$ triangles shown in Fig. 81. The triangles extend from operating point A to the mode curves.

In Fig. 81, mode 1 is the lasing mode for operating point A and the distance Δg (mode 1) has been greatly exaggerated for the purpose of illustration. For the lasing mode, Δg is determined by the optical power and spontaneous emission rate and has the same magnitude as that of a conventional laser (Kazarinov and Henry, 1982). For example, if the facet power $P_1 = 2$ mW, then Δg (mode 1) $\approx 10^{-3}$ cm^{-1}. Since Δg for the lasing mode (mode 1) is very small, Δg characterizing suppression of the nonlasing mode (e.g., mode 2 in Fig. 81) is equal to the value of Δg associated with moving from mode 1 to mode 2 along a $45°$ line. Notice that this value of Δg goes to zero as the crossing point is approached. Under steady-state operation at a facet power of several milliwatts, 1 cm^{-1} of suppression corresponds to nearly three orders of magnitude of mode suppression.

The gain curves for $\phi = 0$ and intermediate coupling $T = 0.99$ are shown in Fig. 81. This occurs at $d = 1.5\lambda$. Maximum selectivity of 15 cm^{-1} occurs for mode 0 when both cavities have equal amplification and $\delta = 0$, corresponding to both cavities being resonant. As we move along the curve of mode 0 to the crossing points where $\delta = \pm 40°$, mode selectivity goes to zero and mode hopping occurs.

As we move to higher-numbered modes in Fig. 81 and as g_2 decreases, the maximum mode selectivity decreases. However, even at mode 7, where $g_2 \approx 0$ and cavity 2 is passive (no stimulated emission), the maximum mode selectivity is about 4 cm^{-1}.

Figures 82 and 83 contrast the gain curves for strong and weak coupling and a gap equal to an integral number of half wavelengths. Figure 82 shows strong coupling with $T = 2.6$ and $d = \lambda/2$. In this case maximum selectivity is nearly the same for each mode and is given by $\Delta g \simeq 5$ cm^{-1}. The case of weak coupling with $T = 0.69$ and $d = 2.5\lambda$ is shown in Fig. 83. Here there is much greater selectivity ($\Delta g \simeq 22$ cm^{-1}) when both cavities have equal amplification than when one cavity is passive (near $g_2 = 0$, $\Delta g \simeq 2.6$ cm^{-1}).

Figures 84 and 85 show $d = 1.4\lambda$ and $d = 1.6\lambda$. The major change from Fig. 81 where $d = 1.5\lambda$ is a decrease in mode selectivity by a factor of $2-3$. Nevertheless, selectivity is still adequate for many applications. For example, $\Delta g \approx 1.8$ cm^{-1} for modes near $g_2 = 0$ in these figures.

Figure 86 shows the gain curves for a narrow gap ($d = 0.06\lambda$). This case is of interest since gap losses are small and the second mechanism in which cavity 2 acts as a resonant reflector for cavity 1 is dominant. Figure 86 shows that there is no selectivity when the sections have equal amplification, but

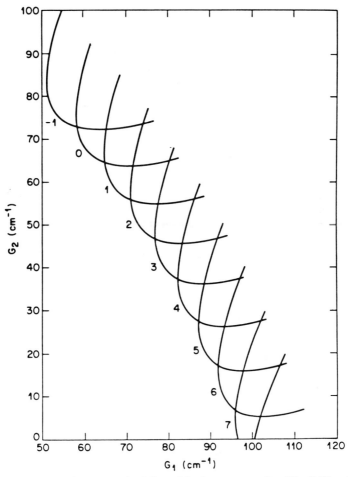

FIG. 82. Gain curves for the case $d = \lambda/2$, resulting in strong coupling ($T = 2.60$) and $\phi = 0$.

there is good selectivity near $g_2 \approx 0$, in which cavity 2 acts as a passive etalon. Here the maximum selectivity is 2.5 cm^{-1}. The intersection of the mode curves at the point ($g_1 = 58$ cm^{-1}, $g_2 = 65$ cm^{-1}) is an illustration that for equal amplification $x = y$, there is no mode selectivity when $\phi \approx 90°$.

Until now, we have been describing laser operation in terms of g_1 and g_2. The parameters under direct experimental control are not the gains, but rather the currents I_1 and I_2. We will now discuss the relationship between gain and current and the conditions necessary to stabilize g_1 and g_2 at the points of high single-mode selectivity.

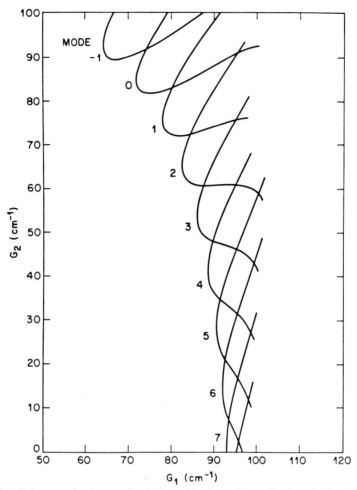

FIG. 83. Gain curves for the case $d = 2.5\lambda$, resulting in weak coupling ($T = 0.69$) and $\phi = 0$.

The current to each cavity generates spontaneous and stimulated emission as

$$I_1 = I_0(L_1/L)[1 + (g_1/\alpha_0)] + g_1 l_1 v_g, \tag{11}$$

$$I_2 = I_0(L_2/L)[1 + (g_2/\sigma_0)] + g_2 l_2 v_g, \tag{12}$$

where I_0 is the current to reach inversion in the uncleaved laser. This current is reduced by L_1/L in cavity 1. Assuming a linear relationship between gain and current in the vicinity of threshold, the current $I_0 L_1/L$ must be increased by $1 + (g_1/\alpha_0)$ to take cavity 1 from the point of inversion ($g_1 = 0$) to

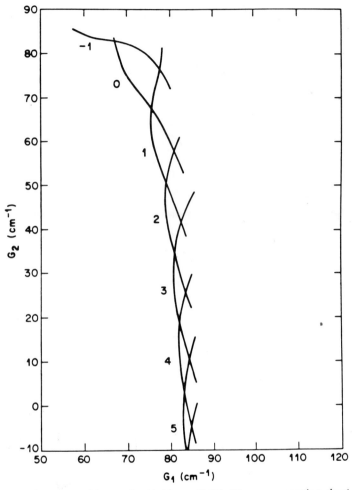

FIG. 84. Gain curves with negative ϕ ($\phi = -58.1°$). The gap separation $d = 1.4\lambda$ and $T = 0.68$.

g_1, where α_0 is the absorption in unpumped material at the laser energy. The second terms in Eqs. (11) and (12) are the rates of stimulated emission in each cavity, where ι_1 and ι_2 are the integrated light intensities in each cavity expressed as the number of photons.

For each mode there is a definite field distribution in each cavity that depends only on the operating point (g_1, g_2). Therefore, ι_1 and ι_2 are not independent, and their ratio can be calculated in terms of g_1 and g_2. The squared fields associated with propagation in each direction are proportional

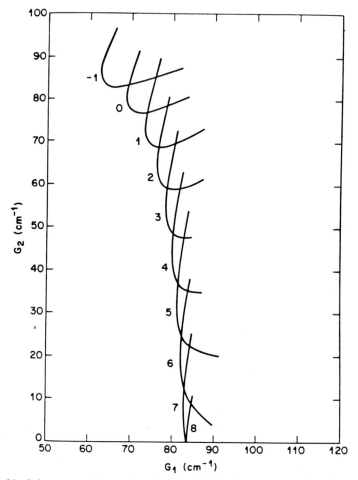

Fig. 85. Gain curves with positive ϕ. ($\phi = 56.6°$). The gap separation $d = 1.6\lambda$ and $T = 0.64$.

to the photon density. It is easily shown by integration of these quantities that

$$\frac{\iota_2}{\iota_1} = \frac{(g_1 - \alpha)(1 - R_1 R_2 y)(1 + R_2 y/R_1)}{(g_2 - \alpha)(1 - R_1 R_2 x)(1 + R_2 x/R_1)} \left| \frac{E_2}{E_1} \right|^2, \tag{13}$$

where E_1/E_2 is determined by Eq. (4) or (5) and x amd y are given by

$$x = \exp[-(g_1 - \alpha)L_1]/R_1 R_2, \tag{14}$$

$$y = \exp[-(g_2 - \alpha)L_2]/R_1 R_2. \tag{15}$$

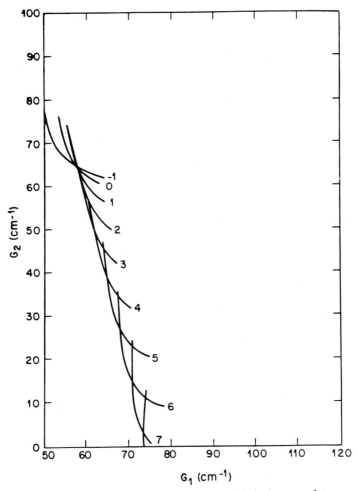

FIG. 86. Gain curves for a narrow gap ($d = 0.06\lambda$). This is the case of strong coupling ($T = 1.69$) and $\phi \approx 90°$. Note that $R_2 = 0.51$ and is substantial even for this small gap.

The rates of stimulated emission in each cavity are linearly related to the facet powers. For cavity 1, a simple calculation leads to

$$P_1 \pm v_g g_1 \iota_1 / \eta, \tag{16}$$

where

$$\eta = \frac{(1 - R_1^2)(g_1 - \alpha)}{g_1(1 - R_1 R_2 x)(1 + (R_1/R_2 x))}. \tag{17}$$

We can eliminate ι_1 and ι_2 from Eqs. (11) and (12) in favor of P_1 and I_2/I_2 as

$$I_1 = I_0(L_1/L)[1 + (g_1/\alpha)] + (P_1/\eta), \tag{18}$$

$$I_2 = I_0(L_2/L)[1 + (g_2/\alpha)] + (g_2\iota_2/g_1\iota_1)(P_1/\eta). \tag{19}$$

Equations (18) and (19) determine the currents required to operate the laser at power P_1 while staying at operating point (g_1, g_2). The threshold currents are determined by setting $P_1 = 0$. To take the laser above threshold and generate power P_1 (without altering g_1 or g_2) additional currents ΔI_1 and ΔI_2 should be supplied to cavities 1 and 2, where

$$\Delta I_1 = P_1/\eta \tag{20}$$

and

$$\Delta I_2/\Delta I_1 = g_2\iota_2/g_1\iota_1. \tag{21}$$

It is clear that additional currents can cause changes in either gain or stimulated emission. In order to prevent changes in g_1 and g_2, the ratio $\Delta I_1/\Delta I_2$ should be equal to the ratio of the rates of stimulated emission in the two cavities.

Figure 87 shows the currents required for threshold and for $P_1 = 5$ mW for each mode on the (I_1, I_2) plane for a laser with $d = 1.5\lambda$. For each mode and operating point (g_1, g_2) there is a linear relationship between ΔI_1 and ΔI_2 corresponding to Eq. (21) that holds g_1 and g_2 fixed. These current changes correspond to straight lines on the (I_1, I_2) plane.

At threshold, the currents of each mode just replicate the gain curves shown in Fig. 81. Only the linear relationships of I_1 and I_2 for the crossing points have been drawn. Along these lines, the intensities of two neighboring modes are equal. These lines divide the (I_1, I_2) plane into zones of single-mode operation. This is qualitatively in agreement with the experimentally observed results. (See, for example, Fig. 26.) The experimental (I_1, I_2) plane shows very distinct regions of very closely spaced zones (regions A and B) and a large single-mode zone (region C). Such features are consistent with the explanation that each cavity behaves as if it maintains its own separate lasing threshold, and when its threshold is reached, its carrier density or gain is more or less pinned. The calculated (I_1, I_2) plane shown in Fig. 87, on the other hand, is consistent with and calculated based on the assumption that there is a strong relationship between g_1 and g_2, and they are not pinned under all current levels. Figures 88 and 89 show plots of P_2 and P_1 versus I_2 for fixed I_1. Both plots show intensity variations along each mode curve and discontinuities in slope where mode hopping occurs. These calculated curves are to be compared with those obtained experimentally shown in Fig. 12. The intensity variations calculated are in qualitative agreement with experi-

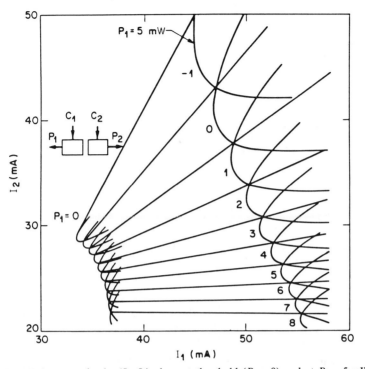

FIG. 87. Mode curves in the (I_1, I_2) plane at threshold ($P_1 = 0$) and at $P_1 = 5$ mW for intermediate coupling ($T = 0.99$), $d = 1.5\lambda$, and $\phi = 0$. The plane divides into zones of single-mode operation.

mental observation. The no-threshold-like behavior of the calculated $(L-I)$ curves in Figs. 88 and 89 (in contrast to the presence of threshold-like behavior of the measured curves shown in Fig. 12) is again due to the assumption in the calculation that g_1 and g_2 are related and are never pinned at all current levels. In spite of these differences, the key features are well explained by their theory.

10. BISTABLE AND UNSTABLE OPERATION

The ratio $\Delta I_1/\Delta I_2$ for each mode necessary to hold (g_1, g_2) fixed is largely determined by $|E_2/E_1|^2$, given by Eq. (4) or (5). For $\phi = 0$ and $\phi = 90°$, $|E_2/E_1|^2$ is the same for two adjacent modes at the point of intersection of the gain curves. Consequently, the same ratio $\Delta I_2/\Delta I_1$ will keep both modes at the point of intersection as the laser is brought above threshold. This results in the division of the (I_1, I_2) plane into zones of single-mode operation above threshold that touch but do not overlap.

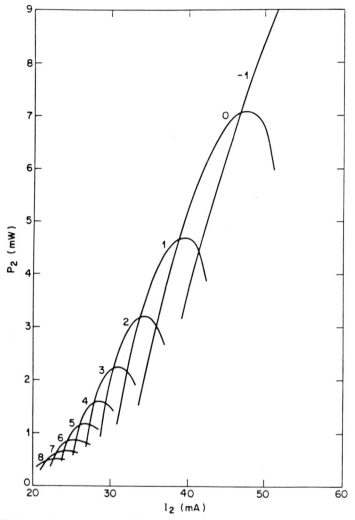

FIG. 88. Power P_2 versus current I_2 for constant $I_1 = 50$ mA, $\phi = 0$, $d = 1.5\lambda$, and intermediate coupling $T = 0.99$.

This simple situation does not hold in general, but only at $\phi = 0$ and $\pm 90°$ in principle. Figure 80 shows the variation of ϕ with gap separation d. We find that for negative values of ϕ, the zones overlap resulting in regions with two possible operating points and bistable operation. For positive ϕ there are voids between the current zones that are regions of unstable operation in which no steady state laser operation can exist. Current settings in these regions should lead to unstable dynamic modes of operation.

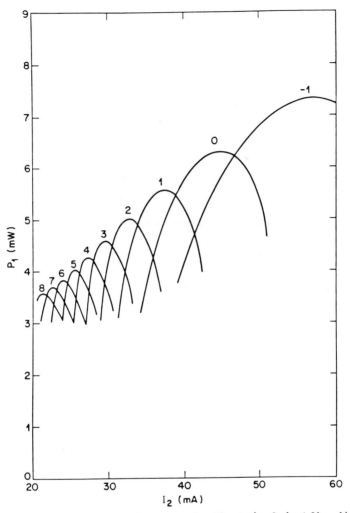

FIG. 89. Power P_1 versus current I_2 for constant $I_1 = 50$ mA, $\phi = 0$, $d = 1.5\lambda$, and intermediate coupling $T = 0.99$.

Experimentally, bistable regions exhibit hysteresis. Each of two different modes can be dominant at a given (I_1, I_2) setting, depending on the history of how (I_1, I_2) was established. These hysteresis effects have been observed by Tsang and Olsson (1983) (Part XI), who have also observed unstable regions in C^3 lasers.

A plot of (I_1, I_2) for $\phi = -58°$ (corresponding to $d = 1.4\lambda$) is shown in Fig. 90. For modes having nearly equal amplification in both cavities, there

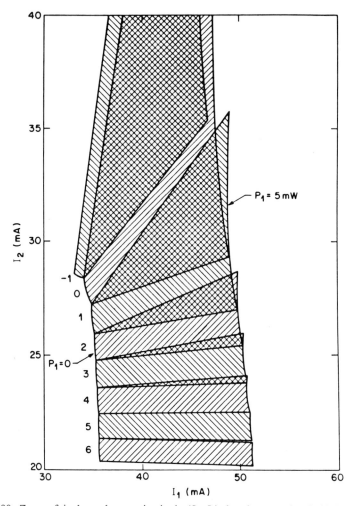

FIG. 90. Zones of single mode operation in the (I_1, I_2) plane between threshold $(P_1 = 0)$ and 5 mW above threshold for $d = 1.4\lambda$, $I = 0.68$, and negative ϕ ($\phi = -58.1°$). Overlapping zones are regions of bistability in which the same currents (I_1, I_2) can excite the laser in two different modes with different cavity gains.

are large overlapping regions of bistable operation. However for modes 4 and 5 (corresponding to $g_2 \approx 0$ and passive operation of cavity 2), the bistable regions are very small. Figure 91 shows a plot of P_2 versus I_2 at fixed current in cavity 1. In the shaded bistable regions, sudden transitions between modes can occur causing sharp drops in P_2. The point of switching can depend

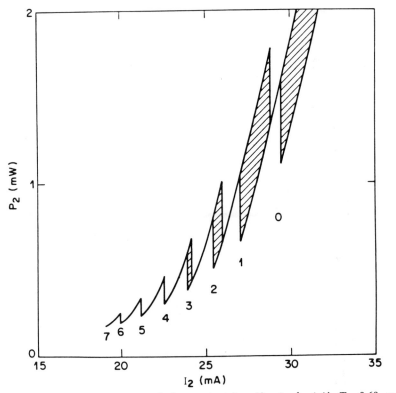

FIG. 91. Power P_2 versus current I_2 for constant $I_1 = 50$ mA, $d = 1.4\lambda$, $T = 0.68$, and $\phi = -58.1°$. Mode switching can occur anywhere in the shaded bistable regions.

upon whether I_2 is increasing or decreasing. These effects have been observed by Tsang and Olsson (1983). (See Part XI.)

The current curves for $\phi = 56.6°$, corresponding to $d = 1.6\lambda$, are shown in Fig. 92. The dashed lines represent constant values of (g_1, g_2). There are voids where zones of stable steady-state operation separate and no stable stationary states exist. There are also points where the current lines for constant (g_1, g_2) cross. These are bistable regions in which the same (I_1, I_2) can excite one mode at several different operating points. Again, we find that near $g_2 = 0$ (modes 6 and 7) the regions of instability are small: again, these curves should be considered qualitative.

Therefore, coupled-cavity lasers formed by cleaving have two selection mechanisms. The stronger mechanism is that different modes have different losses in the gap due to interference that can reduce or enhance transmission

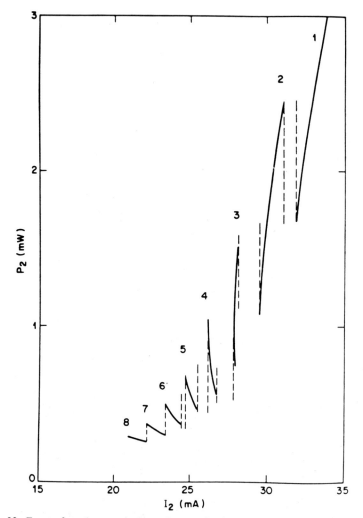

Fig. 92. Zones of steady-state single-mode operation in the (I_1, I_2) plane between threshold $(P_1 = 0)$ and $P_1 = 5$ mW for $d = 1.6\lambda$ and positive ϕ ($\phi = 56.6°$). The dashed lines represent the currents necessary to keep (g_1, g_2) constant. The gaps between zones are regions in which there are no steady-state modes of operation. Points where several dashed lines intersect are points of unstable equilibrium in which the same mode can be excited with different operating points (g_1, g_2).

into the gap. The weaker mechanism relies on one section acting as the laser cavity and the other acting as a resonant reflector that reflects some modes more strongly than others. The above analysis by Henry and Kazarinov (1983) is for steady-state operation. Transient analysis (pulse response of different longitudinal modes) based on numerical calculations of rate equations has also been carried out by Coldren and Koch (1983).

XVIII. Summary

A fabrication technique for forming a new self-aligned coupled-cavity semiconductor laser is described. Unlike prior techniques, this technique is compatible with the present processing employed in mass production. The new C³ semiconductor laser has two standard Fabry–Perot cavity diodes separated by < 5 μm and has active stripes precisely aligned with respect to each other. All four reflecting mirrors are formed by cleaved crystallographic facets and are perfectly parallel to each other for each cavity. Complete electrical isolation (> 50 kΩ) was obtained between the two optically coupled diodes, making it possible to bias each diode independently. This forms a truly three-terminal device with the two cleaved cavities mutually coupled optically.

Essential to the unique operational characteristics of the C³ laser is the realization that the effective optical length of each cavity can be varied separately by controlling the injection current applied to each cavity. This realization makes possible the first, practical, electrically tunable, single-frequency semiconductor laser. In operation, the laser diode, is always operated above threshold. The other diode, the modulator, can either be below or above threshold corresponding to two modes of operation for the C³ laser. With the modulator below threshold, the lasing wavelength of the C³ laser can be controllably tuned for the highest longitudinal mode discrimination ratio or switched among as many as 15 different discrete frequencies. The tuning, in effect, is a result of the shift in the Fabry–Perot modes of the modulator with respect to that of the laser diode resulting from the index change that is controlled by the modulator current via the carrier density. With the modulator above threshold, no tuning is possible since the carrier density is pinned at threshold value.

In this regime, the C³ laser operates in the same single longitudinal mode even when the injection current to both the modulator and laser diodes are varied. Stable single-longitudinal-mode operation can also be achieved by biasing the modulator diode with the optimal dc current for the highest longitudinal mode discrimination ratio and pulse-code-modulating the laser diode. The C³ lasers were demonstrated using 1.3- and 1.5 μm InGaAsP-type buried crescent lasers with separate optical confinement and 1.5-μm ridge-waveguide lasers. Time-resolved spectral and spectrally-resolved pulse response measurements showed that such C³ semiconductor laser are highly single-longitudinally moded with extreme spectral purity and stability even under high-speed direct modulation. Single longituidnal mode power discrimination ratios in excess of 1500:1 have been obtained under direct modulation up to 2 Gbit sec^{-1}. Bit-error-rate-free operation was also obtained under direct modulation up to 2 Gbit sec^{-1} indicating that there are no detectable laser mode-partition events. Using this new single-frequency laser operating at 1.55 μm, unrepeatered transmission through 119 and 161.5 km at 420 Mbit sec^{-1} and 101 km at 1 Gbit sec^{-1} have been demonstrated. The fibers used have a chromatic dispersion of ~ 17 psec km^{-1} nm^{-1} at 1.55 μm. There was no measurable dispersion penalty in these experiments confirming that the C^3 laser indeed operated in a single longitudinal mode even when modulated at 1 Gbit sec^{-1}.

With 1.3- and 1.5-μm C^3 lasers, step-tuning with frequency excursions of 150–300 Å over as many as 15 discrete frequencies and averaged frequency tuning rates of 10–25 Å mA^{-1} were obtained. This makes possible the first demonstration of multilevel multichannel FSK using a 1.3-μm C^3 laser. Further, a new kind of FSK based on the newly observed spectral bistability in a C^3 laser was also demonstrated. The step-tuning capability of the C^3 laser also makes possible an optical switching and routing system. In addition to step-tuning, analog FM can also be obtained by biasing the modulator below threshold with a dc current and a small modulating current superimposed to produce the FM. Compared with direct FM of a conventional semiconductor laser, this results in significantly larger frequency deviation with negligible spurious intensity modulation. Further, the present FM responses are also much more uniform.

The realization that the optical lengths of the cavities can be tuned separately by electrical means allows also for electrical feedback to compensate for drift during operation. The present multicavity scheme can also be operated as a laser–detector module for high coupling efficiency and close tracking of laser output power.

Besides applications in lightwave communications, the use of such multicavity devices in electronically gateable mode-locked semiconductor lasers capable of coding information on the picosecond optical pulses emitted was also demonstrated. Q-switching with an integrated intracavity modulator employing a C^3 laser was also demonstrated. Utilizing the frequency tunability of C^3 lasers, the operation of a new optical scheme that performs the complete set of basic logic operations, AND, OR, EXCLUSIVE OR, and INVERT, was demonstrated. The steady-state theoretical analysis of C^3 lasers carried out by Henry and Kazarinov (1983) was also presented. It was shown that the calculated results agree qualitatively with the measured results.

REFERENCES

Ainslie, B. J., Beales, K. J., Day, C. R., and Rush, J. D. (1981). *IEEE J. Quantum Electron.* **QE-19,** 854.

Allen, L. B. Koenig, H. G., and Rice, R. R. (1978). *Proc. Soc. Photo-Opt. Instrum. Eng.* **157,** 110.

Basov, N. G. (1968). *IEEE J. Quantum Electron.* **QE-4,** 855.

Bowden, C. M., Ciftan, M., and Robl, H. R., eds. (1980). "Optical Bistability." Plenum, New York; also in the special issue of *IEEE J. Quantum Electron.* **QE-17,** (1981).

Bowers, J. E., Bjorkholm, J. E., Burrus, C. A., Coldren, L. A., Hemenway, B. R., and Wilt, D. P. (1983). *Appl. Phys. Lett.* **44,** 821.

Cameron, K. H., Chidgey, P. J., and Preston, K. R. (1982). *Electron. Lett.* **18,** 650.

Campbell, J. C., Dentai, A. G., Holden, W. S., and Kasper, B. L. (1983). *Eur. Conf. Opt. Commun. '83, Geneva* postdeadline paper.

Chang, M. B., and Garmire, E. (1980). *IEEE J. Quantum Electron.* **QE-16,** 997.

Cheung, N. K. (1983). *Eur. Conf. Opt. Commun. '83, Geneva* Pap. 20.

Cheung, N. K., Sandahl, C. R., Lipson, J., Olsson, N. A., Sallada, C. V., and Tsang, W. T. (1983). *Opt. Fiber Commun., New Orleans,* Pap. 30.

Coldren, L. A., and Koch, T. L. (1983). Unpublished data.

Coldren, L. A., Miller, B. I., Iga, K., and Rentschler, J. A. (1981). *Appl. Phys. Lett.* **38,** 315.

Coldren, L. A., Furuya, K., Miller, B. I., and Reutschler, J. A. (1982). *IEEE J. Quantum Electron.* **QE-18,** 1679.

Dandridge, A., and Goldberg, L. (1982). *Electron. Lett.* **18,** 303.

Dutta, N. K., Agrawal, G. P., and Focht, M. W. (1983). Unpublished observations.

Ebling, K. J., and Coldren, L. A. (1983). *J. Appl. Phys.* **54,** 2962.

Henry, C. H. (1982). *IEEE J. Quantum Electron.* **QE-18,** 259.

Henry, C. H., and Kazarinov, R. F. (1983). *IEEE J. Quantum Electron.* **QE-20,** 1000.

Ho, H., Yokoyama, H., and Inoba, H. (1980). *Electron. Lett.* **16,** 620.

Ho, P. T., Glasser, L. A., Ippen, E. P., and Haus, H. A. (1978). *Appl. Phys. Lett.* **33,** 241.

Holbrook, M. B., Sleat, W. E., and Bradley, D. J. (1980). *Appl. Phys. Lett.* **37,** 59.

Ichihashi, Y., Nagai, H., Miya, T., and Miyajima, Y. (1983). *Integr. Opt. Opt. Commun. Conf. '83, Tokyo* postdeadline paper.

Ippen, E. P., Eilenberger, D. J., and Dixon, R. W. (1980). *Appl. Phys. Lett.* **37,** 267.

Iwashita, K., Nakagawa, K., Matsuoka, T., and Nakahara, M. (1982). *Electron. Lett.* **18,** 937.

Kaiser, P. (1983). *Integr. Opt. Opt. Commun. Conf. '83, Tokyo* Pap. 27C2-4.

Kaminow, I. P., Stulz, L. W., Dentin, A. G., Ko, J. S., Nahony, R. E., DeWinter, J. C., and Ballman, A. A. (1983). *Proc.—IEEE Semicond. Laser Conf., Ottawa, 1982* Pap. 35.

Kaminow, I. P., Ko, J.-S., Linke, R. A., and Stalz, L. W. (1983b). Unpublished observation.

Kasper, B. L., Linke, R. A., Campbell, J. C., Dentai, A. G., Vodhanel, R. S., Henry, P. S., Kaminow, I. P., and Ko, J.-S. (1983). *Eur. Conf. Opt. Commun. '83 ECOC '83, Geneva* postdeadline paper.

Kawaguchi, H., and Kawakami, T. (1977). *IEEE J. Quantum Electron.* **QE-13,** 556.

Kazarinov, R. F., and Henry, C. H. (1982). *J. Appl. Phys.* **53,** 4631, Appendix A.

Kleinman, D. A., and Kislink, P. P. (1962). *Bell Syst. Tech. J.* **41,** 453.

Kobayashi, S., Tamamoto, Y., Ito, M., and Kimura, T. (1982). *IEEE J. Quantum Electron.* **QE-18,** 582.

Kogelnik, H., and Patel, C. K. N. (1962). *Proc. IRE* **50,** 11.

Kosnocky, W. F., and Connely, R. H. (1968). *IEEE J. Quantum Electron.* **QE-4,** 125.

Lang, R., and Kobayashi, K. (1976). U.S. Patent 3,999,146.

Lazay, P. D., and Pearson, A. D. (1982). *IEEE J. Quantum Electron.* **QE-18,** 504.

Lee, T. P., and Roldan, R. H. R. (1970). *IEEE J. Quantum Electron.* **QE-6,** 339.

Lee, T. P., Burrus, C. A., Copeland, J. A., Deutai, A. G., and Marcuse, D. (1982). *IEEE J. Quantum Electron.* **QE-18,** 1101.

Lee, T. P., Burrus, C. A., Sessa, W. B., and Besomi, P. (1984). *Electron. Lett.* **20,** 625.

Lee, T. P., Burrus, C. A., Eisenstein, G., Sessa, W. B., and Besomi, P. (1984). *Electron. Lett.* **20,** 625.

Linke, R. A., Kasper, B. L., Ko, J.-S., Kaminow, I. P., and Vodhanel, R. S. (1983). *Electron. Lett.* **19,** 775.

Linke, R. A., Kasper, B. L., Campbell, J. C., Dentai, A. G., and Kaminow, I. P. *Electron. Lett.* **20,** 489.

Lipson, J., and Harvey, G. (1983). *J. Lightwave Technol.* **LT-1,** 387.

Logan, R. A., van deer Ziel, J. P., Temkin, H., and Henry, C. H. (1982). *Electron. Lett.* **18,** 95.

Malyon, D. J., and McComa, A. P. (1982). *Electron. Lett.* **18,** 445.

Manning, J., Olshansky, R., and Su, C. B. (1983). *IEEE J. Quantum Electron.* **QE-19,** 1525.

Marcuse, D., and Lee, T. P. (1984). *IEEE J. Quantum Electron.* **QE-20,** 166.

Matsuoka, S. T., Nagai, H., Itaya, Y., Noguchi, Y., Suzuki, Y., and Ikegami, T. (1982). *Electron Lett.* **18,** 27.

Merz, J. L., Logan, R. A., and Sergent, M. (1979). *IEEE J. Quantum Electron.* **QE-15,** 72.

Olsson, N. A., and Tang, C. L. (1979). *IEEE J. Quantum Electron.* **QE-15,** 1085.

Olsson, N. A., and Tsang, W. T. (1983a). *IEEE J. Quantum Electron.* **QE-19,** 1621.

Olsson, N. A., and Tsang, W. T. (1983b). *Electron. Lett.* **19,** 808.

Olsson, N. A., and Tsang, W. T. (1983c). *IEEE J. Lightwave Commun.* **LT-2,** 49.

Olsson, N. A., Tang, C. L., and Green, E. L. (1980). *Appl. Opt.* **19,** 1897.

Olsson, N. A., Dutta, N. K., and Besomi, P. (1983a). Unpublished observations.

Olsson, N. A., Tsang, W. T., Logan, R. A., and Patel, C. K. N. (1983b). *Appl. Phys. Lett.* **43,** 1091.

Pearson, A. D., Lazay, P. D., Reed, W. A., and Saunders, M. J. (1982). *Eur. Conf. Opt. Commun., 8th, Cannes, Fr.* Pap. AIV-3.

Preston, K. R. (1982). *Electron. Lett.* **18,** 1092.

Preston, K. R., Woollard, K. C., and Cameron, K. H. (1981). *Electron. Lett.* **17,** 931.

Reinhart, F. K., and Logan, R. A. (1975). *Appl. Phys. Lett.* **27,** 532.

Reinhart, F. K., and Logan, R. A. (1980). *Appl. Phys. Lett.* **36,** 954.

Runge, P. K. (1983). *Tech. Dig.—Opt. Fiber. Commun. '83, New Orleans.* Pap. MD2.

Saito, S., Yamamoto, Y., and Kimura, T. (1980). *Electron. Lett.* **16,** 826.

Saito, S., Yamamoto, Y., and Kimura, T. (1981). *IEEE J. Quantum Electron.* **QE-17,** 935.

Saito, S., Yamamoto, Y., and Kimura, T. (1982). *Electron. Lett.* **18,** 468.

Sakai, K., Utaka, K., Akiba, S., and Matushima, Y. (1982). *IEEE J. Quantum Electron.* **QE-18,** 1272.

Sakakibara, Y., Furuya, K., Utaka, K., and Suematsu, Y. (1980). *Electron Lett.* **16,** 456.

Salathe, R. P. (1979). *Appl. Phys.* **20,** 11.

Shumate, P. W., Jr., Chen, F. S., and Dorman, P. W. (1978). *Bell Syst. Tech. J.* **57,** 1823.

Smith, P. W. (1972). *Proc. IEEE* **60,** 422.

Suematsu, Y., Arai, S., and Kishino, K. (1983). *J. Lightwave Technol.* **LT-1,** 161.

Tang, C. L., Kreismanis, V., and Ballantyne, J. (1976). *Appl. Phys. Lett.* **30,** 113.

Tsang, W. T. (1983). Unpublished observations.

Tsang, W. T., and Logan, R. (1978). *J. Appl. Phys.* **49,** 2629.

Tsang, W. T., and Olsson, N. A. (1983). *Appl. Phys. Lett.* **43,** 527.

Tsang, W. T., Olsson, N. A., and Logan, R. A. (1983a). *Appl. Phys. Lett.* **42,** 650.

Tsang, W. T., Logan, R. A., Olsson, N. A., Temkin, H., van de Ziel, J. P., Kasper, B. L., Linke, R. A., Mazurczyk, V. J., Wagner, R. E., Kaminow, I. P., and Miller, B. I. (1983b). *Tech. Dig.—Opt. Fiber Commun. '83, New Orleans.* postdeadline section.

Tsang, W. T., Olsson, N. A., Logan, R. A., and Linke, R. A. (1983c). *Electron. Lett.* **19,** 341.

Tsang, W. T., Olsson, N. A., and Ditzenbeyer, J. A. (1983d). *Appl. Phys. Lett.* **43,** 1003.

Tsang, W. T., Olsson, N. A., and Logan, R. A. (1983e). *Appl. Phys. Lett.* **43,** 339.

Tsang, W. T., Olsson, N. A., and Logan, R. A. (1983f). *IEEE J. Quantum Electron.* **QE-19**, 1621.

Tsang, W. T., Olsson, N. A., and Logan, R. A. (1983g). Unpublished observations.

Tsang, W. T., Olsson, N. A., and Logan, R. A. (1983h). Unpublished observations.

Tsang, W. T., Olsson, N. A., and Logan, R. A. (1983i). Unpublished observations.

Tsang, W. T., Olsson, N. A., and Logan, R. A. (1983j). *Electron. Lett.* **19**, 488.

Tsang, W. T., Olsson, N. A., Linke, R. A., and Logan, R. A. (1983k). *Electron. Lett.* **19**, 415.

Tsang, W. T., Olsson, N. A., Logan, R. A. (1983l). *Appl. Phys. Lett.* **42**, 1003.

Utaka, K., Suematsu, Y., Kobayashi, K., and Kawanishi, H. (1980). *Jpn. J. Appl. Phys.* **19**, L137.

van der Ziel, J. P. (1981). *J. Appl. Phys.* **52**, 4435.

van der Ziel, J. P., Tsang, W. T., Logan, R. A., Mikulyak, R. M., and Augustyniak, W. M. (1981). *Appl. Phys. Lett.* **39**, 525.

Vodhanel, R. S. (1983). Unpublished observations.

Yamada, J., Kawana, A., Nagai, H., and Kimura, T. (1982). *Electron. Lett.* **18**, 98.

Yamamoto, Y., and Kimura, T. (1981). *IEEE J. Quantum Electron.* **QE-17**, 919.

Index

Contents of Volume 22

Part E

Contents of Previous Volumes